D1825684

ALCOHOL AND INJURIES

Emergency Department Studies in an International Perspective

Editors: Cheryl J. Cherpitel, Guilherme Borges, Norman Giesbrecht, Daniel Hungerford,

Margie Peden, Vladimir Poznyak, Robin Room, Tim tockwell

World Health Organization

WHO Library Cataloguing-in-Publication Data:

Alcohol and injuries: emergency department studies in an international perspective.

1.Alcohol drinking - adverse effects. 2.Alcoholic intoxication - diagnosis. 3.Wounds and injuries - etiology. 4.Emergency service, Hospital. 5.Multicenter studies. I.World Health Organization.

ISBN 978 92 4 154784 0   (NLM classification: WM 274)

TABLE OF CONTENTS

FOREWORD

The harmful use of alcohol is one of the main risk factors to health. It is responsible each year for about 2.3 million premature deaths worldwide. Injuries – both unintentional and intentional – account for more than a third of the burden of disease attributable to alcohol consumption. These include injuries from road traffic crashes, burns, poisoning, falls and drowning as well as violence against oneself or others. The impact of alcohol-related injuries affects not only those who are intoxicated at the time of injury occurrence, but also those who fall victim to their behavior. These include the pedestrian or cyclist knocked over by a drunk driver or the woman or children beaten by a drunk husband or father.

Alcohol-attributable injuries and violence are of growing concern to the World Health Organization. Alcohol-related injuries are especially evident in hospital emergency rooms and trauma centers. The clinical encounters in these settings present a one-time chance for health professionals to get access to a population often difficult to reach. Reduction of the burden of such injuries can be achieved by implementing evidence-based public health strategies, policy measures and effective interventions on a broad scale.

This publication draws together the current state of knowledge on research, practice and policy issues on the association of alcohol with injuries. It synthesizes the results of studies from a number of hospital emergency departments conducted in different cultural settings, including the World Health Organization's Collaborative Study on Alcohol and Injuries. The book provides an introduction to the epidemiology of alcohol-related injuries and refers to methodological issues of studies conducted in emergency departments. It also addresses public policy implications and equips the reader with practical information on interventions that can be implemented in emergency departments such as screening and brief interventions for hazardous and harmful drinking.

Alcohol and Injuries: Emergency Department Studies in an International Perspective will be a useful and important source for researchers, service providers and policy makers on international and national levels as well as for all those who are concerned with alcohol-related injuries and violence and the reduction of public health problems caused by the harmful use of alcohol.

Dr Benedetto Saraceno
Director
Department of Mental Health and Substance Abuse
World Health Organization

Dr Etienne Krug
Director
Department of Injuries and
Violence Prevention and Disability
World Health Organization

PREFACE

The Government of Valencian Community, Spain, is pleased to support through the Valencian Health Agency and its General Directorate for Drug Dependence the publishing of this book on alcohol and injuries in emergency departments. This publication is a highly relevant and timely product. The health burden of injuries is significant, and reducing alcohol-related injuries is a priority area for public health interventions which makes this book highly relevant to public health professionals. The book presents a synthesis of the latest international findings drawing on a number of studies conducted in high-, low- and middle-income countries, including the WHO Collaborative Study on Alcohol and Injuries and the Emergency Room Collaborative Alcohol Analysis Project (ERCAAP). The reader will benefit from the latest experiences and knowledge summarized and presented in this book, which also embraces conceptual issues in the association between alcohol and injuries, methods of identifying alcohol involvement in injuries in emergency departments as well as specific examples of interventions from culturally diverse and different health care settings. This publication is recommended to a wide audience of health professionals, scientists and policy makers interested in epidemiology and monitoring of alcohol involvement in injuries and developing effective interventions for reducing alcohol-related injuries in emergency departments and trauma centers.

Dr Sofía Tomás Dols
Director General
General Directorate for Drug Dependence
Health Department
Government of Valencian Community

ACKNOWLEDGEMENTS

The publication of this book grew, in part, out of the *WHO Collaborative Study on Alcohol and Injuries and the Emergency Room Collaborative Alcohol Analysis Project (ERCAAP)* and, in part, out of the *International Conference on Alcohol and Injuries: New Knowledge from Emergency Room Studies*, held in Berkeley, California, October 3-6, 2005, sponsored by the U.S. National Institute of Alcohol Abuse and Alcoholism, and co-sponsored by the World Health Organization and the U.S. Centers for Disease Control and Prevention.

Publication of this book would not have been possible without Dr Cheryl Cherpitel who provided the overall coordination for the preparation of the book and made invaluable contribution to individual chapters and organization of the above-mentioned conference.

A special acknowledgement is also extended to all contributors and editors of this book as well as to all of those who have contributed so greatly to the field of alcohol and injury in emergency room studies over the last 25 years.

A number of colleagues at WHO assisted in different capacities during the preparation of this book for publication, including Dr Alexandra Fleischman, Daniela Fuhr, Mylène Schreiber and Tess Narciso under the supervision of Dr Vladimir Poznyak, WHO Department of Mental Health and Substance Abuse.

WHO wishes to acknowledge the financial contribution of the Health Department, Government of Valencian Community (Spain), to the publication of this book. Special thanks are extended to Dr Sofía Tomás Dols, Director General, General Directorate for Drug Dependence, Health Department, Government of Valencian Community (Spain), for her contribution to this project.

OVERVIEW :
ALCOHOL AND INJURIES: EMERGENCY DEPARTMENT STUDIES IN AN INTERNATIONAL PERSPECTIVE

Cheryl J. Cherpitel - Alcohol Research Group | Emeryville, CA

This edited volume on alcohol and injuries provides a research, practice and policy perspective of the current state of knowledge of alcohol's association with casualties based on hospital emergency department (ED) studies in a number of countries and settings. While a multitude of studies of ED patients now exists which addresses this association, many of these studies have serious limitations regarding generalizability of findings. These studies also have largely been restricted to Western cultures. This volume focuses on the most recent evidence regarding alcohol's association with injury, and the scope, magnitude and diversity of association across countries and cultures, based on epidemiological studies in EDs around the world. It also addresses issues related to conducting studies in the ED, identifying patients with alcohol-related injuries and practical aspects of implementing brief intervention in this setting, as well as implications and application of findings for alcohol policy and public health.

This is the first such book which focuses on alcohol and injuries exclusively in the emergency department setting, and draws on contributions from researchers and emergency service clinicians involved in ED studies and practice from a number of countries. Some over-lap of issues is apparent across sections of the book; for example, alcohol attributable fraction of injury has been touched on in a number of chapters. Many of the contributors are not new to the field of alcohol and injury in the emergency department. Furthermore, it is not uncommon to any field which is continuing to grow and evolve, that differing viewpoints are evident across chapters. Terminology to designate the emergency services setting has not been restricted to a single term, but uses "emergency room" (ER) and "emergency department" interchangeably. These terms are also inclusive of Accident and Emergency (A&E) as used in the United Kingdom. While the term "trauma center" is sometimes used interchangeably with ED or ER in the U.S., trauma centers have a distinct designation and this is discussed. Some countries, as well, do not have designated emergency services as such, because emergency services are not a defined medical specialty.

The first section of the book focuses on the epidemiology of alcohol and injuries, and includes latest findings across all injuries, as well as variation by type and cause, from a large number of international ED studies comprising the Emergency Room Collaborative Alcohol Analysis Project (ERCAAP) and the World Health Organization Collaborative Study of Alcohol and Injuries, both of which used comparable methodology and instrumentation. A distinction is made between the risk of injury associated with drinking, and factors associated with an alcohol-related injury. Especially important in findings presented is the culturally-specific contextual variables (socio-cultural factors and organizational/administrative factors related to obtaining treatment in the ED) that explain, to some extent, the variation in associations of alcohol and injury found across ED studies, countries and cultures.

Causality is an important issue related to a better understanding of the alcohol-injury nexus. While alcohol may demonstrate a positive association with injury, it may not be a sufficient cause for injury occurrence and this issue is explored. A related issue is the patient's perception as to whether the injury would have occurred if he or she had not been drinking, and this 'casual attribution' of alcohol to injury is also explored in relation to deriving estimates of alcohol attributable fractions. Finally, this section concludes by putting these data into a larger perspective in relation to the global burden of disease based on comparative risk analysis across sub-regions of the world.

In the second section a number of issues are explore that are related to the study of alcohol and injury based on ED populations, including limitations of these studies, and the quality and usefulness of the resulting data. An overview of ED studies is provided, and includes the scope and quality of data. Estimates of the alcohol attributable fraction obtained from individual-level data and from aggregate-level data are compared, and new approaches are presented, with the potential of ED data for improving these estimates. Methodologies which have been used for conducting epidemiological studies of alcohol and injury in the ED and their utility are discussed. Conceptual issues related to explaining the magnitude of the association of alcohol and injury are explored, including the interaction of usual drinking patterns with drinking in the injury event, and choice of the best control population.

The third section examines issues related to identifying alcohol-related injuries in the ED. Assessment of alcohol intoxication of ED patients based on estimated BAC (ICD-10 Y90 codes) and clinical assessment (ICD-10 Y91 codes) is explored, including lack of correspondence between the two measures, and suggestions made related to the development and utility of an international standard for intoxication assessment. This section also explores methods related to the surveillance of alcohol-related injury and monitoring options, including the use of "surrogate" measures for identifying alcohol-related injuries in the ED.

Brief intervention for alcohol problems in the ED is the focus of the fourth section. Evidence of the effectiveness of brief interventions for ED patients and methodologies which have been used are reviewed. Practical issues related to implementation of brief intervention in the ED setting are explored in a series of five papers, and include clinician training, barriers to implementation, and adaptation of brief intervention across various types of emergency service settings including trauma centers. Finally, the potential impact of implementing a brief intervention among ED patients in relation to cost-effectiveness and policy implications is discussed.

The last section of the book focuses on application and implications of findings from ED studies. Practical experiences with conducting ED studies in diverse cultural circumstances (Mexico, Argentina, Poland, Czech Republic, and India) are presented in a series of papers, and include lessons learned in adapting and implementing an ED study design in specific national circumstances. The community context of findings from ED studies and policy implications are discussed, including the use of ED data in community prevention trials. The book concludes with three papers on alcohol policy and public health implications in the U.S. and European context, and in the global perspective.

The themes addressed in this book provide state-of-the-art knowledge regarding alcohol and injury based on studies of ED patients, globally; practical issues including surveillance of alcohol-related injuries in the ED, and implementation of brief intervention in this setting; and the implication and importance of these findings for public health policy formulation in an international perspective.

SECTION I : EPIDEMIOLOGY OF ALCOHOL AND INJURY IN EMERGENCY DEPARTMENT STUDIES
INTRODUCTION

Guilherme Borges - National Institute of Psychiatry, Metropolitan Autonomous University |
Mexico City, MEXICO

Epidemiology as the study of disease occurrence in populations has long contributed to the field of alcohol and injuries. A privileged scenario for such inquiry is the emergency department (ED). In modern societies, most injuries of consequence will come to the attention of the ED. Even if a person does not have economic resources to pay for medical attention, in many countries it is mandatory for EDs to provide initial treatment to any person experiencing a life-threatening medical emergency. While it thus appears that the ED is a virtual gold mine for studies of alcohol and injuries, prior to 1984, at which time a series of ED studies was initiated, all using a similar rigorous methodology (Cherpitel, 1989), few sound epidemiologic studies had been conducted in the ED. A general claim made by ED personnel treating these patients was that it was a common situation to have patients "under the influence" of alcohol at the time of ER admission, a situation which further highlighted the important need to develop and conduct research in this setting.

The four chapters presented in this section represent the latest approaches to move forward this field of study. A great deal of new data is presented together with challenging ideas for the future in the study of the epidemiology of alcohol and injury.

The first chapter by Ye and Cherpitel goes beyond the link of alcohol and injury to address the issue of variables associated with an alcohol-related injury which, as the authors underscore, has great significance for injury prevention in society. The authors make an important contribution in bringing the problem of identification of factors associated with alcohol-related injury to the forefront, with implications for calculations of attributable fractions of alcohol and injury.

The work of Borges et al. in the following next chapter, is an example of the use of modern epidemiological techniques of study design and analyses to face a pervasive problem in several ED studies: the lack of adequate control groups. The authors use the case-crossover study methodology to obtain relative risk estimates of alcohol and injury by injury type and cause, taking advantage of a large dataset in this field.

The contribution of Bond and Macdonald in the next chapter, is a novelty in this area. The authors provide an alternative approach to objective measures of calculating estimates of the relative risk of injury due to alcohol, using the subjective measure of "causal attribution", or the belief on the part of the patient that drinking prior to the event contributed to injury occurrence. How this combination of objective and subjective risk estimates and resulting alcohol attributable fractions will be used will certainly be a matter of great discussion for future work in the field.

The last chapter by Rehm, et al. quantifies alcohol-attributable injuries in a global perspective across sub-regions of the world for the year 2002. It shows in a clear and comparative fashion how alcohol consumption contributes to the burden of injury mortality and morbidity and how changes in consumption can lead to a substantial improvement in the health status of a population. Such exercise has proven to be a powerful argument for scientists and policy makers alike.

References

Cherpitel CJ (1989). Study of alcohol use and injuries among emergency room patients. In: *Drinking and casualties: Accidents, poisonings and violence in an international perspective*. Giesbrecht N et al., eds, pp. 288-299. Routledge, New York.

CHAPTER 1 :
RISK OF INJURY ASSOCIATED WITH ALCOHOL AND ALCOHOL-RELATED INJURY

Yu Ye, Cheryl J. Cherpitel - Alcohol Research Group | Emeryville, CA USA

Introduction

Alcohol consumption is a leading risk factor for mortality and morbidity related to both intentional and unintentional injury. In 2000, 16.2% of deaths and 13.2% of disability-adjusted life years (DALYs) from injuries, globally, were estimated to be attributed to alcohol (see Chapter 4 in this volume). Alcohol affects individual psychomotor skills involving brain-eye-hand-foot coordination. The effect on visual focus, reaction time and judgment leads to injuries from causes like motor vehicle accidents and falls. Alcohol intoxication also has an effect on an individual's cognitive skill. Persons exposed to alcohol may place themselves in dangerous situations, be more aggressive and less averse to risk taking, leading to both unintentional injuries such as drowning and burns, and to intentional injuries as either perpetrators or victims.

Alcohol-Related Injury and Risk of Injury Associated with Alcohol

Historically, alcohol involvement among fatal and nonfatal injuries has been reported in a large number of studies. For example, in a meta-analysis of 65 articles published between 1975 and 1995, Smith et al. (1999) reported that the aggregate percentage of intoxicated cases (BAC≥.10%) was 31.5% among homicide deaths, 22.7% among suicide cases and 31.0% among non-traffic unintentional deaths. While such moderate to high percentages of alcohol involvement in injury events might suggest alcohol is a risk factor for injury, prevalence rates of exposure to alcohol do not provide information to evaluate the actual risk of injury due to alcohol consumption.

In epidemiology, the exposure-disease association is normally assessed either in cohort studies through exposure-based sampling or by case-control design through disease-based sampling. One important aspect of the alcohol-injury association is that injury is presumably caused by episodic, acute alcohol consumption, in which the effect of alcohol exposure is transient, rather than by the accumulative effect of alcohol as found in such chronic diseases as liver cirrhosis. Because of the transient nature of the causal mechanism, the traditional cohort study is hard to implement given the difficulty in measuring the baseline acute exposure to alcohol. For this reason, risk of injury associated with alcohol consumption is normally assessed using the case-control design in which drinking prior to the injury event among cases is compared with drinking at a given time among the non-injury control group. More recently, the case-crossover design has also been used for this purpose, in which alcohol consumption before the injury event for each case is compared with his or her own drinking in a pre-determined prior time period or with usual alcohol consumption (Borges et al., 2004; Vinson et al., 1995).

While alcohol involvement in both case and control groups is required to evaluate risk of injury associated with drinking, once risk of injury due to alcohol is established, alcohol involvement among injury cases only, or *alcohol-related injury* is an outcome of interest in itself. Alcohol is the third leading cause of total mortality in the United States (Mokdad et al., 2004) and a major risk factor for fatal and non-fatal injury, especially in some subpopulation groups such as young males. The identification of alcohol-related injury, thus, has great significance for injury prevention in society.

While injuries with alcohol involvement might not all be causally linked to drinking, rate of alcohol-related injury is an important measure of the burden of injury attributable to alcohol in a society, i.e. alcohol-attributable injury. One formula to calculate Alcohol-Attributable Fraction of injury (AAF), which is particularly used in case studies, is $AAF = P(Alcohol|injury) \times (1 - \frac{1}{RR})$ (Jewell, 2004). Thus, both risk of injury related to alcohol (the relative risk or RR) and the rate of alcohol-related injury (probability of drinking among injured patients) contribute to alcohol-attributable injury. The rate of alcohol-related injury is positively associated with the proportion of injury attributed to alcohol, especially when the relative risk of injury due to alcohol is high. On the other hand, a low rate of alcohol-related injury implies a relative low attributable risk of injury to alcohol, even if the relative risk of a given type of injury associated with drinking is high.

Risk of Injury Associated with Drinking from ED Studies

Emergency department controls

To study the risk of injury associated with drinking, hospital emergency departments (EDs) have been an ideal place to sample injury patients seeking treatment, and many studies have used medical condition (non-injury) patients as control subjects. Emergency department controls are assumed to be from the same geographic area as cases and to possibly share some other similar characteristics such as socioeconomic status. The case-control design is used to estimate the risk of injury associated with acute alcohol use, measured by blood alcohol concentration (BAC) taken immediately on arrival at the ED or by the patient's self-reported alcohol consumption during a period (e.g. six hours) prior to the injury event or, for the non-injured patient, to first noticing the medical condition which brought him or her to the ED.

In a meta-analysis of the alcohol and injury relationship, Cherpitel et al. (2003a) combined 12 ED studies, all using ED controls and similar methodology, in the Emergency Room Collaborative Alcohol Analysis Project (ERCAAP). These data were re-analyzed with three additional studies (two in Poland and one in Argentina), and excluded from analysis those medical patients primarily admitted to the ED for alcohol intoxication or for alcohol withdrawal. These results are presented in Table 1. When analysis was restricted to patients arriving at the ED within six hours following the event, to provide a more valid BAC measure, the odds ratio of injury risk associated with positive BAC (≥.01%) ranged from 0.9 to 7.2. The overall pooled odds ratio across studies was 2.4 (95% CI = 1.9-3.0). When different BAC levels were evaluated, a dose-response relationship between increasing BAC levels and risk of injury was found to vary across studies. An elevated risk was observed at a modest BAC level (.01-.049%) but then leveled off (Contra Costa, California and Mar del Plata, Argentina), or was only observed when BAC was higher than .10% (San Francisco, California). The overall pooled effect of BAC levels at .01-.049%, .05-.099% and ≥.1% was 2.0, 1.5 and 2.9, respectively, compared with a negative BAC (<.01%). Since BAC level predicted cases perfectly in some of the studies, the pooled estimates should be viewed with caution as those studies with positive infinite odds ratios were excluded in pooled estimate calculation. Also shown in Table 1 (last column) is the risk of injury associated with self-reports of any drinking within six hours prior to the injury or medical condition, evaluated on the total ED sample regardless of time of arrival. Estimates varied significantly across studies, and the pooled odds ratio across all studies was 2.1 (95% CI = 1.6-2.7). A dose-response relationship was not possible to evaluate for self-reported drinking since medical patients were not asked the amount they consumed during the six-hour period prior to the medical condition in most of the ERCAAP studies.

Table 1. Odds Ratios (OR) and 95% CI for BAC and drinking within 6 hours prior to the event for risk of injury vs medical condition[1]

Studies	Year	N (total/6 hr arrival)	Positive BAC (≥ .01%)[2]	BAC levels[2]			Any drinking 6 hour prior
				.01-.049%	.05-.099%	≥ .10%	
1 San Francisco, CA	1984-85	1896/658	1.6 (1.0, 2.5)	1.3 (.5, 3.5)	.6 (.2, 1.4)	2.5 (1.4, 4.3)	2.9 (2.3, 3.7)
2 Contra Costa, CA	1985	2400/1247	1.9 (1.1, 3.3)	2.1 (.9, 4.9)	2.0 (.5, 8.2)	1.8 (.9, 3.8)	1.1 (.9, 1.5)
3 Kaiser (Contra Costa) Co., CA	1989	961/447	.9 (.3, 3.2)	.3 (.04, 3.1)	.7 (.04, 12.7)	3.6 (.3, 38.8)	1.1 (.6, 1.8)
4 Jackson, MS	1992	1017/276	7.2 (2.0, 25.9)	4.8 (1.0, 23.4)	∞	9.1 (1.1, 78.1)	3.0 (2.0, 4.7)
5 Santa Clara, CA	1995-96	1334/516	2.5 (1.0, 6.1)	.4 (.1, 3.5)	5.3 (.7, 38.0)	3.4 (1.1, 11.1)	3.0 (1.9, 4.6)
6 Mexico City, Mexico	1986	2188/1210	2.9 (1.5, 5.6)	2.7 (.9, 8.6)	4.2 (.9, 20.2)	2.6 (1.1, 6.4)	2.9 (2.0, 4.1)
7 Acapulco, Mexico	1987	640/354	2.7 (1.1, 6.8)	1.4 (.5, 4.0)	∞	5.6 (.7, 45.6)	2.9 (1.8, 4.6)
8 Pachuca, Mexico	1996-97	1417/777	4.9 (2.0, 11.6)	6.6 (1.8, 25.1)	2.5 (.7, 8.7)	11.5 (.7, 206)	3.3 (1.9, 5.8)
9 Edmonton (Alberta), Canada	1989	842/424	6.1 (2.6, 14.3)	3.0 (.8, 11.0)	3.8 (.3, 54.4)	9.9 (3.2, 30.3)	3.7 (2.5, 5.6)
10 Quebec City, Canada	1989	655/344	1.5 (.5, 4.6)	.9 (.1, 5.9)	.7 (.1, 4.1)	∞	.9 (.6, 1.5)
11 Barcelona, Spain	1987	2363/1014	2.4 (.9, 6.2)	1.7 (.5, 5.8)	∞	2.3 (.5, 10.1)	1.2 (.9, 1.6)
12 Trieste, Italy	1990	476/NA	NA[2]	NA[2]	NA[2]	NA[2]	1.9 (1.2, 3.1)
13 Mar del Plata, Argentina	2001	800/354	3.2 (1.0, 10.0)	3.2 (.4, 28.8)	4.3 (.5, 37.6)	2.6 (.5, 13.6)	1.9 (1.2, 3.0)
14 Warsaw, Poland	2002	714/311	∞	∞	∞	∞	2.4 (1.0, 5.7)
15 Sosnowiec, Poland	2002-03	735/373	6.8 (.9, 54.5)	∞	∞	3.0 (.4, 25.1)	2.7 (1.5, 4.7)
Test of homogeneity			Q=18(df=12) p=.117	Q=12 (df=11) p=.381	Q=11(df=8) p=.228	Q=9 (df=11) p=.603	Q=85(df=14) p<.001
Pooled effect			2.4 (1.9, 3.0)	2.0 (1.3, 2.8)	1.5 (.9, 2.6)	2.9 (2.1, 4.0)	2.1 (1.6, 2.7)

[1] Adjusted ORs from logistic regressions controlling for gender, age, education, martial and employment status
[2] Analysis on BAC was restricted to injury or medical condition patients who came to the ED within 6 hours after event; OR estimates of BAC were not available for Trieste, Italy since ED arrival time was missing
∞ refers to positive infinite ORs when positive BAC or BAC levels predicted the cases perfectly

Community controls

Population samples from the communities where the ED injury patients reside have also been used in case-control studies of injury risk associated with drinking (Borges et al., 1998). A study was conducted in Western Australia sampling 797 ED injury cases and 797 controls paired to cases in relation to suburb of residence (Mcleod et al., 1999). Acute alcohol use among controls was evaluated by questions related to exactly the same time period during which the case was injured. Drinking any alcohol six hours prior to the injury approximately doubled the risk of injury.

However, injury risk increased significantly only after consuming more than 60g of alcohol (OR=3.4) and elevated risk from modest drinking (<60g) was small and non-significant. In another Australian study, Watt and colleagues (2004) sampled 488 ED injury patients in Queensland and matched them to 488 community controls on gender, age group, neighborhood, day and time of injury. Drinking any alcohol within six hours prior to injury increased the risk of injury by an odds ratio of 2.13 (95% CI = 1.2-3.9). A moderate dose-response relationship was also observed, with "low-risk" quantities of alcohol (≤40grams for females and ≤60grams for males) associated with an odds ratio of 1.71 (95% CI = .9-3.3) while "risky/high risk" quantities (>40g for females and >60g for males) were associated with an odds ratio of 2.40 (95% CI = 1.1-5.2). In another study in the United States, Vinson and colleagues (2003) sampled 2,161 injury patients from three EDs in Boone County, Missouri, who were matched to 1,856 community controls based on sex, age, and day of week and hour of injury. In case-control analysis an odds ratio for any drinking within the six hours prior to injury versus none was 3.1 (95% CI: 2.3-4.2). A dose-response relationship was also observed with odds ratios for 1-2, 3-4, 5-6 and 7+ drinks of 1.5, 3.7, 13.5 and 8.8, respectively.

Cases as controls in case-crossover analysis

A case-crossover analysis (in which each patient serves as his or her own control) (Maclure, 1991) was also performed in parallel to the case-control study by Vinson et al. (2003) described above. The odds ratio for any drinking within six hours prior to injury compared with drinking in the same six-hour interval the previous day was 3.2 (95% CI: 2.4-4.3). A dose-response relationship was also detected in case-crossover analysis, with odds ratios for 1-2, 3-4, 5-6 and 7+ drinks of 1.8, 6.2, 9.5 and 17, respectively. A similar analysis was performed on data from the World Health Organization (WHO) Collaborative Study on Alcohol and Injuries, which were collected in 2001-02 from 10 EDs around the world (Borges et al., 2006b). When data from the 10 studies were combined, the odds ratio for any drinking within six hours prior to injury compared with the same time period the previous week was 5.7 (95% CI: 4.5-7.3). The odds ratios across levels of acute alcohol consumption were 3.3, 3.9, 6.5 and 10.1 for 1, 2-3, 4-5 and 6+ drinks, respectively, versus no drinking, again suggesting a dose-response relationship.

In addition to this pair-matched method using a pre-determined time period as described above, another method of case-crossover analysis compares cases' drinking prior to the injury with their usual frequency of drinking. This usual frequency method was utilized by Borges et al (2006a) on combined data from the ERCAAP and WHO collaborative study. The relative risk of any drinking prior to injury was evaluated for the 28 studies separately, with a pooled estimate of 5.7 (95% CI: 4.5-7.3). Large variations in relative risk estimates were observed across studies, ranging from 1.1 to 35.

In order to compare estimates from different methods, pooled estimates are evaluated here separately for the ERCAAP and WHO projects. For the WHO studies, the pooled estimate of relative risk from the usual frequency method of the case-crossover design was 6.8, not greatly different from the odds ratio of 5.7 from the pair-matched method on the same data. For the ERCAAP studies, the pooled relative risk estimate from the case-crossover usual frequency method was 5.2, which was larger than estimates from the case-control design observed in Table 1 (2.4 for a positive BAC and 2.1 for any drinking in the six-hour period preceding injury). While there is the possibility of under-estimation of the relative risk of injury due to alcohol, using medical patients in the same ED as controls since they may come to the ED for conditions related to their use of alcohol or be heavier drinkers than those in general population (Cherpitel, 1993), on the other hand, it is possible that relative risk estimates from the case-crossover design may be subject to upward bias, when the usual frequency method is used, due to potential recall bias. Furthermore, although both the case-control and case-crossover analyses found large variations across studies, estimates from the usual frequency method appeared to show even greater variation. Therefore, further studies comparing different methodologies are needed to derive more valid estimates

Alcohol-injury relationship confounded and modified by usual consumption pattern

In the above described Australian study by Watt and colleagues (2004), when usual alcohol consumption, together with risk taking disposition and drug use, were controlled, the odds ratio of injury associated with any drinking six hours prior to the event increased from 2.13 to 3.73 (95% CI = 1.5-9.5). Observing the confounding of usual drinking patterns, the authors argued that this should be considered in analysis of the alcohol-injury relationship.

By way of an example, this confounding effect of usual drinking patterns is best demonstrated in analysis of data from the Grand Rapids Study (Borkenstein et al., 1964; Hurst et al., 1994). In estimating the risk of an accident for drinking drivers at increasing BACs, Borkenstein et al. found the relative risk for a BAC of .01- .04% dropped below one compared to a negative BAC (as shown in Table 2, Model 1). This seemingly protective effect of modest alcohol exposure was then attributed to a disproportionate distribution of usual drinking patterns across acute BAC levels, and is an example of Simpson's Paradox (Simpson, 1951). This explanation was confirmed by Hurst et al. Re-analyzing the data here by fitting logistic regressions, we derived the same results. As shown in Model 2 of Table 2, the odds ratio of a car accident for a BAC of .01-.04% was 1.17 compared with a negative BAC, controlling for usual frequency pattern. Furthermore, compared to risk across BAC levels when usual frequency is not controlled (Model 1), in Model 2, risk of injury is greater across all BAC levels.

In both the Watt and Borkenstein/Hurst analyses above, usual frequency of drinking was positively associated with acute drinking but negatively associated with risk of injury. After controlling for usual frequency pattern, at a given BAC level, the risk of injury was higher among less frequent drinkers. While it is tempting to interpret this finding in relation to a higher tolerance among frequent drinkers, such an interpretation should be viewed with caution, since exploring only the main effects of usual drinking and acute drinking on risk of injury simultaneously assumes that risk of injury for usual frequency pattern is fixed across BAC levels. In Model 2 of Table 2, higher usual frequency drinkers had a smaller risk of injury than lower usual frequency drinkers at all positive BAC levels, as well as when BAC was zero. While the former might be explained by tolerance, the latter cannot (that is, when BAC was zero), since drinking was not involved, and may be related to other risk factors not observed. In Model 3 of Table 2, when the risk for each usual frequency pattern across BAC levels was analyzed separately, the odds ratio at a given BAC level was not necessarily smaller for more frequent drinkers. Therefore, controlling for the usual frequency of drinking without exploring its interaction with acute drinking is not sufficient for determining the possibility of a tolerance effect.

Table 2.   Odds Ratios of driving accidents for BAC intervals, re-analyzing data of Borkenstein et al. (1964)

| | Model 1 | Model 2 | Model 3: within usual frequency groups | | | | |
			≤yearly	monthly	weekly	3x per week	daily
BAC (ref .00)							
.01-.04%	.94	1.17*	1.20	1.04	1.08	1.13	1.46*
.05-.07%	1.14	1.71***			1.84**	1.64	1.61
.08-.10%	3.11***	3.93***			4.30***	2.52**	3.16***
.11%+	6.79***	10.68***			11.59***	7.25***	13.22***
Usual frequency (ref ≤yearly)							
Monthly		1.03					
Weekly		.78***					
3 times per week		.59***					
Daily		.37***					

* p<.05, ** p<.01, *** p<.001,

The modifying effects of usual consumption pattern on the relationship of acute drinking and injury was examined in multilevel analysis of the ERCAAP data (Cherpitel et al., 2004). A negative interaction was observed for the risk of injury between having five or more drinks on an occasion at least weekly (5+ weekly) or 5+ at least monthly and a positive BAC, and similarly between 5+ weekly/5+ monthly or any drinking at least weekly, and drinking within six hours prior to the injury event. That is, among frequent heavy drinkers, the risk of injury for any drinking in the event versus no drinking was less than that among infrequent heavy drinkers. Thus, evidence for tolerance among frequent heavy drinkers was suggested by the negative interaction term. Gmel and his colleagues (Gmel et al., 2006) also explored the interaction of usual and acute consumption on injury among 8,736 ED patients in Lausanne, Switzerland using ED controls. Patients were categorized into three levels of drinking within 24 hours prior to injury (no drinking, <4 drinks for women and 5 drinks for men, 4+ for woment/5+ for men) and five usual drinking patterns: abstainers, low risk (usual volume less than 7 drinks a week for women or 14 drinks for men and never had any 4+ occasions for women or 5+ occasions for men during the last month), chronic high-volume (usual volume more than 7/14 drinks a week, but never 4+/5+ last month), risky single-occasion drinkers (usual volume less than 7/14 drinks a week, but at least 4+/5+ once in the last month) and risk accumulators (usual volume more than 7/14 drinks a week, and at least 4+/5+ once in the last month). The cross-tabulation of acute drinking in the event and usual drinking levels resulted in 11 drinking groups since some combinations were logically not possible. This analytic approach found that at a given level of acute alcohol consumption, risky single-occasion drinkers were at a higher risk of injury than chronic high-volume drinkers or risk accumulators.

In summary, a strong association between acute alcohol consumption and injury has been consistently shown. However, the magnitude of risk from alcohol varies across ED samples and also across different study designs, with case-crossover analysis tending to have higher estimates than studies using case-control designs. Dose-response relationships were also found to vary across ED studies, which may have two possible explanations. First, the alcohol and injury relationship may be confounded and modified by demographic characteristics, risk-taking dispositions and societal drinking cultures, including usual drinking patterns and places of drinking, for example; all of which may vary across studies. Second, in relation to the causal mechanism of alcohol's effect on injury occurrence, it may be alcohol in combination with other behaviors or activities, rather than drinking itself, which increases risk of injury. Therefore, an accurate measure of the alcohol and injury relationship also depends on the joint distribution of drinking and particular activities, which are difficult to measure and quantify.

Risk of Alcohol-Related Injury from ED Studies

While acute alcohol consumption is believed to be associated with injury morbidity and mortality, the likelihood and risk of alcohol-related injury is related to one's usual drinking patterns. In meta-analysis of the ERCAAP data, Cherpitel et al. (2003b) reported that among injury patients, the risk of alcohol-related injury (defined by a positive BAC or any drinking within six hours prior to the event) for those reporting 5 or more drinks on an occasion at least monthly was about 4 times greater than for other drinkers. Alcohol-related injury was also found to be associated with alcohol-related problems.

To evaluate how risk of alcohol-related injury is associated with drinking pattern defined by risky drinking levels, we re-analyzed the combined ERCAAP/WHO data reported in Cherpitel (2005) by adopting the usual drinking pattern typology used by Gmel, et al. (2006). As described above, drinkers were grouped into four usual drinking categories: low-risk, chronic high-volume, risky single-occasion drinkers and risk accumulators. Rather than using 4+ drinks for women and 5+ for men during the last month to define heavy episodic drinking, 5+ at least once a month was used for both genders. The analyses were restricted to injury patients arriving at the ED within six hours following the event across 27 studies. Abstainers in the last 12 months were excluded from analysis as they are presumably free from risk of alcohol-related injury. As shown in Table 3, in most studies, compared to low-risk drinkers, chronic high-volume drinkers, risky single-occasion drinkers and risk accumulators all reported a higher risk of alcohol-related injury (based on reporting drinking prior to injury or a positive BAC). When the effects from all studies were pooled together, although risk accumulators had the highest risk of alcohol-related injury (OR 7.5 and 6.1 based on a positive BAC and self-reported drinking, respectively), both chronic high-volume drinkers (OR 3.9 and 3.5) and risky single-occasion drinkers (OR 2.9 and 2.7) also showed an elevated risk compared with low-risk drinkers. These findings suggest that risk of alcohol-related injury is not only related to heavy episodic drinking, but also to chronic high-volume drinking, and highest when both conditions are met.

Table 3. Odds Ratios (OR) and 95% CI for risk of alcohol-related injury, adjusting for gender and age

	Predicting injury with positive BAC (ref. low-risk drinkers)			Predicting injury with drinking any alcohol within 6 hours prior (ref. low-risk drinkers)		
	Chronic high-volume	Risky single-occasion	Risk accumulator	Chronic high-volume	Risky single-occasion	Risk accumulator
ERCAAP						
San Francisco, CA	6.2 (1.2, 30.9)	3.6 (1.6, 7.9)	10.0 (4.7, 21.3)	7.4 (1.2, 40.9)	4.0 (2.0, 8.1)	6.2 (3.1, 12.3)
Contra Costa, CA	7.7 (1.9, 31.7)	1.6 (.8, 3.3)	3.3 (1.7, 6.4)	3.8 (1.0, 14.7)	1.7 (1.0, 3.1)	2.6 (1.4, 4.6)
Martinez, CA	3.5 (.7, 17.0)	.7 (.3, 1.8)	3.3 (1.7, 6.6)	2.8 (.6, 12.9)	1.7 (.9, 3.5)	3.4 (1.8, 6.6)
Kaiser (Contra Costa) Co., CA	6.3 (.5, 79.7)	1.3 (.1, 12.4)	2.7 (.5, 15.2)	4.5 (.4, 53.8)	2.5 (.5, 11.3)	2.7 (.6, 11.6)
Jackson, MS	13.8 (.6, 331)	4.4 (1.2, 15.9)	5.1 (1.2, 21.2)	9.3 (.6, 139)	13.0 (3.4, 48.8)	26.6 (4.9, 143)
Santa Clara, CA	NA[1]	4.9 (1.1, 22.5)	8.6 (2.2, 34.0)	NA[1]	1.4 (.3, 6.2)	6.2 (2.0, 19.8)
Mexico City, Mexico	3.6 (1.4, 9.7)	2.6 (1.7, 4.0)	7.8 (4.6, 13.0)	4.7 (1.7, 12.9)	2.3 (1.5, 3.5)	5.6 (3.4, 9.2)
Pachuca, Mexico	1.9 (.2, 20.4)	5.6 (2.7, 11.4)	12.7 (3.3, 49.3)	1.0 (.1, 17.2)	3.6 (1.8, 7.4)	16.6 (4.3, 64.1)
Alberta, Canada	NA[1]	.9 (.3, 3.0)	4.5 (.9, 23.6)	NA[1]	3.0 (1.2, 7.7)	10.4 (3.0, 35.9)
Quebec, Canada	NA[1]	1.9 (.3, 11.9)	-∞	NA[1]	1.5 (.4, 5.6)	2.3 (.2, 28.1)
Fremantle, Australia	3.1 (.6, 15.5)	3.6 (1.7, 7.9)	9.5 (4.9, 18.3)	3.2 (1.1, 9.6)	2.2 (1.2, 3.9)	4.2 (2.6, 6.9)
Barcelona, Spain	3.3 (1.2, 9.1)	3.5 (1.6, 7.5)	14.9 (3.9, 57.2)	4.2 (1.7, 10.7)	3.8 (1.8, 8.0)	9.3 (2.4, 35.8)
Mar Del Plata, Argentina	8.8 (1.2, 66.7)	3.6 (1.0, 12.6)	7.4 (2.4, 23.5)	3.4 (.7, 15.3)	1.9 (.8, 4.7)	3.7 (1.5, 8.9)
Warsaw, Poland	NA[2]	NA[2]	NA[2]	NA[1]	3.4 (.6, 18.9)	11.6 (2.1, 64.8)
Sosnowiec, Poland	-∞	1.6 (.4, 6.4)	4.1 (1.2, 14.6)	1.2 (.1, 11.3)	3.6 (1.4, 9.1)	4.9 (1.9, 12.7)
WHO						
Mar Del Plata, Argentina	2.8 (.5, 14.8)	4.5 (2.1, 9.6)	8.0 (3.6, 17.6)	4.5 (1.2, 16.5)	4.8 (2.4, 9.6)	6.8 (3.3, 13.9)
Minsk, Belarus	.5 (.1, 4.3)	3.3 (1.9, 5.5)	11.5 (4.4, 30.3)	.5 (.1, 4.4)	2.2 (1.3, 3.7)	8.0 (3.3, 19.5)
São Paulo, Brazil	5.4 (1.4, 20.0)	1.8 (.8, 4.3)	5.7 (2.3, 14.2)	3.2 (.8, 12.8)	4.0 (1.9, 8.7)	5.6 (2.3, 13.6)
Ontario, Canada	NA[2]	NA[2]	NA[2]	4.3 (.6, 28.7)	3.5 (.5, 23.1)	26.7 (5.1, 140)
Hunan, China	16.6 (1.5, 180)	2.2 (.8, 6.5)	20.3 (7.4, 55.9)	∞	2.1 (.9, 4.7)	13.7 (5.4, 35.0)
Prague, Czech Republic	11.7 (2.5, 55.1)	19.2 (4.4, 84.5)	14.5 (2.9, 73.5)	4.9 (1.8, 13.2)	3.8 (1.3, 11.4)	8.3 (3.0, 22.8)
Bangalore, India	10.5 (2.2, 49.9)	8.3 (1.7, 41.1)	35.5 (8.0, 158)	13.8 (2.5, 75.5)	20.5 (2.9, 143)	47.1 (9.0, 146)
Mexico City, Mexico	3.5 (.2, 58.8)	2.3 (1.2, 4.6)	6.2 (2.0, 19.1)	2.8 (.2, 45.8)	2.6 (1.4, 4.8)	4.2 (1.5, 11.8)
Maputo, Mozambique	-∞	3.8 (1.5, 9.3)	16.2 (5.3, 52.1)	-∞	4.8 (1.9, 11.9)	12.1 (4.2, 35.4)
Auckland, New Zealand	1.2 (.1, 9.4)	1.7 (.6, 4.8)	4.8 (1.5, 15.6)	1.4 (.2, 9.1)	2.3 (.9, 5.5)	2.5 (.9, 7.1)
Cape Town, South Africa	.7 (.2, 2.5)	2.4 (.9, 6.6)	3.0 (1.4, 6.4)	1.3 (.3, 4.9)	2.6 (.9, 7.1)	4.0 (1.8, 8.9)
Malmö, Sweden	5.1 (1.2, 21.2)	5.1 (2.1, 12.5)	17.0 (6.1, 47.2)	3.4 (1.1, 10.3)	2.9 (1.3, 6.2)	10.2 (3.9, 26.6)
Test of homogeneity	Q=20.4 (df=19) p=.369	Q=36.6 (df=24) p=.048	Q=36.8 (df=23) p=.034	Q=12.8 (df=20) p=.887	Q=26.7 (df=26) p=.425	Q=42.3 (df=26) p=.023
Pooled effect	3.9 (2.7, 5.5)	2.9 (2.3, 3.6)	7.5 (5.8, 9.7)	3.5 (2.6, 4.8)	2.7 (2.4, 3.2)	6.1 (4.8, 7.7)

[1] No chronic high-volume drinkers were sampled among injury patients

[2] Reference category (low-risk drinkers) all had negative BAC

∞/-∞ refers to positive/negative infinite ORs when alcohol-related injury was perfectly predicted positively or negatively

Study-Level Risk Estimates and Variations

Given the large number of cross-national ED studies conducted around the world, all using a similar methodology, collaborations among researchers such as the Emergency Room Collaborative Alcohol Analysis Project and the WHO Collaborative Study on Alcohol and Injuries have provided unique opportunities to examine variations in estimates of injury risk associated with drinking and risk of alcohol-related injury across different studies. These collaborations have also provided the opportunity to examine how variations in findings can be explained by aggregate-level predictors such as organizational variables affecting the ED as well as regional/country socio-cultural characteristics. In a multi-level analysis of the ERCAAP data by Cherpitel et al. (2004), cross-ED variations in risk of injury associated with both a positive BAC and any drinking within six hours prior to injury were strongly related to detrimental drinking pattern score. The pattern score, developed for the comparative risk assessment project of the WHO's Global Burden of Disease study (Rehm et al., 2004) measures the detrimental impact of a country's drinking pattern, given the same level of per capita alcohol consumption, for a large number of countries, and includes such indicators as heavy drinking occasions, drinking with meals and drinking in public places (Rehm et al., 2001). The findings suggest that EDs in countries with higher detrimental drinking pattern scores are more likely to have a higher risk of injury related to drinking than in countries with lower detrimental pattern scores. In a separate study, rates of alcohol-related injury among drinkers were examined (Cherpitel et al., 2005) from the combined ERCAAP/WHO ED studies. ED studies with a higher average volume of alcohol consumption and studies in countries with higher detrimental pattern scores and higher legal intoxication levels for driving reported higher rates of alcohol-related injury.

In summary, injury risk associated with drinking and risk of alcohol-related injury are different but related concepts. Risk of injury is normally assessed from acute drinking in the event using case-control or case-crossover analysis. Once risk of injury related to drinking is established, drinking prior to the injury event or alcohol-related injury becomes a concept of interest that may be explained, in part, by variations in individuals' usual drinking patterns. Additionally, both concepts are necessary to determine the attributable fraction of injury to alcohol, which reflects the burden of injury in a given society that is attributable to alcohol. The rate of alcohol-related injury, and the risk of injury associated with drinking, are both positively related to the attributable fraction, and are both important in the prevention of morbidity and mortality related to injury.

References

Borges G et al. (1998). Alcohol consumption in emergency room patients and the general population: a population based study. *Alcoholism: Clinical and Experimental Research* **22**(9), 1986-1991.

Borges G, Cherpitel CJ, Mittleman M (2004). The risk of injury after alcohol consumption: a case-crossover study in the emergency room. *Social Science & Medicine* **58**(6), 1191-1200.

Borges G et al. (2006a). Acute alcohol use and the risk of non-fatal injury in sixteen countries. *Addiction* **101**(7), 993-1002.

Borges G et al. (2006b). A multicenter study of acute alcohol and nonfatal injuries: data from the WHO Collaborative Study on Alcohol and Injuries. *Bulletin of the World Health Organization* **84**(6), 453-460.

Borkenstein RF et al. (1964). *The Role of the Drinking Driver in Traffic Accidents*. Department of Police Administration, Indiana University, Bloomington, IN.

Cherpitel CJ (1993). Alcohol consumption among emergency room patients: Comparison of county/community hospitals and an HMO. *Journal of Studies on Alcohol* **54**, 432-440.

Cherpitel CJ et al. (2003a). A cross-national meta-analysis of alcohol and injury: data from the Emergency Room Collaborative Alcohol Analysis Project (ERCAAP). *Addiction* **98**, 1277-1286.

Cherpitel CJ et al. (2003b). Alcohol-related injury in the ER: a cross-national meta-analysis from the Emergency Room Collaborative Alcohol Analysis Project (ERCAAP). *Journal of Studies on Alcohol* **64**, 641-649.

Cherpitel CJ, Ye Y, Bond J (2004). Alcohol and injury: multi-level analysis from the Emergency Room Collaborative Alcohol Analysis Project (ERCAAP). *Alcohol and Alcoholism* **39**(6), 552-558.

Cherpitel CJ et al. (2005). Multi-level analysis of alcohol-related injury among emergency department patients: a cross-national study. *Addiction* **100**, 1840-1850.

Gmel G et al. (2006). Alcohol-attributable injuries in admissions to a Swiss Emergency Room – an analysis of the link between volume of drinking, drinking patterns, and preattendance drinking. *Alcoholism: Clinical & Experimental Research* **30**(3), 501-509.

Jewell NP (2004). Statistics for Epidemiology. Chapman and Hall, Boca Raton, FL.

Maclure M (1991). The case-crossover design: a method for studying transient effect on the risk of acute events. *American Journal of Epidemiology* **133**(2), 144-53.

Mcleod R et al. (1999). The relationship between alcohol consumption patterns and injury. *Addiction* **94**(*1*), 1719-1734.

Mokdad AH et al. (2004). Actual causes of death in the United States, 2000. *Journal of the American Medical Association* **291**(10), 1238-1245.

Rehm J et al. (2001). Steps towards constructing a global comparative risk analysis for alcohol consumption: determining indicators and empirical weights for patterns of drinking, deciding about theoretical minimum, and dealing with different consequences. *European Addiction Research* **7**(3), 138-147.

Rehm J et al. (2004). Alcohol use. In: *Comparative Quantification of Health Risks: Global and regional burden of disease attributable to selected major risk factors*. Vol. 1. Ezzati M et al., eds, pp. 959-1108. World Health Organization, Geneva, Switzerland.

Simpson EH (1951). The interpretation of interaction in contingency tables. *Journal of the Royal Statistical Society, Series B* **13**, 238-241.

Smith GS, Branas CC, Miller TR (1999). Fatal nontraffic injuries involving alcohol: a meta-analysis. *Annals of Emergency Medicine* **33**, 659-668.

Vinson DC et al. (1995). Alcohol and injury. A case-crossover study. *Archives of Family Medicine* **4**, 505-511.

Vinson DC et al. (2003). A population-based case-crossover and case-control study of alcohol and the risk of injury. *Journal of Studies on Alcohol* **64**(3), 358-366.

Watt K et al. (2004). Risk of injury from acute alcohol consumption and the influence of confounders. *Addiction* **99**, 1262-1273.

CHAPTER 2 :
VARIATION IN ALCOHOL-RELATED INJURY BY TYPE AND CAUSE OF INJURY

Guilherme Borges - National Institute of Psychiatry, Metropolitan Autonomous University | Mexico City, MEXICO

Scott Macdonald - Centre for Addictions Research of BC | Victoria, BC CANADA

Cheryl J. Cherpitel - Alcohol Research Group | Emeryville, CA USA

Ricardo Orozco - National Institute of Psychiatry | Mexico City, MEXICO

Margie Peden - Injuries and Violence Prevention | World Health Organization, Geneva, SWITZERLAND

Alcohol and Mode of Injury

Other chapters in this volume have documented the importance of injury in the epidemiological context of many countries in the world, and have demonstrated the importance of alcohol consumption as a risk factor for all injuries combined and opportunities for prevention in relation to alcohol policy development. However, the role of alcohol may vary considerably, depending on the cause or mode of injury (eg., intentional, unintentional, traffic crashes, falls, etc.). The current chapter examines how risk of injury for different modes is related to acute alcohol consumption.

It is worthwhile to briefly address some of the mechanisms by which we expect alcohol to be directly related to different modes of injury. Alcohol creates profound decrements in psychomotor capabilities and may increase the risk of simple falls (Honkanen et al., 1983). By the same mechanism, we would expect that these decrements would increase the likelihood of injuries related to traffic fatalities (Ogden et al., 2004) and other injuries related to actions and activities that require fine motor performance, such as boat driving (Driscoll et al., 2004), skiing (Cherpitel et al., 1998), bicycle (Olkkonen et al., 1990), as well as job related activities, even though the latter is a matter of debate (Webb et al., 1994). Other explanations of increased risk for specific modes of injury have also been postulated. For example, disinhibition, emotional effects and power issues have been suggested as key concepts contributing to alcohol related violence (Hoaken and Stewart, 2003). The strong relationship between smoking and alcohol consumption likely contributes to findings of studies that indicate alcohol impairment is related to fire fatalities (Hingson and Howland, 1993). Suicide and suicidal behaviour (i.e. suicide ideation, plan and attempt) is a complex topic in which alcohol may play a role as an acute (Cherpitel et al., 2004), as well as a chronic, exposure (Skog, 1991).

In fact, there is no reason why we should expect alcohol to be related in the same way with the same mechanism, and with a similar strength for all types of injuries. It is not possible at present, however, to develop a ranking of alcohol's association with injury across even major causes of injury. A closer look at the literature shows that the body of scientific information on the causal role of alcohol varies depending on the mode of injury. For example, studies are conclusive that alcohol is a major cause of traffic crashes and fatalities (Connor et al., 2004), and the preponderance of evidence suggests that alcohol is a risk factor for other types of injuries, such as those resulting from violence (Macdonald et al., 2005) and falls (Hingson and Howland, 1987), even though the exact mechanisms for such associations are not completely understood. When judging the differential impact that alcohol may have, say, on falls compared to homicides, a major difficulty is that usually there are a number of important differences across studies, including the research methodology used, the population studied, measurement of exposure (to alcohol), secondary (confounding) variables associated with both alcohol use and injury, and the time frame for alcohol use in relation to injury. For example, how can we compare the risk associated with self-reported alcohol consumption prior to a fall, measured in a case-control study of primarily older male pedestrians crossing a city street, with risks associated with alcohol's presence in urine samples of burn victims among cases of primarily younger female patients admitted to a trauma unit in a large city hospital? A limitation of most prior research in this area is that sample sizes are usually small, clinical populations are subjected to several sources of selection bias (see Chapter 8), and studies are conducted in limited jurisdictions, making comparisons and generalizations of findings inconclusive.

Studies conducted in representative samples of patients attending the emergency department (ED) (Cherpitel, 1989) have usually collected data across a variety of injury causes, and results from these studies on alcohol consumption and mode of injury are more comparable. Indeed, some of these ED studies have produced comparative estimates of risk for violent and non-violent injuries (Cherpitel, 1993) as well as for other causes of injury (Borges et al., 1994). But even in these larger, and for the most part, comparable ED studies, sample sizes for specific modes of injury diminish rapidly in analysis of alcohol and injury. Another pervasive limitation when comparing estimates of the relative risk (RR) of injury due to alcohol across modes of injury is the presence of possible sources of confounding. For example, studies on suicide attempt should control for chronic depression, while studies on traffic accidents should control for the driver's experience behind the wheel. Rarely have ED studies attempted to collect data on such a large number of diverse confounding variables, and the typical analyses carried out in these studies do not allow for control of a wide assortment of possible sources of confounding within various modes of injury, making it difficult to compare RR estimates across injuries from various causes.

Case-crossover Studies on Alcohol and Mode of Injury

Estimates of the relative risk (RR) of the relationship between acute alcohol use and non-fatal injuries in the emergency department setting, that take into account within-person, stable sources of confounding, are possible using a case-crossover analytic design (Maclure, 1991). Prior studies on alcohol as a risk factor for non-fatal injury in ED settings have used the case-crossover design (Vinson et al., 1995; Borges et al., 2004b) and have, to some extent, presented estimates of the relationship between alcohol and injury type.

ED studies comprising the Emergency Room Collaborative Alcohol Analysis Project (ERCAAP) (Cherpitel et al., 2003a) and the WHO Collaborative Study on Alcohol and Injuries (Cherpitel et al., 2005) project, all using a similar design and study methodology, have recently provided a sufficiently large data set for examining the relative risk of non-fatal injury associated with alcohol consumption by mode of injury. The ERCAAP data include 16 ED studies (1984-2002) across seven countries

(Argentina, Australia, Canada, Mexico, Poland, Spain and the U.S.), with the number of ED's per study ranging from one in most countries to eight in the Mexico City study. The WHO Collaborative Study on Alcohol and Injuries include 12 ED studies (2001-2002) in 12 countries (Argentina, Belarus, Brazil, Canada, China, Czech Republic, India, Mexico, Mozambique, South Africa, Sweden, New Zealand). The WHO study collected data only from those patients arriving at the ED within six hours of the injury event. The ERCAAP data analyzed here were selected to include only those meeting this six-hour arrival criterion. Excluding the New Zealand study that did not include questions defining mode of injury, the combined ERCAAP and WHO datasets provide a final sample of 11 383 injured patients in 27 studies.

In these studies, mode of injury was assessed with a single question. This variable was categorized as unintentional injury (non-violence related injury) or intentional by someone else (violence-related injury). Injuries that were unintentional were further classified according to the following categories: a) traffic accident; b) blunt force injury, struck against or caught between; c) stab, cut, bite or acciden-tal gunshot; d) fall, trip e) a mixed group of other injuries that included choking, hanging, drowning, poisoning, burn with fire, hot liquid, other and not specified.

The ERCAAP and WHO studies elicited data on the quantity and frequency of usual drinking that allow case-crossover analyses using the "usual frequency method" (Maclure, 1991), which was modified to take into account the amount of alcohol consumed during the six hours prior to injury, with each drink consumed assuming to have an effect period of one hour (Tables 1-2). This approach compares, for each patient, their reported use of alcohol during the six-hour period prior to the injury with their usual frequency and quantity of alcohol during the last 12 months. The "pair match-ing strategy" (Maclure, 1991) was also used with data from the WHO project (Table 3) and compares, for each patient, their reported use of alcohol during the six-hour period prior to the injury with their use of alcohol during the same time period the same day the previous week. (The ERCAAP studies did not collect this latter data).

Intentional and unintentional injury cases

In Table 1 the RR estimates are reported for alcohol in violence-related and other injuries for each study site. For the total sample, the total (crude) estimate of RR was 21.50 for a violence-related injury compared with 3.37 for a non-violence-related injury. In all sites violence- related cases had increased RRs, which ranged from 4.84 in Warsaw to a high of 903.78 in Alberta (where 82% of the violence-related injury cases reported drinking prior to the event). The test of homogeneity suggested that the RR estimates were not homogenous across studies. Based on this a random estimate for pooled RR was calculated (RR=20.25). In contrast, for five sites among those with non-violence-related injuries, either non-elevated (WHO-Canada and the Czech Republic) or non-significant RR estimates (Kaiser, Warsaw, Sosnowiec) were observed. RR estimates across study sites for a non-violence-related injury also showed heterogeneity, with a pooled random RR estimate of 3.67. In all circumstances (crude, fixed and random estimates), the homogeneity test for the comparison of the pooled estimates for violence and non-violence cases (the p values in the middle column) showed differences. Among study sites, only in Mar del Plata, China, India and Mozambique were no differences found between the RR estimates for cases of violence-related and non-violence-related injuries.

Table 1.   WHO-ERCAAP Relative Risk and 95% Confidence Intervals for alcohol use six hours prior and the risk of injury, by study level and selected characteristics

Site	Violence-related			Non Violence-related	
	RR	95% CI	p value †	RR	95% CI
1 San Francisco, CA	10.68	(6.57 - 17.36)	0.003	4.24	(2.98 - 6.03)
2 Contra Costa, CA	16.06	(8.51 - 30.31)	<0.001	2.35	(1.84 - 3.01)
3 Martinez, CA	8.98	(5.84 - 13.80)	<0.001	2.62	(1.94 - 3.53)
4 Kaiser (Contra Costa, Co. CA)	21.02	(2.66 - 166.24)	0.013	1.43	(0.85 - 2.40)
5 Jackson MS	51.65	(15.34 - 173.88)	0.007	7.21	(3.32 - 15.68)
6 Santa Clara, CA	28.55	(4.11 - 198.51)	0.006	1.75	(1.03 - 2.95)
7 Mexico City, Mexico	56.77	(39.99 - 80.58)	<0.001	9.16	(7.36 - 11.40)
8 Acapulco, Mexico	135.01	(55.61 - 327.75)	<0.001	12.57	(7.91 - 19.97)
9 Pachuca, Mexico	58.96	(30.63 - 113.47)	<0.001	11.55	(7.81 - 17.08)
10 Alberta, Canada	903.78	(9.17 - 89092.27)	0.028	5.12	(3.39 - 7.72)
11 Quebec, Canada ‡				2.80	(1.58 - 4.97)
12 Fremantle, Australia	15.42	(8.64 - 27.51)	<0.001	2.19	(1.73 - 2.78)
13 Barcelona, Spain	7.18	(3.61 - 14.27)	0.001	2.05	(1.71 - 2.47)
14 Mar del Plata, Argentina	5.32	(3.17 - 8.93)	0.353	3.92	(2.67 - 5.76)
15 Warsaw, Poland	4.84	(1.56 - 14.99)	0.022	1.09	(0.61 - 1.96)
16 Sosnowiec, Poland	8.38	(5.87 - 11.98)	<0.001	1.13	(0.76 - 1.68)
17 WHO – Argentina	21.29	(10.56 - 42.89)	<0.001	3.53	(2.70 - 4.61)
18 WHO – Belarus	56.76	(25.20 - 127.85)	<0.001	9.96	(7.74 - 12.8)
19 WHO – Brazil	5.36	(3.20 - 8.98)	0.001	2.00	(1.47 - 2.73)
20 WHO – Canada ‡				0.95	(0.52 - 1.76)
21 WHO – China	51.34	(14.86 - 177.38)	0.139	18.58	(10.95 - 31.51)
22 WHO – Czech	7.44	(1.34 - 41.14)	0.023	0.98	(0.68 - 1.42)
23 WHO – India	36.61	(15.46 - 86.72)	0.757	30.37	(13.50 - 68.30)
24 WHO – Mexico	27.23	(17.49 - 42.38)	<0.001	8.56	(6.08 - 12.06)
25 WHO – Mozambique	9.26	(5.79 - 14.81)	0.065	5.22	(3.55 - 7.67)
26 WHO – South Africa	45.99	(29.18 - 72.49)	0.030	21.09	(12.34 - 36.04)
27 WHO – Sweden	44.88	(13.03 - 154.57)	<0.001	3.04	(2.27 - 4.09)

	Violence-related			Non Violence-related	
Test of Homogeneity across studies	X^2=210.32 (df=24), p < 0.001			X^2=553.10 (df=26), p < 0.001	

		Violence-related			Non Violence-related	
Crude		21.50	(19.04 - 24.27)	<0.001	3.37	(3.17 - 3.58)
Pooled effect Size	Fixed effect	17.63	(15.61 - 19.92)	<0.001	3.67	(3.44 - 3.92)
	Random effect	20.25	(13.69 - 29.96)	<0.001	3.97	(2.92 - 5.41)

† By the chi-square test of homogeneity. Contrasts are "Violence vs. Non violence-related injury" at each study site.

‡ Could not calculate RR because unexposed cases (did not drink six hours prior) reported no habitual alcohol use in the last 12 months

RR=relative risk; CI= Confidence interval

Types of unintentional injuries

Table 2 shows estimates by mode of injury, separating unintentional injuries into sub-categories, and by number of injuries sustained. Traffic accidents had the highest RR among unintentional injuries, followed by a fall/trip accident. These estimates were not homogenous across injury mode. Also, among the five types of unintentional injuries, considerable heterogeneity across RR estimates was found (chi-square= 73.735 with 4 degrees of freedom, p< 0.001). Those sustaining two or more injuries were found to have higher RR estimates than those attending the ED with only one injury (6.53 vs. 4.94).

Table 2. Relative Risk and 95% Confidence Intervals for alcohol use six hours prior and the risk of injury, by mode of injury and number of injuries (ERCAAP-WHO usual frequency*)

	Relative Risk	95% Confidence Interval
Mode of injury		
Intentional by someone else (n=1,943)	21.50	(19.04 - 24.27)
Traffic accident (n=1,851)	5.24	(4.62 - 5.95)
Blunt force inj/ Struck against/ Caught between (n=1,064)	2.60	(2.17 - 3.11)
Stab, cut, bite/ Gunshot (n=1,462)	3.10	(2.68 - 3.58)
Fall, trip (n=2,981)	3.39	(3.02 - 3.80)
Other/ Choking, hanging/ Drowning/ Poisoning/ Burn with fire, hot liquid/ Taking drugs, other substances/ Something in lungs/ Don't know (n=1,802)	2.30	(1.96 - 2.70)
TOTAL	**5.15**	**(4.89 - 5.42)**

Test of Homogeneity of the relative risk across mode of injury	X^2=782.50 (df=5) p<0.001	
Pooled effect size — Fixed effect	5.00	(4.72 - 5.28)
Pooled effect size — Random effect	4.39	(2.17 - 8.86)

Severity of injury*		
One injury (n=9747)	4.94	(4.67 - 5.23)
Two or more (n=1750)	6.53	(5.81 - 7.34)
TOTAL	**5.18**	**(4.92 - 5.44)**

Test of Homogeneity of the relative risk across severity	Q=17.66 (df=1) p<0.001	
Pooled effect size — Fixed effect	5.20	(4.95 - 5.47)
Pooled effect size — Random effect	5.65	(4.30 - 7.42)

* Data weighted

Dose response estimates

In Table 3 odds ratio (OR) estimates for the WHO data only are shown based on the matched-pair analyses. For any alcohol use prior to injury, the total OR estimates for violent-related injuries (OR=20.71) and non-violent-related injuries (OR=4.07) are roughly similar to the ones based on the usual frequency analyses in Table 1. Both violence and non-violence cases showed increasing ORs with increasing amount of drinking (a dose-response relationship), as measured by the number of drinks consumed during the six hours prior to injury. Elevated risks even at low levels of consumption (1-2 drinks) are evident for both intentional and unintentional injury. Interestingly, at low levels of consumption there are no differences in risk across injury mode (by the homogeneity test), but at middle and high levels of consumption those with a violence-related injury had much higher risks at similar levels of consumption than cases of non-violence-related injuries.

Table 3. WHO Alcohol use six hours prior and one week prior to the injury.
Matched Pair analyzes by violence-related injury and number of drinks (n=4,167).

	Violence-related (n=570)		Non Violence-related (n=3,425)		
	OR	CI (95%)	OR	CI (95%)	
Any alcohol	20.71	(9.41 - 45.58)	4.07	(3.14 - 5.26)	Hom $X^2(1)$=14.77; p<0.001
0	1.00	-	1.00	-	
1-2	11.14	(3.96 - 31.36)	3.86	(2.53 - 5.89)	Hom $X^2(1)$=3.45; p=0.063
3-4	13.76	(5.22 - 36.31)	2.19	(1.44 - 3.33)	Hom $X^2(1)$=11.62; p<0.001
5+	35.57	(13.97 - 90.59)	6.40	(4.31 - 9.50)	Hom $X^2(1)$=10.98; p=0.001
dose response trend stat.	$X^2(3)$=150.08 p<0.001		$X^2(3)$=160.37 p<0.001		

Discussion

The use of a case-crossover methodology allowed us to make comparable multiple RR estimates across a large number of injuries, assuring that basic demographic characteristics such as age and gender were controlled in these comparisons. This large sample size is unprecedented in the literature and allowed us to obtain stable estimates of the association of alcohol and injury by mode for the pooled sample. Large differences were found for acute alcohol use between intentional and unintentional injuries, with larger risks for intentional injuries. This finding was present in 24 out of the 27 sites included here. Within study sites considerable heterogeneity of effects was also found, with a large range of RR estimates across sites for both intentional and unintentional injuries. Increases in risks were also apparent among unintentional injuries, with traffic accidents showing the largest RR within this category. Risk was also greater for those with more than one injury. Strong evidence was also found for a dose-response relationship between the amount of alcohol consumed within six hours prior to injury and increase in the OR for both types of injury; however, at larger amounts of drinking, the OR increased more sharply for intentional than for non-intentional injury. The relationship between alcohol and different modes of injury was replicated in practically all the sites and in different cultural contexts. Although the strength of these relationships was heterogeneous, based on statistical tests, this is not surprising considering the large sample size, number of sites and different cultures where data were collected. Overall, the direction and significance of relationships are most important. The significant dose-response relationship found between increased alcohol consumption and likelihood of an injury (Borges et al., 2006a) is consistent with a causal interpretation of the data. Similar dose-response relationships that have been observed from case-control studies of alcohol and traffic crashes have provided convincing support that alcohol impairment is a cause of crashes (Connor et al., 2004). Lastly, while severity of injury data were not available from these datasets, the number of injuries sustained could be taken as a marker for injury severity, suggesting that risk of injury related to drinking is elevated for those with more serious injury.

The stronger relationship found between alcohol and violence compared to other modes of injury is suggestive that additional mechanisms than simply reduced psychomotor capabilities might explain the elevated RR found. One possible causal explanation is that alcohol reduces inhibitions and increases the sense of power for some individuals (Hoaken and Stewart, 2003). Drinking may also lead to more aggressive behaviour on the part of some individuals, resulting in elevated risk for violence-related injury. Additionally, people tend to drink in social situations, some of which may be more likely to spawn a violent-related event than others. For example, it is more likely that drinking in a bar or other public place may lead to such an event than drinking at home with family members. More research is needed to better understand competing hypotheses for the relationship between acute alcohol consumption and violence.

Variations in alcohol involvement across other modes of injury were found as well, with traffic crashes having higher RRs than other non-intentional injuries. Again, the situation in which people drink is an important consideration in the likelihood of injury occurrence. Driving a vehicle requires certain skills and can be a dangerous activity, which might explain the elevated risk found for drinking and traffic crashes.

These findings showed important differences in RR estimates across sites, and considerable heterogeneity of effects. Much of our current knowledge on social factors related to differences in the effects of alcohol and injury across cultures is based on ecological studies that examined changes in per-capita alcohol consumption and mortality. Such studies across several countries in northern, southern and central Europe (Skog, 2001a; Skog, 2001b) found that differences in the effects of alcohol on mortality varied by accident and injury type, depending on drinking culture in relation to alcohol consumption levels, drinking patterns and social norms. Findings here of the variation in RR estimates associated with drinking by injury mode across societies may be related to such factors as risky consumption patterns (Borges et al., 2006a; Room et al., 2005) and this should be further explored.

Study limitations

This study is limited to an analysis of data from patients with non-fatal injuries who attended specific emergency departments. Although the study design provides a representative sample of patients from each facility, patients may not be representative of other facilities in the city or the country. Additionally, as is common with other studies conducted in emergency departments, cases cannot be assumed to be representative of other individuals who were injured but did not seek medical attention. All analyses reported here are based on the patient's reported alcohol consumption across different times, and it is possible that participants were more likely to recall their consumption more accurately immediately before an injury than during a previous period, including the last year. Differential recall may lead to an overestimate of the association between alcohol and injury if patients are more likely to remember and report alcohol use in the short term. Prior case-crossover research on alcohol consumption and injury has used other control periods (Vinson et al., 1995; Vinson et al., 2003; Borges et al., 2004a; Borges et al., 2004b) ranging from a day to a year, and findings of relative risk have been found to vary. Legal or other issues, however, may encourage patients to minimize their reports of drinking prior to an injury, as in the case of drivers in motor vehicle accidents. On the other hand, it is also possible that patients may overestimate their drinking; for example, those with violence-related injuries may over-report alcohol consumption to excuse behavior that would otherwise be viewed as socially unacceptable (MacAndrew and Edgerton, 1969). Clearly, more research on the validity of methods for eliciting alcohol use in case-crossover analyses is needed.

Despite the fact that case-crossover studies are well suited to control for between-person confounders, they do not remove the possibility that within-person confounders exist. For example, in our study it is possible that a patient may have been suffering from a transient depressive episode that gave rise to an increase in alcohol consumption. This co-occurrence of depression and alcohol use, or any other psychiatric disorder or substance use, could confound RR estimates, especially for suicide attempt, for example. Because we lack measures of other acute variables that vary over time and that could be considered possible confounders of the relationship between acute alcohol use and injury, we cannot quantify this bias or adjust our results accordingly.

Finally, when examining the association of drinking with specific types of injuries, it is important to control the activity in which the individual was engaged at the time of injury, and which may have contributed to injury occurrence. For example, for drivers injured in traffic-related accidents, it is important to know whether they were also driving in the control period. These data were not available for analysis here, and this also is an important topic for future research on the risk at which alcohol places the individual for injury. It is likely that none of these limitations, however, fully explain the elevated relative risk estimates reported in these analyses.

Conclusions

Within the scope of these limitations, this report on a large and representative sample of patients attending emergency departments suggests that even one drink may increase the risk of injury. A dose-response relationship for the number of drinks and risk for both a violence and non-violence-related injury was evident, with increasing risks even at low levels (1-2 drinks) of alcohol use. Higher levels of drinking were associated with much higher elevation in the OR for violence-related injuries than for non-violent injuries. If subjects decided to drink, increasing amounts may have pronounced consequences in their risk of triggering an injury, specially a violence-related injury. Future research should include an examination of a dose-response relationship for other injuries by cause or mode.

References

Borges G et al. (1994). Casualties in Acapulco: results of a study on alcohol use and emergency room care. *Drug and Alcohol Dependence.* **36**, 1-7.

Borges G, Cherpitel C, Mittleman M (2004a). Risk of injury after alcohol consumption: a case-crossover study in the emergency department. *Social Science and Medicine* **58**, 1191-1200.

Borges G et al. (2004b). Episodic alcohol use and risk of nonfatal injury. *American Journal of Epidemiology* **159**, 565-571.

Borges G et al. (2006a). Acute alcohol use and the risk of non-fatal injury in sixteen countries. *Addiction.* **101**, 993-1002

Borges G et al. (2006b). Multicentre study of acute alcohol use and non-fatal injuries: data from de WHO collaborative study on alcohol and injuries. *Bulletin of the World Health Organization.* **84**, 453-60.

Cherpitel C et al. (2005). Clinical assessment compared with breathalyser readings in the emergency room: concordance of ICD-10 Y90 and Y91 codes. *Emergency Medicine Journal.* **22**,689-95.

Cherpitel CJ, Borges GL, Wilcox HC (2004). Acute alcohol use and suicidal behaviour: a review of the literature. *Alcoholism Clinical and Experimental Research.* **28**, 18S-28S.

Cherpitel CJ et al. (2003). A cross-national meta-analysis of alcohol and injury: data from the Emergency Room Collaborative Alcohol Analysis Project (ERCAAP). *Addiction* **98**, 1277-1286.

Cherpitel CJ, Meyers AR, Perrine MW (1998). Alcohol consumption, sensation seeking and ski injury: a case-control study. *Journal of Studies on Alcohol.* **59**, 216-21.

Cherpitel CJ (1993). Alcohol and violence-related injuries: an emergency room study. *Addiction.* **88**, 79-88.

Cherpitel CJ (1989). Study of alcohol use and injuries among emergency room patients. In: *Drinking and casualties: Accidents, poisonings and violence in an international perspective.* Giesbrecht N et al., eds, pp. 288-299. Routledge, New York.

Cherpitel CJ et al. (2006). Multi-level analysis of causal attribution of injury to alcohol and modifying effects: Data from two international emergency room projects. *Drug and Alcohol Dependence* **82**, 258-268.

Connor J et al. (2004). The contribution of alcohol to serious car crash injuries. *Epidemiology* **15**, 337-344.

Driscoll TR, Harrison JA, Steenkamp M (2004). Review of the role of alcohol in drowning associated with recreational aquatic activity. *Injury Prevention.* **10**,107-13.

Hingson R, Howland J (1987). Alcohol as a risk factor for injury or death resulting from accidental falls: a review of the literature. *Journal of Studies on Alcohol* **48**, 212-219.

Hingson R, Howland J (1993). Alcohol and non-traffic unintended injuries. *Addiction* **88**, 877-883.

Hoaken PN, Stewart SH (2003). Drugs of abuse and the elicitation of human aggressive behavior. *Addictive behaviors* **28**, 1533-1554.

Honkanen R et al. (1983). The role of alcohol in accidental falls. *Journal of Studies on Alcohol.* **44**, 231-45.

MacAndrew C, Edgerton RB (1969). *Drunken comportment: A social explanation.* Aldine Publishing Co., Chicago.

Macdonald S et al. (2005). The criteria for causation of alcohol in violent injuries based on emergency room data from six countries. *Addictive Behaviors* **30**, 103-113.

Maclure M (1991). The case-crossover design: a method for studying transient effects on the risk of acute events. *American Journal of Epidemiology* **133**, 144-153

Ogden EJ, Moskowitz H (2004). Effects of alcohol and other drugs on driver performance. *Traffic Injury Prevention.* **5**:185-98.

Olkkonen S, Honkanen R (1990). The role of alcohol in nonfatal bicycle injuries. *Accident; analysis and prevention.* **22**, 89-96.

Room R, Babor T, Rehm J (2005). Alcohol and public health. *Lancet* **365**, 519-530.

Skog OJ. Alcohol And Suicide – Durkheim Revisited (1991) *Acta Sociologica.* **34**, 193-206.

Skog OJ (2001a). Alcohol consumption and mortality rates from traffic accidents, accidental falls, and other accidents in 14 European countries. *Addiction* **96**, S49-S58.

Skog OJ (2001b). Alcohol consumption and overall accident mortality in 14 European countries. *Addiction* **96**, S35-S47.

Vinson DC et al. (1995). Alcohol and injury. A case-crossover study. *Archives of Family Medicine* **4**, 505-511.

Vinson DC et al. (2003). A population-based case-crossover and case-control study of alcohol and the risk of injury. *Journal of Studies on Alcohol* **64**, 358-366.

Webb GR et al. (1994). The relationships between high-risk and problem drinking and the occurrence of work injuries and related absences. *Journal of Studies on Alcohol.* **55**, 434-46.

CHAPTER 3 :
CAUSALITY AND CAUSAL ATTRIBUTION
OF ALCOHOL IN INJURIES

Jason Bond - Alcohol Research Group | Emeryville, CA USA
Scott Macdonald - Centre for Addictions Research of BC | Victoria, BC CANADA

Introduction

Emergency Department (ED) studies, in which data are collected from injured patients on factors related to their injury and alcohol use, provide a unique opportunity to examine the causal role of alcohol in injuries. Acute alcohol involvement among those injured can be compared to a control group to determine the relative risks for injury associated with alcohol. Although experimental studies clearly indicate that psychomotor performance declines considerably with ingestion of alcohol, ED studies more accurately reveal real world conditions. These studies allow us to better determine whether people are equally likely to use alcohol in a variety of activities and to what degree alcohol ingestion increases likelihood of injury for these activities. As well, there may be other personal factors than decrements of psychomotor capabilities that may act to increase the likelihood of certain types of injuries. For example, alcohol can reduce inhibitions or increase sense of power for some, which could contribute to the likelihood of violence (Hoaken and Stewart, 2003).

In social science research, three conditions must exist to show causality: (1) the suspected cause (i.e., acute alcohol consumption) must precede the effect (i.e., injury) in time, (2) a statistical association must be found between the two variables, and (3) the relationship cannot be better explained by a third variable (MacMahon and Pugh, 1970). ED studies are ideal for assessing the first requirement of causality and the second requirement of a statistical association (assuming an unbiased control group). The third requirement is the most difficult to assess. The list of possible variables that might better explain the relationships seen can be large, and multi-causal phenomena often best explain relationships, making it problematic to sort out different causal pathways.

There are a number of methodological obstacles when assessing a determination of the precise role of alcohol in various types of injuries. A major challenge for ED studies is the selection of an appropriate control group. Ideally, the control group would be composed of those without injuries who are as similar as possible to the injury group in terms of age, sex and other demographic characteristics that might be confounders. As well, each control subject would ideally be sampled from the same type of setting, time of day, and day of week as the cases. With the exception of Honkanen et al. (1983) who investigated falls, to date most injury studies that meet these criteria are ones conducted on alcohol involvement in traffic crashes (see Compton et al., 2002, for a recent example), which have some unique advantages over other types of injuries – crashes occur in public places and there are laws prohibiting driving while impaired by alcohol, which facilitates the collection of breath alcohol samples from control drivers. Other types of injuries occur in a variety of settings, such as private residences or drinking establishments, where response rates may be lower (as there may be fewer reasons that subjects would wish to participate), resulting in biases in the selection of both case and control subjects and risk estimates.

In ED studies, several sources of control groups have been used. These include: (1) external controls from the general population (McLeod et al., 2003) (2) medical patients (Cherpitel, 1994), (3) patients with different types of injuries (Macdonald et al., 2005; Watt et al., 2005), (4) injured patients as their own controls in case-crossover designs (Borges et al., 2004a) and (5) injured patients who attributed a causal association of their injury to alcohol use. This fifth approach is the focus of this paper, and refers to one where the patient believes the injury event would not have happened if he or she had not been drinking at the time. Each of these approaches can introduce conservative or liberal bias (i.e., under or overestimate risk).

Large data sets with multiple sites allow for a better assessment of causality than smaller datasets from a single location, and permit an assessment of whether relationships found are consistent across regions and countries. Such consistency increases confidence that the observed relationships are rooted in real processes and helps us assess factors that best explain variation in findings (Macdonald et al., 2005).

Subjective Causal Relationship Between Alcohol and Injury

In considering the role of alcohol in injuries, the actual mechanism linking the two can be complex and is thought to be associated with many factors in combination (one of which is alcohol), with time also playing an important role (Romelsjö, 1995). Although alcohol consumption, along with other factors, may contribute to the injury, much can be learned regarding the alcohol-injury link from patients' own attribution of their drinking to injury, and is potentially important in tailoring effective intervention and prevention strategies. Prior studies of subjective assessments in emergency department populations have found a substantial proportion of injured patients who reported feeling drunk at the time of the event themselves attributed a causal association of their drinking with the injury (Cherpitel, 1996). Although rates of attribution of the injury to alcohol are high among those reporting they were feeling drunk (63%), alcohol may be implicated in injuries even for those who did not feel drunk at the time. For example, one study found that 34% of those drinking in the six hours prior to injury, but did not feel drunk, also attributed a similar causal association (Stephens, 1987). Causal attribution of injury to drinking in ED studies has been found to vary, based on whether the injury was related to their own or someone else's drinking (Macdonald et al., 1999) and region in the U.S. where the data were collected (Cherpitel, 1997), as well as across other cultures and countries (Cherpitel, 1993; Cherpitel, 1999). Other factors, such as usual drinking patterns or frequency of drunkenness may also play an important role in subjective causal association (MacAndrew and Edgerton, 1969).

Two analyses of the subjective assessment of the causal role of alcohol in injuries are summarized here. The first is a study of the variation in rates of causal attribution using the combined Emergency Room Collaborative Alcohol Analysis Project (ERCAAP) and the World Health

Organization (WHO) Collaborative Study on Alcohol and Injury. The second considers the role of causal attribution in the formation of aggregate epidemiologic estimates of the proportion of injuries that are attributable to alcohol.

Causal attribution of injury to the use of alcohol in the ERCAAP and WHO studies

In order to explore the role of subjective causal assessments, results from a prior multilevel analysis (Cherpitel et al., 2006) of data from 35 ED sites in 24 ED studies across 15 countries comprising the combined ERCAAP and WHO Collaborative Study are summarized here. Individual-level, or patient-level, variables included in the analysis were age, gender, *volume* consumed in the six hours prior to the event, whether or not the respondent reported *feeling drunk* at the time of the injury, and whether the injury resulted from violence or was traffic-related. Study-level variables, which typically were measured as characteristics of the EDs themselves, were used to explain the between-study variation of the relationship between the amount of alcohol consumed in the six hours prior to injury (log volume), as well as whether or not the patient reported feeling drunk, and the likelihood of attributing a causal association of the injury to drinking. Study-level variables included both (1) aggregations of individual-level variables (log volume of consumption six-hours prior to the event) within a study and, (2) socio-cultural contextual variables pertaining to a region or country, including detrimental drinking pattern (DDP) (Rehm et al., 2001). The latter is a variable scored with a range from 1 to 4, with a higher score indicating a higher postulated detrimental effect of the same per capita consumption of alcohol (Rehm et al., 2003a; Rehm et al., 2003b). In addition, possible modifying influences of the relationship between volume and causal attribution and between feeling drunk and casual attribution were also considered. These include individuals' usual drinking patterns as well as characteristics of the injury type (whether the injury resulted from a violence-related or traffic-related event).

In order to explore whether the relationships varied across EDs, a multilevel model was estimated (Raudenbush et al., 2004). Such a model predicts the relationship between individual-level variables while allowing this relationship to vary as a function of study-level variables. Random effects were included for the intercept (in order to examine whether the overall rate of causal attribution varied across studies, controlling for volume consumed in-the-event) as well as the slope (to examine whether the relationship between volume consumed and causal attribution, or the relationship between volume and feeling drunk, varied across studies) (Raudenbush and Bryk, 2002). Self-reported volume was included as a logged variable to help reduce skewness and stabilize variance of the outcome variable. Separate multilevel logistic regression models were estimated for self-reported volume and whether the respondent felt drunk at the time of injury. In each model, gender and age were used as individual-level control variables. In the first model only volume was entered and included as a random effect. In the second model, both volume and feeling drunk were entered, but here only feeling drunk was included as a random effect. Table 1 shows results for these two models.

Neither gender nor age was associated with the likelihood of causal attribution in either model, so estimates for these coefficients are not shown. However, both self-reported volume and whether or not the patient felt drunk (controlling for volume) were significant predictors of attribution of injury to alcohol (p<.001 for both). Significant variation was found for the rate of causal attribution across studies (p<.05) in both models, as seen in the variance estimate for the intercept; however, variation was not found for the relationship of volume or feeling drunk with causal attribution (p<.10 for both).

Table 1. Parameter estimates and standard errors (se) for multilevel generalized linear models predicting causal attribution

Predictor	Model 1	Model 2
	Fixed effect: β(se)	
Log volume	.528(.103)***	.254(.074)***
Feeling drunk	N.A.	.901(.199)***
	Random effect: τ^2(se)	
Intercept[a]	.810(.368)*	.421(.176)*
Slope[c] of log volume	.117(.069) [+]	N.A. [b]
Slope[c] of Feeling drunk	N.A.	.455(.238) [+]

[+] Each of the models included an intercept and controlled for Gender and Age as fixed effects (neither were significant)

[a] The intercept indicates the rate of causal attribution at low consumption levels and for those not feeling drunk (when included)

[b] Forced to be fixed effect

[c] This slope term indicates the relationship between log volume (feeling drunk) and log odds of causal attribution

*** p<.001, ** p<.01, * p<.05, [+] p<.10 (Wald test)

Volume Consumed

Table 2 shows the role of study-level aggregated and contextual variables in predicting variation in both the rate of causal attribution (the intercept) and in the relationship between volume and reporting causal attribution (the slope). In the model, gender and age were included as fixed effects predictors. Although variation in th e volume slope was marginally significant (p<.10), study-level variables were still entered to predict their variances. Aggregate and socio-cultural study-level variables were first entered in separate models (results not shown) for volume. Significant predictors in these separate models for either the intercept or slope were then entered simultaneously.

The DDP measure was a highly significant predictor for both the intercept and slope of volume (Table 2). A higher rate of causal attribution at low levels of alcohol consumption was found in regions with low DDP (as seen in the intercept model). In contrast, regions with low DDP were less likely than high DDP regions to attribute a causal association of drinking and injury at low consumption levels, but more likely to attribute an association at higher consumption levels.

Feeling Drunk

Study-level variables were also used to explain the variation in the relationship between feeling drunk and the probability of casual attribution when gender, age and volume of alcohol consumed were controlled, along with the rates of causal attribution at low levels of consumption (the intercept). Although a number of study variables, when entered separately, significantly predicted variation of either the rate of attribution for those not feeling drunk at the time (the intercept) or the relationship between feeling drunk and attribution (the slope), the only significant predictors in the simultaneous models were aggregated volume of consumption (for the intercept) and DDP (for the slope). In those areas with higher DDP, patients who were feeling drunk were less likely to attribute their injury to drinking.

Table 2. Parameter estimates and standard errors of contextual variables predicting the variation of intercept and slope, built on model 2 (for log volume) and model 3 (for feeling drunk)

	Significant predictors from marginal models entered simultaneously			
	Log Volume		Feeling drunk	
Modifying Study Level Variable:	**Intercept**	**Slope of log volume**	**Intercept**	**Slope of feeling drunk**
Group mean of log volume	.837(.348)*	Not entered	1.148(.526)*	-.028(.523)
Detrimental drinking pattern	586(.183)**	-.275(.088)**	.107(.156)	-.483(.179)**
Random effect: τ^2(se)	.250(.194)	.058(.050)	.138(.083)	.019(.082)

*** $p<.001$, ** $p<.01$, * $p<.05$ (Wald test)

Modifying effects of injury event characteristics and usual drinking patterns

Several important individual and event characteristics were also considered, including drinking pattern and the type of injury (violence vs. traffic-related vs. other), as modifiers of the relationship between volume as well as feeling drunk and the likelihood of causal attribution. Candidate drinking pattern variables included weekly drinking of any amount, 5+ drinks a day at least monthly and heavy drinking, defined as whether the respondent reported usually having 5 or more drinks on an occasion when they drank. Of these, only weekly drinking was a moderator ($p<.001$), with those who drank at least weekly less likely to attribute their injury to drinking at low volume levels compared with those who drank less often (Table 3).

An additional exploratory analysis considered the role of cause of injury (violence vs. traffic-related vs. other) as a modifier in the relationship between alcohol and causal attribution. This analysis approached the role of alcohol in injuries in a slightly different way than that of more traditional drunken comportment theories (MacAndrew and Edgerton, 1969), which suggest that drinking may be used as an excuse for explaining otherwise socially unacceptable behavior. While the drunken comportment argument implicitly suggests that the respondent has admitted responsibility for his or her behavior and may be looking for an excuse on which to place blame, the act of acknowledging causal attribution occurs cognitively at a stage before responsibility has been assigned. If acknowledgment of causal attribution is interpreted as an admission of responsibility for the injury regardless of other potential contributing factors, then for injuries that are traffic or violence-related, where legal or financial repercussions may ensue, such responsibility may be less likely to be accepted.

Table 3 shows that patients sustaining injuries related to either traffic crashes or violence were no more likely than those with non-violence and non-traffic-related injuries to attribute a causal association of their injury to drinking *at low consumption levels*. As the level of drinking increased, however, those with violence-related injuries were significantly less likely to causally attribute their injury to drinking than those with non-violence-related/non traffic-related injuries, as seen in the interaction term shown in the last row of Table 3 (p=.01). Interestingly, these effects were found only for violence, and not for traffic-related injuries. One possible explanation for this phenomenon is that violence-related injuries result from aggression, either on the part of the injured patient, another person, or both, and individuals who engage in such behavior may be less likely to willingly accept blame (which could easily be placed on the other party) for the incident. Although empirical results here suggest that those with violence-related injuries may be relatively less willing to accept responsibility for their injuries than those with other non-violence/non-traffic types of injuries at higher volumes of consumption, these results should be considered exploratory.

Table 3. Individual usual drinking patterns and drinking and injury event characteristics on the probability of causal attribution, controlling for gender and age

Sub-models:	β(se)
1) Any WD+	-.668(.255)*
Log volume • Any WD	447(.123)***
2) Traffic Injury	.430 (.357)
Violence-Related Injury	.463 (.292)
Log Volume	.694 (.124)***
Log Volume • Traffic Injury	-.307 (.198)
Log Volume • Violence-Related Injury	-.327 (.141)*

+ WD: Weekly Drinking

*$p<.05$, **$p<.01$, ***$p<.001$ (t-test)

Subjective Causality in Estimates of the Attributable Fraction of Injuries to Alcohol

The nature of subjective causal attribution of injury to alcohol (as described above) makes it a natural candidate to consider for incorporation into epidemiologic estimates of the fraction of injuries that are caused by alcohol (the attributable fraction or AF). In order to assess alcohol's true role in injury for those drinking prior to the event, an unbiased assessment of whether the injury would have happened had the respondent not been drinking is necessary. This information is rarely available, however. As a result, in order to estimate the fraction of all injuries in a population attributable to drinking, one method which has been used has examined drinking in the event among injured patients compared to non-injured patients in ED settings. Data from the ERCAAP studies, for example, used as a control group, patients with medical (non-injury) conditions who entered the EDs during the same times as injured patients. The underlying assumption is that an appropriate study design using a well-defined case and control group will generate data that will be generalizable to a desired target population. The use of medical patients as a control group has some disadvantages, however, giving rise to two issues that likely result in biased estimates of the AF: (1) the rate of injury in the target population in relation to the relative representation of injury patients compared to controls in the ED population; and (2) the drinking characteristics of the control sample.

Even in situations where the study characteristics and design may be considered unbiased, it would still be of interest to explore the subjective assessment of the role of alcohol in injury in estimates of the fraction of injury which can be attributed to alcohol. Additionally, one characteristic of many studies that makes subjective assessment even more relevant is the definition of the exposure period of drinking prior to injury. Some might argue that one drink as long as six hours prior to the event (the time frame used in the ERCAAP studies) should not result in sufficient decrements in psychomotor capacities as to lead to an injury. Subjective assessments, then, may provide a more realistic evaluation of the true role of alcohol in injury compared with self-reported drinking prior to the event. Of course, patients' judgments of whether an act that might be considered socially undesirable (such as drinking) resulted in injury may be somewhat biased.

There are two components of the AF that are required for its estimation, as will be seen in the specific formulation of the AF that is used in the following section. The first is the rate of exposure to alcohol 6 hours prior to injury among only injured patients and the second is the relative risk (RR) of injury due to alcohol exposure. An estimate of this first quantity is readily available from the injured patients in the ED samples that are available (ignoring arguments of the appropriateness of the ED as a venue for sampling those with injuries). However, no "gold" standard is available for the estimate of the second component, the RR of injury due to alcohol, although several different approaches have been taken (Borges et al., 2004b; Cherpitel et al., 2005). The use of subjective assessments of the causal association of alcohol and injury simply presents yet another method for the estimation of the RR and, therefore, is worthwhile to investigate.

Comparison of subjective and objective AF of injury due to alcohol use

Table 4 shows the notation for the distribution of drinking within six hours prior to the event leading to the ED visit by injury status of the patient. It should be noted that non-injured patients in these studies were asked about alcohol use within the six hours prior to noticing the condition for which they were now seeking ED treatment. Also shown in the table is, among the injured patients, the split of those who admitted that alcohol indeed played a causal role in the event (denoted as a_c) and those who responded that the injury would have taken place regardless of whether or not they had been drinking (denoted as a_{nc}).

Table 4. Distribution of Injury and Non-Injury Patients by Drinking Status

		Injured	
		Yes	No
Drinking 6-hours prior to the event	Yes	$a = a_c + a_{nc}$	b
	No	c	d

a_c # of injured reporting drinking within the six hours who attributed a causal association of injury to drinking

a_{nc} # of injured reporting drinking within the six hours who did not attribute a causal association of injury to drinking

The definition of the "objective" estimate of the AF of injury due to alcohol use within six hours prior to the event, when drinking during the same period is also available for a group of non-injured control subjects, is: $AF = P(Drinking \mid Injury) \cdot \left(1 - \frac{1}{RR}\right)$, where RR (relative risk) is an estimate of the risk of injury due to drinking relative to the risk of non-injury due to drinking. In comparison, given a split of the injury patients into those attributing a causal association of alcohol and injury and those not making such attribution, an estimate of the "subjective" attribution fraction (AFs) of injury due to the use of alcohol within six hours prior to the event is $\left(AF_s = \frac{a_c}{a+c}\right)$ (see Table 4). The main advantage of direct subjective assessments here is readily apparent: a sample from a control population is not required. Therefore, any relevant differences between the target and sample population, stemming from appropriateness of the control sample used do not affect the estimates. Neither the relative representation of injury patients compared to non-injured in the population nor the rate of exposure in the control population has any effect on the estimates derived. Of course, these advantages need to be weighed against concerns of any bias that may result from an individual's reporting of a causal attribution of injury to drinking.

Results for AF of injury due to alcohol use in the ERCAAP study

Table 5 shows the relevant data needed for estimating the subjective attributable fraction. Several of the ERCAAP studies are not represented because either there were too many missing observations for the causal attribution variable, or the variable was only asked of those who reported feeling drunk at the time of injury. Column 2 displays the number of people within each study that did not think that their injury would have happened had they not been drinking (a_c). Column 3 shows the number of respondents that thought the injury would have occurred regardless of their drinking (a_{nc}) and column 4 represents those who were not sure. Column 6 shows the subjective AF produced from averaging two ways of estimating the "not sure" responses; one in which all of these respondents attributed their injury to drinking and, conversely, none of the respondents attribute their injury to drinking. The subjective AF shown in column 6 can then be compared to the objective AF as shown in column 5 (corresponding to the equation above), estimated using the standard definition of relative risk.

Table 5. Objective and Subjective Estimates of Attributable Fraction of Alcohol Use

(1) Studies	Would your injury have happened if you didn't drink?*			Objective and Subjective AF of Injury		Objective and Subjective AF of Violence-Related Injury**	
	(2) No a_c	(3) Yes a_{nc}	(4) Not sure	(5) AF_o	(6) $\overline{AF}_s = \frac{\bar{a}_c}{a+c}$	(7) AF_o	(8) $\overline{AF}_s = \frac{\bar{a}_c}{a+c}$
San Francisco, CA	60	105	25	15.8	13.4	41.6	19.4
Jackson, MS	13	38	7	**13.1**	**6.4**	50.7	15.4
Santa Clara, CA	16	14	7	6.4	9	35.4	32.3
Mexico City, Mexico	239	154	22	**4.3**	**15.5**	37.4	30
Acapulco, Mexico	68	24	2	**10.2**	**21.4**	43.1	35.1
Pachuca, Mexico	45	21	32	4.3	6.6	32	22.6
Edmonton (Alberta), Canada	28	29	2	13.6	14.1	81	46.7
Quebec City, Canada	5	11	3	1.1	5.3	83.3	28.6
Barcelona, Spain	39	172	22	0.9	3.4	24.9	16.7
Mar del Plata, Argentina	12	58	4	7	**4.3**	36.9	5.9
Warsaw, Poland	9	14	19	1.8	4	36.2	22.6
Sosnowiec, Poland	16	35	17	4.7	5.9	**47.2**	17.9

* Numbers reported are Unweighted Ns of those reporting the Injury would not have happened if they hadn't drank been drinking during the 6-hours prior to the event

** All non-violent injury patients were grouped with medical patients for calculation of violence-related injury AF estimates

$\bar{a}_c$ is defined as a_c + (# not sure / 2)

Some of the larger differences between subjective and objective AF estimates are indicated in bold.

The subjective AF provides larger estimates, generally, than does the objective AF measure. This finding is consistent with prior indications that the AFs derived from the standard objective epidemiologic estimates have often been found to be lower than would be expected (Cherpitel et al., 2005). For several studies, such as those from Mexico City and Acapulco, the subjective AF is dramatically larger than the corresponding objective estimate. Among those studies for which the AF declines, the Mississippi and Argentina studies show the largest decreases. (Note that in Table 5 the largest differences between the subjective and objective estimates are bolded.) One variable that may be of anecdotal interest in discussing differences found between objective and subjective assessments is whether or not the use of alcohol is stigmatized in the population that the ED serves. This variable was obtained from key informants at each of the study sites that assessed how acceptable they believed alcohol use was in the culture (on a four-point scale from very acceptable to not very acceptable) and the likelihood that patients who had been drinking prior to injury might deny the use of alcohol. Interestingly, for the Mexico City and Acapulco studies, in which the subjective AF is much larger than the objective estimate, stigmatization of alcohol use is perceived to be low, whereas for those studies in which the subjective AF decreases relative to the objective estimate, stigmatization is perceived as being high (Mississippi and Argentina). Not only is the stigma variable consistent for these largest differences, the average paired difference between the subjective and objective AFs was found to be lower ($p<.05$) for those studies reporting high compared to low stigma, suggesting the possibility of a societal influence on individuals' willingness to report attribution of their injury to alcohol.

Results of AF for violence-related injury due to alcohol use

Estimates of the objective and subjective AF are reported for violence-related injury in columns 7 and 8 of Table 5, respectively, where a dramatic reduction in the subjective measures is seen compared to the objective measures. For each study, the subjective AF was smaller than the objective AF (and was only half the objective measure in 5 of the 12 studies), although the subjective AF for violence-related injury was larger than the subjective AF across all injuries in each study. These differences between objective and subjective measures are due to the fact that, along with elevated rates of drinking within six hours prior to the event among those with violence-related injuries, the relative risks of injury derived from the objective assessment were quite large. In effect, from the equation above, the objective AF estimate closely mirrored the exposure estimate. For the two Canadian studies, the objective AF estimates were both above 80% and much larger than that in other studies. This elevated AF estimate in the Canadian studies was due to the very high rate of drinking in the violence-related event (which was over 80% compared to, at most, 56% in the other studies) compared to drinking in the non-violence related event, for which the rates were similar between the Canadian and the other studies.

In comparing the two estimates for violence-related injuries, a reasonable question is whether the objective AF is an overestimation or the subjective AF is an underestimation of the true population parameter. Although it may be difficult to directly answer this question, analyses discussed in Cherpitel et al. (2006) indicated that those reporting violence-related injuries were no less likely to report a causal association between drinking and their injury, when controlling for the amount consumed, than other injured patients who also drank within the six hours before the event. Causal attribution was also found to be significantly lower across the range of volume consumed for those with violence-related injuries. As a result, patients with violence-related injuries may be more likely to have consumed alcohol and in larger quantities, but no more likely to report causal attribution. In addition, the average paired differences between objective and subjective AF measures were not related to stigmatization, as was found for analyses using all injuries. While this suggests there is no evidence that supports subjective AF estimates necessarily underestimating their population values for violence-related injury, this is an area that requires further research.

As noted above, the population values for the parameters required for estimation of AF (prevalence of alcohol exposure and relative risk of injury related to alcohol use) are simply not known. Given the absence of these data and the true AF, it is difficult to discuss the differences found from estimating AF using different techniques. Instead, it is reasonable to discuss differences between estimates in terms of the assumptions underlying each of the estimates, and to explore whether these differences can be explained by other societal-level variables that may be associated with biases in reporting.

Using the standard AF estimator, a control group is required. Given that the control group for each study was chosen within the same ED, and not from the broader population that the emergency department serves, differences in both the drinking patterns between the two populations (emergency room and general population) as well as in the rate of occurrence of injury in a similar six hour period may contribute to differences between estimates produced using other methods. The advantage of using causal assessments here is that the need for a control sample is removed. Other considerations, however, such as individuals' willingness to respond truthfully may be a factor in assessing the unbiasedness of reports of causal attribution.

References

Borges G et al. (2004a). A case-crossover study of acute alcohol use and suicide attempt. *Journal of Studies on Alcohol* **65**, 708-714.

Borges G, Cherpitel CJ, Mittleman M (2004b). The risk of injury after alcohol consumption: a case-crossover study in the emergency room. *Social Science & Medicine* **58**(6), 1191-1200.

Cherpitel CJ (1993). Alcohol and injuries: A review of international emergency room studies. *Addiction* **88**, 923-937.

Cherpitel CJ (1994). Alcohol and injuries resulting from violence: a review of emergency room studies. *Addiction* **89**, 157-165.

Cherpitel CJ (1996). Drinking patterns and problems and drinking in the event: An analysis of injury by cause among ER patients. *Alcoholism, Clinical and Experimental Research* **20**(6), 1130-1137.

Cherpitel CJ (1997). Comparison of screening instruments for alcohol problems between Black and White emergency room patients from two regions of the county. *Alcoholism, Clinical and Experimental Research* **21**(8), 1391-1397.

Cherpitel CJ (1999). Drinking patterns and problems: A comparison of primary care with the emergency room. *Substance Abuse* **20**(2), 85-95.

Cherpitel CJ et al. (2006). Multi-level analysis of causal attribution of injury to alcohol and modifying effects: data from two international emergency room projects. A research report from the Emergency Room Collaborative Alcohol Analysis Project (ERCAAP) and the WHO Collaborative Study on Alcohol and Injuries. *Drug and Alcohol Dependence* **82**(3), 258-268.

Cherpitel CJ, Ye Y, Bond J (2005). Attributable risk of injury associated with alcohol use: a cross-national meta-analysis from the Emergency Room Collaborative Alcohol Analysis Project. *American Journal of Public Health* **95**(2), 266-272.

Compton RP et al. (2002). Crash risk of alcohol impaired driving. In: Mayhew S, Dussault C, eds. *Proceedings of Alcohol, Drugs and Traffic Safety – T2002. 16th International Conference*, August 4–9, Montreal, Quebec, Canada: International Council on Alconol, Drugs and Traffic Safety, 2002, pp. 39-44.

Hoaken PNS, Stewart S (2003). Drugs of abuse and elicitation of human aggressive behaviour. *Addictive Behaviors* **28**, 1533-1554.

Honkanen R et al. (1983). The role of alcohol in accidental falls. *Journal of Studies on Alcohol* **44**, 231-245.

MacAndrew C, Edgerton RB (1969). *Drunken comportment: A social explanation*, Chicago: Aldine Publishing Co.

Macdonald S et al. (2005). The criteria for causation of alcohol in violent injuries based on emergency room data from six countries. *Addictive Behaviors* **30**, 103-113.

Macdonald S et al. (1999). Demographic and substance use factors related to violent and accidental injuries: results from an emergency room study. *Drug and Alcohol Dependence* **55**, 53-61.

MacMahon B, Pugh TF (1970). *Epidemiology: Principles and methods*, Boston, MA: Little, Brown and Company.

Mcleod R et al. (2003). The influence of extrinsic and intrinsic risk factors on the probablility of sustaining an injury. *Accident, Analysis and Prevention* **35**(*1*), 71-80.

Raudenbush SW, Bryk AS (2002). *Hierarchical Linear Models: Applications and data analysis methods* (2nd Edition), Thousand Oaks, CA: SAGE Publications.

Raudenbush SW, Bryk AS and Congdon RT Jr (2004). HLM 6. *Hierarchical Linear and Nonlinear Modeling*, Lincolnwood, IL: Scientific Software International.

Rehm J et al. (2001). Steps towards constructing a global comparative risk analysis for alcohol consumption: determining indicators and empirical weights for patterns of drinking, deciding about theoretical minimum, and dealing with different consequences. *European Addiction Research* **7**(3), 138-147.

Rehm J et al. (2003a). The global distribution of average volume of alcohol consumption and patterns of drinking. *European Addiction Research* **9**(4), 147-156.

Rehm et al. (2003b). Alcohol as a risk factor for global burden of disease. *European Addiction Research* **9**, 157-164.

Romelsjö A (1995). Alcohol consumption and unintentional injury, suicide, violence, work performance, and inter-generational effects. In: Holder HD, Edwards G, eds. *Alcohol and public policy: evidence and issues*, New York, NY: Oxford University Press, pp. 114-142.

Stephens C J (1987). Alcohol consumption and casualties: drinking in the event. *Drug and Alcohol Dependence* **20**, 115-127.

Watt K et al. (2005). The relationship between acute alcohol consumption and consequent injury type. *Alcohol and Alcoholism* **40**(4), 263-268.

CHAPTER 4 :
ALCOHOL-ATTRIBUTABLE INJURY
IN A GLOBAL PERSPECTIVE

Jürgen Rehm - Centre for Addiction and Mental Health | Toronto, ON CANADA

Svetlana Popova - Centre for Addiction and Mental Health | Toronto, ON CANADA

Jayadeep Patra - Centre for Addiction and Mental Health | Toronto, ON CANADA

Global Extent and Trends of Alcohol-Attributable Injury

Alcohol is a major risk factor for burden of mortality and disability of disease and injury (Rehm et al., 2003a; 2004). In total, 3.2% of deaths and 4.0% of the burden of disease as measured in disability adjusted life-years (DALYs; see Lopez et al., 2006) in the year 2000 was attributable to alcohol (see also WHO, 2002). Relatively, the impact of alcohol on acute conditions is more pronounced than on chronic disease. This contribution will try to quantify the amount of alcohol-attributable injuries in a global perspective for the year 2002.

Methodology and Data Sources

The methodology of estimating injury has been summarized in Chapter 6 of this volume. The methodology specified in the Comparative Risk Assessment (CRA) of the Global Burden of Disease (GBD) study was followed, and estimates were updated for 2002 (see also Rehm et al., 2004; 2006a).

As a very short overview:

- Criteria of causality used were the standard epidemiological criteria (Hill, 1965; see Rothman and Greenland, 1998). The determination of alcohol as a causal influencing factor of injury is discussed in Rehm et al. (2003b, c).

- Exposure data were from the Global Alcohol Database (www3.who.int/whosis). To obtain stable estimates, the mean of adult per capita consumption for the years 2001 to 2003 was used.

- Risk relations were taken from the CRA (Rehm et al., 2004).

- For a categorization of alcohol-attributable injuries, GBD categories were used. Thus, there were five categories of alcohol-attributable unintentional injuries and three categories of alcohol-attributable intentional injuries. Table 1 gives an overview of the relevant ICD codes.

Table 1. ICD-10 codes for alcohol-attributable injury categories used

Injuries	ICD-10 Code
Unintentional injuries	
Road traffic accidents	*
Poisonings	X40 - X49
Falls	W00 - W19
Drowning	W65-W74
Other unintentional injuries	** Rest of V & W20 –W64, W75 - W99, X10 -X39, X50 - 59, Y40 -Y86, Y88, Y89
Intentional injuries	
Self-inflicted injuries	X60 - X84, Y87.0
Violence	X85 -Y09, Y87.1
Other intentional injuries	Y35

*V021-V029, V031-V039, V041-V049, V092, V093, V123-V129, V133-V139, V143-V149, V194-V196, V203-V209, V213-V219, V223-V229, V233-V239, V243-V249,V253-V259, V263-V269, V273-V279, V283-V289, V294-V299, V304-V309, V314-V319, V324-V329, V334-V339, V344-V349, V354-V359, V364-V369, V374-V379, V384-V389, V394-V399, V404-V409, V414-V419, V424-V429, V434-V439, V444-V449, V454-V459, V464-V469, V474-V479, V484-V489, V494-V499, V504-V509, V514-V519, V524-V529, V534-V539, V544-V549, V554-V559, V564-V569, V574-V579, V584-V589, V594-V599, V604-V609, V614-V619, V624-V629, V634-V639, V644-V649, V654-V659, V664-V669, V674-V679, V684-V689, V694-V699, V704-V709, V714-V719, V724-V729, V734-V739, V744-V749, V754-V759, V764-V769, V774-V779, V784-V789, V794-V799, V803-V805, V811, V821, V830-V833, V840-V843, V850-V853, V860-V863, V870-V878, V892.
**Rest of V = V-series MINUS*.

- The outcome data, mortality and burden of disease as measured in DALYs were taken from the GBD study for 2002 (source: Dr Colin Mathers, WHO; for a general description of the methodology see Lopez et al., 2006). To be comparable with the CRA, a discount rate of 3% and age-weighting was applied. That is, in the calculation of years of life lost and for DALYs, future events were discounted by 3% per year, and years of life lost were given slightly different weights in earlier years in the life course than in later years (eg., a life year lost at age 30 was weighted higher than a life year lost at age 70; see Lopez et al., 2006, for further explanations).

- For regional distributions, the WHO sub-regions were used. These sub-regions were created based on levels of adult and infant mortality (WHO, 2000). The regional groups are organized as follows: A denotes very low child and low adult mortality, B is low child and low adult mortality, C is low child and high adult mortality, D is high child and high adult mortality, and E is high child and very high adult mortality. There were 14 sub-regions defined in total. Table 2 gives a classification of countries to the different sub-regions.

Table 2. Classification of countries into WHO sub-regions

Africa D	Algeria, Angola, Benin, Burkina Faso, Cameroon, Cape Verde, Chad, Comoros, Equatorial Guinea, Gabon, Gambia, Ghana, Guinea, Guinea-Bissau, Liberia, Madagascar, Mali, Mauritania, Mauritius, Niger, Nigeria, Sao Tome and Principe, Senegal, Seychelles, Sierra Leone, Togo
Africa E	Botswana, Burundi, Central African Republic, Congo, Côte d'Ivoire, Democratic Republic of the Congo, Eritrea, Ethiopia, Kenya, Lesotho, Malawi, Mozambique, Namibia, Rwanda, South Africa, Swaziland, Uganda, United Republic of Tanzania, Zambia, Zimbabwe
Americas A	Canada, Cuba, United States of America
Americas B	Antigua and Barbuda, Argentina, Bahamas, Barbados, Belize, Brazil, Chile, Colombia, Costa Rica, Dominica, Dominican Republic, El Salvador, Grenada, Guyana, Honduras, Jamaica, Mexico, Panama, Paraguay, Saint Kitts and Nevis, Saint Lucia, Saint Vincent and the Grenadines, Suriname, Trinidad and Tobago, Uruguay, Venezuela
Americas D	Bolivia, Ecuador, Guatemala, Haiti, Nicaragua, Peru
Eastern Mediterranean B	Bahrain, Iran (Islamic Republic of), Jordan, Kuwait, Lebanon, Libyan Arab Jamahiriya, Oman, Qatar, Saudi Arabia, Syrian Arab Republic, Tunisia, United Arab Emirates
Eastern Mediterranean D	Afghanistan, Djibouti, Egypt, Iraq, Morocco, Pakistan, Somalia, Sudan, Yemen
Europe A	Andorra, Austria, Belgium, Croatia, Cyprus, Czech Republic, Denmark, Finland, France, Germany, Greece, Iceland, Ireland, Israel, Italy, Luxembourg, Malta, Monaco, Netherlands, Norway, Portugal, San Marino, Slovenia, Spain, Sweden, Switzerland, United Kingdom
Europe B	Albania, Armenia, Azerbaijan, Bosnia and Herzegovina, Bulgaria, Georgia, Kyrgyzstan, Poland, Romania, Slovakia, The Former Yugoslav Republic Of Macedonia, Tajikistan, Turkmenistan, Turkey, Uzbekistan, Yugoslavia
Europe C	Belarus, Estonia, Hungary, Kazakhstan, Latvia, Lithuania, Republic of Moldova, Russian Federation, Ukraine
South East Asia B	Indonesia, Sri Lanka, Thailand
South East Asia D	Bangladesh, Bhutan, Democratic People's Republic of Korea, India, Maldives, Myanmar, Nepal
Western Pacific A	Australia, Brunei Darussalam, Japan, New Zealand, Singapore
Western Pacific B	Cambodia, China, Cook Islands, Fiji, Kiribati, Lao People's Democratic Republic, Malaysia, Marshall Islands, Micronesia (Federated States of), Mongolia, Nauru, Niue, Palau, Papua New Guinea, Philippines, Republic of Korea, Samoa, Solomon Islands, Tonga, Tuvalu, Vanuatu, Viet Nam

Source: WHO (2000)

An Overview of Injuries in Different Sub-regions of the World

Table 3 gives an overview of alcohol-attributable injury deaths in different parts of the world. There was a great deal of variation in alcohol-attributable injury death categories across the different sub-regions. The highest proportion of total deaths attributable to alcohol for both unintentional and intentional injuries was found in Europe (EUR) C followed by the Americas (AMR) B. The third highest was EUR B for alcohol-attributable deaths due to unintentional injuries, while for alcohol-attributable deaths due to intentional injuries Africa (AFR) E was the third highest. The lowest proportion of total deaths attributable to alcohol for both unintentional and intentional injuries was found in the Eastern Mediterranean (EMR) D followed by EMR B and South East Asia (SEAR) D.

Two main factors explain at least two-thirds of both unintentional and intentional injury deaths as evidenced in multivariate regression models:

- Exposure, as measured by adult per capita consumption, explains most of the variation in injury. With each increase of one litre of pure alcohol consumed, there were 2.9% more alcohol-attributable unintentional injury deaths, and 2.7% more alcohol-attributable intentional injury deaths.

- The level of economic development was inversely related to injury deaths. The higher the economic development, the lower the proportion of alcohol-attributable injury deaths, controlling for exposure. This effect had a much lower effect size than exposure.

In terms of different categories of alcohol-attributable injuries, road traffic injury accidents were clearly the largest contributor to alcohol-attributable unintentional injury deaths, whereas violence was the largest contributor to alcohol-attributable intentional injury deaths in the majority of sub-regions and globally. In relation to absolute level, alcohol-attributable unintentional injury deaths were more than twice as prevalent as alcohol-attributable intentional injury deaths. In terms of relative level with respect to alcohol involvement, there was little difference. In many of the poorest sub-regions, however, there was a relatively higher proportion of alcohol-attributable intentional injury deaths compared to alcohol-attributable unintentional injury deaths (i.e. AFR D, AFR E, AMR B, and AMR D), whereas the opposite was true for all other sub-regions.

Table 3. Deaths* due to alcohol-attributable unintentional and intentional injuries in WHO sub-regions and world in 2002

Injuries	WHO Subregions														World		
Unintentional Injuries	AFR D	AFR E	AMR A	AMR B	AMR D	EMR B	EMR D	EUR A	EUR B	EUR C	SEAR B	SEAR D	WPR A	WPR B	M**	W**	TOTAL
Road traffic accidents	13.5	25.7	12.9	30.9	2.7	2.8	1.5	15.2	7.9	31.5	15.3	30.8	3.0	49.1	213.6	29.3	242.9
Poisonings	1.6	2.9	2.7	0.5	0.1	0.1	0.1	1.3	1.4	41.6	0.7	5.7	0.3	6.8	52.3	13.4	65.7
Falls	0.7	1.3	2.0	2.9	0.2	0.2	0.1	5.9	1.8	10.0	1.4	6.2	0.9	9.5	36.0	7.1	43.0
Drowning	4.5	6.7	1.0	4.9	0.6	0.2	0.2	1.3	1.9	15.5	2.0	7.5	1.5	11.2	47.9	11.0	59.0
Other unintentional injuries	10.4	14.9	7.2	19.9	4.2	0.7	0.7	9.7	10.5	51.0	3.4	25.7	4.0	23.4	151.2	34.7	185.9
All alcohol-attributable deaths due to unintentional injuries	30.7	51.6	25.8	59.1	7.8	3.9	2.7	33.4	23.5	149.7	22.8	75.9	9.6	100.0	501.0	95.5	596.5
Total deaths due to unintentional injuries	258.3	210.2	119.9	167.5	33.4	98.0	195.8	136.1	76.2	320.3	148.9	931.4	48.5	779.0	2,307.0	1,243.6	3,550.6
% of all alcohol-attribuable deaths due to unintentional injuries	11.9%	24.6%	21.5%	35.3%	23.3%	4.0%	1.4%	24.6%	30.8%	46.7%	15.3%	8.1%	19.9%	12.8%	21.7%	7.7%	16.8%
Intentional Injuries	AFR D	AFR E	AMR A	AMR B	AMR D	EMR B	EMR D	EUR A	EUR B	EUR C	SEAR B	SEAR D	WPR A	WPR B	M**	W**	TOTAL
Self-inflicted injuries	1.1	3.3	4.4	5.6	0.3	0.2	0.2	7.3	4.5	31.8	3.0	10.9	3.6	18.5	77.2	17.4	94.6
Violence	9.5	27.0	4.7	51.7	3.1	0.2	0.5	1.4	3.0	35.1	5.5	11.0	0.3	11.5	141.6	22.8	164.3
Other intentional injuries	0.0	0.0	0.1	0.3	0.0	0.0	0.0	0.0	0.2	0.0	0.1	0.5	0.0	0.3	1.2	0.2	1.4
All alcohol-attributable deaths due to intentional injuries	10.6	30.3	9.1	57.5	3.4	0.4	0.7	8.7	7.6	66.9	8.6	22.4	3.9	30.2	220.0	40.3	260.3
Total deaths due to intentional injuries	85.8	96.8	52.6	153.6	12.6	14.6	83.2	52.9	33.2	170.2	75.8	310.3	35.7	366.2	1.156,7	461.0	1,617.7
% of all alcohol-attribuable deaths due to intentional injuries	12.3%	31.3%	17.4%	37.4%	26.9%	2.9%	0.8%	16.5%	23.0%	39.3%	11.3%	7.2%	11.1%	8.3%	19.0%	8.7%	16.1%
Total alcohol-attribuable deaths due to unintentional and Intentional injuries	41.3	81.9	34.9	116.6	11.2	4.4	3.4	42.2	31.1	216.5	31.4	98.2	13.6	130.2	721.0	135.8	856.8

*Numbers are rounded to the nearest thousand. Zero (0) indicates fewer than 500 alcohol-attributable deaths in the injury category

** M=Men, W=Women

AFR=Africa, AMR=American, EMR=Eastern Mediterranean, EUR=Europe, SEAR=South East Asia, WPR=Western Pacific

Fire was excluded, as we could not find reliable estimators on global scale to quantify the impact of alcohol. However, there are national data and indicators of several causal relationships on country levels (see Rehm et al., 2006b for Canada).

War as an injury category was excluded, as we did not have information that alcohol could be a cause of war.

Alcohol-attributable DALYs for the same categories of injuries and sub-regions are shown in Table 4. Overall, the picture was very similar to deaths; the highest proportion of alcohol-attributable DALYs due to both unintentional and intentional injuries was found in EUR C followed by AMR B. The third highest for alcohol-attributable DALYs of unintentional injuries was EUR A, while for alcohol-attributable DALYs of intentional injuries AMR D was the third highest. The lowest proportion of alcohol-attributable injury DALYs was found in EMR D, followed by EMR B and SEAR D.

Similarly to mortality, road traffic accidents were the largest contributor to alcohol-attributable DALYs among unintentional injuries globally and in the majority of sub-regions (excluding AMR D, EUR B and EUR C). Violence was the largest contributor to alcohol-attributable DALYs among intentional injuries in the world and in the majority of the sub-regions. In EUR A, EUR B, SEAR D, Western Pacific (WPR) A, and WPR D self-inflicted injuries were the leading cause of DALYs attributable to alcohol among intentional injuries. In terms of absolute level, alcohol-attributable DALYs due to unintentional injuries were more than twice as great as those due to intentional injuries. Similarly to mortality, the relative level of alcohol-attributable DALYs due to intentional injuries was higher than due to unintentional injuries in most of the poorest sub-regions (i.e. AFR D, AMR B, AMR D), whereas an opposite trend was observed in all other sub-regions.

The distribution of deaths due to alcohol-attributable unintentional and intentional injuries by gender and age in the world for 2002 is presented in Table 5. Men had a slightly higher mortality from unintentional injuries attributable to alcohol than from intentional injuries attributable to alcohol (21.7% vs. 19.0%, respectively). In contrast, women had higher mortality from intentional injuries attributable to alcohol than from unintentional injuries (8.7% vs. 7.7%, respectively), mainly due to self-inflicted injuries. Overall, when genders were combined, the proportion of deaths attributable to alcohol was similar for unintentional and intentional injuries (16.8% vs. 16.1%, respectively).

Alcohol was proportionally most important as a risk factor associated with injury mortality among men in the age group of 15-44 years, for whom about two-thirds of intentional injuries and more than one-fifth of unintentional injuries were attributable to alcohol. Among women, the proportion of unintentional injuries attributable to alcohol was largest between the ages of 30-44 and 45-59 (10.9%), while the proportion of intentional injuries attributable to alcohol was highest among those 30-44 (10.1%). Road traffic accidents were the leading cause of death for alcohol-attributable unintentional injuries for both genders among those 0 to 44 years. In terms of alcohol-attributable intentional injuries, violence was a leading cause of death in all age categories for both men and women. The only exception was in the age group of 60-69 years for men where the leading cause of death attributable to alcohol was self-inflicted injuries.

The distribution of DALYs due to alcohol-attributable unintentional and intentional injuries by gender and age in the world for 2002 is presented in Table 6. Overall, when genders were combined, the proportion of alcohol-attributable DALYs was also higher for intentional compared to unintentional injuries attributable to alcohol (15.2% vs. 12.9%, respectively).

Table 4. DALYs* due to alcohol-attributable unintentional and intentional injuries in WHO sub-regions and world in 2002

| | WHO Subregions | | | | | | | | | | | | | | World | | |
	AFR D	AFR E	AMR A	AMR B	AMR D	EMR B	EMR D	EUR A	EUR B	EUR C	SEAR B	SEAR D	WPR A	WPR B	M**	W**	TOTAL
Unintentional Injuries																	
Road traffic accidents	475.6	831.1	373.0	979.4	86.3	90.5	50.8	435.8	228.4	907.3	481.4	958.5	79.6	1,469.6	6,547.8	899.4	7,447.2
Poisonings	42.1	74.5	65.7	13.5	2.7	2.0	3.2	32.4	31.2	836.8	15.6	122.0	7.0	151.0	1,121.8	278.0	1,399.8
Falls	21.0	35.3	41.2	90.2	9.9	6.0	4.8	96.0	60.1	269.7	37.9	163.9	18.0	263.2	942.8	174.5	1,117.3
Drowning	110.7	167.9	22.6	129.8	15.0	4.8	6.4	22.5	43.4	346.3	47.2	179.7	18.0	260.1	1,142.0	232.4	1,374.3
Other unintentional injuries	439.0	619.0	156.4	677.8	135.0	33.2	36.4	209.7	363.1	1,200.0	134.5	935.4	64.1	803.8	4,744.4	1,063.1	5,807.5
All alcohol-attributable DALYs due to unintentional injuries	1,088.4	1,727.8	659.0	1,890.7	248.9	136.6	101.6	796.4	726.1	3,560.1	716.6	2,359.4	186.7	2,947.7	14,498.7	2,647.4	17,146.1
Total DALYs due to unintentional injuries	11,620.9	9,920.8	3,132.6	6,813.4	1,451.4	4,224.7	9,767.9	3,042.1	3,123.3	8,317.2	5,865.9	37,871.6	971.2	26,732.4	84,118.3	48,993.4	133,111.7
% of all alcohol-attributable DALYs due to unintentional injuries	9.4%	17.4%	21.0%	27.8%	17.2%	3.2%	1.0%	26.2%	23.2%	42.8%	12.2%	6.2%	19.2%	11.0%	17.2%	5.4%	12.9%
Intentional Injuries																	
Self-inflicted injuries	28.3	84.7	102.9	142.7	9.1	5.1	5.7	151.8	107.1	720.5	79.7	309.0	72.5	413.6	1,825.0	407.7	2232.7
Violence	316.8	852.6	148.0	1,882.9	91.9	8.4	16.9	39.1	82.4	918.2	154.8	291.2	7.4	335.9	4,507.2	639.3	5146.5
Other intentional injuries	0.0	0.0	2.0	8.8	0.0	0.6	0.6	0.1	4.6	0.5	2.0	10.8	0.2	7.2	33.6	3.8	37.5
All alcohol-attributable DALYs due to intentional injuries	345.2	937.4	253.0	2,034.5	101.0	14.1	23.1	191.1	194.2	1,639.2	236.4	610.9	80.1	756.7	6,365.8	1,050.8	7,416.7
Total DALYs due to intentional injuries	3,353.1	5,908.2	1,412.2	6,616.8	437.4	542.7	2,946.4	1,038.9	934.8	4,488.5	2,350.8	9,458.8	655.9	8,614.2	35,767.1	13,112.4	48,879.5
% of all alcohol-attributable DALYs due to intentional injuries	10.3%	15.9%	17.9%	30.7%	23.1%	2.6%	0.8%	18.4%	20.8%	36.5%	10.1%	6.5%	12.2%	8.8%	17.8%	8.0%	15.2%
Total alcohol-attributable DALYs due to unintentional and Intentional injuries	1,433.6	2,665.1	911.9	3,925.2	349.9	150.7	124.7	987.4	920.3	5,199.2	953.1	2,970.4	266.9	3,704.5	20,864.5	3,698.2	24,562.8

*Numbers are rounded to the nearest thousand. Zero (0) indicates fewer than 500 alcohol-attributable deaths in the injury category

** M=Men, W=Women

AFR=Africa, AMR=American, EMR=Eastern Mediterranean, EUR=Europe, SEAR=South East Asia, WPR=Western Pacific

Fire was excluded, as we could not find reliable estimators on global scale to quantify the impact of alcohol. However, there are national data and indicators of several causal relationships on country levels (see Rehm et al., 2006b for Canada).

War as an injury category was excluded, as we did not have information that alcohol could be a cause of war.

Table 5. Deaths due to alcohol-attributable unintentional and intentional injuries by sex and age in world in 2002

Injuries	Men							Women							Total
Unintentional Injuries	0-14	15-29	30-44	45-59	60-69	70+	Total Men	0-14	15-29	30-44	45-59	60-69	70+	Total Women	All ages
Road traffic accidents	11,077	78,319	77,086	28,614	9,683	8,785	213,564	4,413	5,649	8,135	5,907	2,175	3,036	29,315	242,879
Poisonings	0	11,051	15,311	18,060	6,590	1,324	52,337	0	2,684	3,215	4,218	2,411	868	13,395	65,732
Falls	0	5,340	7,609	9,782	5,116	8,113	35,959	0	666	840	1,460	894	3,215	7,076	43,036
Drowning	0	15,996	14,918	10,876	3,356	2,802	47,947	0	2,953	2,560	2,301	1,109	2,098	11,022	58,969
Other unintentional injuries	8,981	39,531	38,839	32,232	15,486	16,141	151,210	2,374	5,426	6,100	6,050	3,636	11,111	34,697	185,907
All alcohol-attributable deaths due to unintentional injuries	20,058	150,239	153,762	99,564	40,230	37,166	501,018	6,786	17,379	20,851	19,935	10,225	20,328	95,504	596,522
Total deaths due to unintentional injuries	370,790	546,011	516,665	434,796	198,882	239,826	2,306,970	288,911	220,212	190,902	183,582	109,819	250,177	1,243,603	3,550,573
% of all alcohol-attributable deaths due to unintentional injuries	5.4%	27.5%	29.8%	22.9%	20.2%	15.5%	21.7%	2.3%	7.9%	10.9%	10.9%	9.3%	8.1%	7.7%	16.8%

Intentional Injuries	0-14	15-29	30-44	45-59	60-69	70+	Total Men	0-14	15-29	30-44	45-59	60-69	70+	Total Women	All ages
Self-inflicted injuries	0	23,296	25,598	17,172	7,730	3,415	77,210	0	4,856	5,244	3,684	1,876	1,706	17,367	94,577
Violence	1,505	60,776	44,632	23,827	6,690	4,123	141,554	1,233	6,208	6,569	4,683	1,979	2,112	22,784	164,339
Other intentional injuries	0	515	399	180	83	39	1,216	0	37	49	58	20	9	174	1,390
All alcohol-attributable deaths due to intentional injuries	1,505	84,586	70,629	41,179	14,503	7,577	219,980	1,233	11,101	11,863	8,426	3,874	3,828	40,325	260,305
Total deaths due to intentional injuries	27,442	408,528	342,496	211,965	87,451	78,819	1,156,701	22,385	133,254	117,903	92,926	41,428	53,144	461,041	1,617,742
% of all alcohol-attributable deaths due to intentional injuries	5.5%	20.7%	20.6%	19.4%	16.6%	9.6%	19.0%	5.5%	8.3%	10.1%	9.1%	9.4%	7.2%	8.7%	16.1%

| Total alcohol-attributable deaths due to unintentional and intentional injuries | 21,563 | 234,825 | 224,391 | 140,743 | 54,733 | 44,743 | 720,999 | 8,019 | 28,480 | 32,714 | 28,361 | 14,099 | 24,156 | 135,829 | 856,828 |

Fire was excluded, as we could not find reliable estimators on global scale to quantify the impact of alcohol. However, there are national data and indicators of several causal relationships on country levels (see Rehm et al., 2006b for Canada).
War as an injury category was excluded, as we did not have information that alcohol could be a cause of war.

The highest proportion of alcohol-attributable DALYs due to unintentional injuries among men was in those 30-44 (25.2%) and 70 years and older (23.4%). Among women, the highest proportion of alcohol-attributable DALYs due to unintentional injuries was among those 30-44 (8.7%), 45-59 (9%) and 70 years and older (13%).

In terms of unintentional injuries, the main cause of alcohol-attributable DALYs among men (15-29 and 30-44 years of age) and women (45-59 years of age) was road traffic accidents. The category of "other unintentional injuries" (for definition see Table 1) was the leading cause of alcohol-attributable DALYs in the remainder of the age groups for both genders.

The proportion of alcohol-attributable DALYs due to intentional injuries for men was highest between ages 15-59, and for women between ages 30-44 and over 70. Similar to mortality, violence was the leading cause of alcohol-attributable DALYs in all age categories for both men and women (with the exception of men 60-69 years of age, for whom the leading cause of alcohol-attributable DALYs was due to self-inflicted injuries).

Discussion

These findings showed that alcohol is an important risk factor for injury deaths and DALYs, especially in the age group of young adults 15 – 44 in the majority of the sub-regions and globally. Accidents and the resulting adverse effects (irrespective of being attributable to alcohol), are the leading cause of death for people under 35 years of age in many Western market economies of the world (as examples, for the USA see National Center for Health Statistics, 2004; for Canada see Statistics Canada, 2003). On a global scale, injuries, and unintentional injuries in particular, play a similar role. When infant mortality is excluded, injuries were the most important cause of death for age groups up to age 45 in the year 2002. Ninety percent, 88% and 62% of all male deaths in age groups 5-14, 15-29 and 30-44, respectively, were caused by injuries described in the categories above. For women, the respective proportions were 84%, 76% and 47%, with proportions larger for unintentional injury than for intentional injury.

As injuries are responsible for the overwhelming proportion of premature deaths in adolescence and early adulthood, and as alcohol is one of the most important causes of injuries, it is not surprising that alcohol is the most important risk factor for premature death in these age groups on a global level (Rehm, 2004; Rehm et al., 2006). Similar relationships can be seen for premature disability and premature burden of disease. Given that alcohol-attributable injuries are in principle completely avoidable, and could be avoided in a relatively short time frame (Babor et al., 2003), pragmatically effective and cost effective policies to reduce the impact of alcohol on injury must be promoted (Babor et al., 2003; Chisholm et al., 2004).

Table 6. DALYs due to alcohol-attributable unintentional and intentional injuries by sex and age in world in 2002

Injuries	Men							Women							Total
Unintentional Injuries	0-14	15-29	30-44	45-59	60-69	70+	Total Men	0-14	15-29	30-44	45-59	60-69	70+	Total Women	All
Road traffic accidents	505,753	3,167,860	2,209,085	509,790	94,454	60,826	6,547,768	226,596	256,359	253,993	114,843	23,867	23,778	899,437	7,447,205
Poisonings	0	379,499	382,537	290,166	60,739	8,887	1,121,828	0	94,089	82,292	70,844	24,242	6,518	277,985	1,399,814
Falls	0	371,503	268,242	187,186	55,829	60,014	942,775	0	61,520	39,552	33,870	12,500	27,070	174,512	1,117,286
Drowning	0	546,583	372,136	173,566	30,910	18,770	1,141,965	0	102,172	65,091	38,232	11,149	15,731	232,376	1,374,341
Other unintentional injuries	609,957	1,967,742	1,259,650	617,191	171,527	118,314	4,744,380	134,435	390,441	262,425	138,371	48,596	88,805	1,063,074	5,807,454
All alcohol-attributable DALYs due to unintentional injuries	1,115,710	6,433,188	4,491,649	1,777,899	413,459	266,812	14,498,716	361,031	904,582	703,353	396,161	120,354	161,902	2,647,384	17,146,100
Total DALYs due to unintentional injuries	24,958,411	29,022,961	17,857,390	8,884,658	2,256,223	1,138,622	84,118,265	19,904,070	13,937,172	8,039,556	4,388,947	1,476,882	1,246,738	48,993,364	133,111,628
% of all alcohol-attributable DALYs due to unintentional injuries	4.5%	22.2%	25.2%	20.0%	18.3%	23.4%	17.2%	1.8%	6.5%	8.7%	9.0%	8.1%	13.0%	5.4%	12.9%

Intentional Injuries	0-14	15-29	30-44	45-59	60-69	70+	Total Men	0-14	15-29	30-44	45-59	60-69	70+	Total Women	All
Self-inflicted injuries	0	809,009	645,563	275,851	71,546	23,008	1,824,977	0	177,010	136,475	62,109	19,102	12,978	407,674	2,232,651
Violence	84,557	2,591,602	1,325,720	411,695	64,612	29,001	4,507,188	54,806	269,693	194,856	83,105	20,498	16,363	639,321	5,146,509
Other intentional injuries	0	19,024	10,589	2,977	772	279	33,641	0	1,305	1,270	1,006	197	71	3,849	37,491
All alcohol-attributable DALYs due to intentional injuries	84,557	3,419,635	1,981,872	690,523	136,930	52,289	6,365,806	54,806	448,008	332,602	146,220	39,797	29,412	1,050,845	7,416,651
Total DALYs due to intentional injuries	1,978,065	17,989,708	10,848,764	3,765,257	836,537	348,729	35,767,060	1,616,270	5,728,173	3,411,062	1,677,330	421,555	258,041	13,112,431	48,879,491
% of all alcohol-attributable DALYs due to intentional injuries	4.3%	19.0%	18.3%	18.3%	16.4%	15.0%	17.8%	3.4%	7.8%	9.8%	8.7%	9.4%	11.4%	8.0%	15.2%
Total alcohol-attributable DALYs due to unintentional and intentional injuries	1,200,266	9,852,824	6,473,521	2,468,422	550,389	319,100	20,864,522	415,837	1,352,591	1,035,955	542,381	160,151	191,314	3,698,229	24,562,751

Fire was excluded, as we could not find reliable estimators on global scale to quantify the impact of alcohol. However, there are national data and indicators of several causal relationships on country levels (see Rehm et al., 2006b for Canada).

War as an injury category was excluded, as we did not have information that alcohol could be a cause of war.

References

Babor T et al. (2003). *Alcohol: no ordinary commodity. Research and public policy*. Oxford University Press, Oxford and London.

Chisholm D et al. (2004). Reducing the global burden of hazardous alcohol use: a comparative cost-effectiveness analysis. *Journal of Studies on Alcohol* **65**, 782-793.

Hill AB (1965). The environment and disease: association or causation? *Proceedings of the Royal Society of Medicine*, **58**, 295-300.

Lopez AD et al. (2006) *Global burden of disease and risk factors*. Oxford University Press and the World Bank, New York and Washington.

National Center for Health Statistics (2004). *Health, United States, 2004 with chartbook on trends in the health of Americans*. Hyattsville, Maryland.

Rehm J et al. (2006b). *The costs of substance abuse in Canada 2002*. ISBN no. 1-897321-10-4. Canadian Centre on Substance Abuse, Ottawa.

Rehm J et al. (2003b). Alcohol-related mortality and morbidity. *Alcohol Research and Health* **27**(1), 39-51.

Rehm J et al. (2006a). *Alcohol consumption and global burden of disease 2002*. Final report to World Health Organization. World Health Organization, Geneva.

Rehm J, Patra J, Popova S (2006c). Alcohol-attributable mortality and potential years of life lost in Canada 2001: implications for prevention and policy. *Addiction* **101**(3), 373-384.

Rehm J et al. (2003c). The relationship of average volume of alcohol consumption and patterns of drinking to burden of disease – an overview. *Addiction* **98**, 1209-1228.

Rehm J et al. (2003a). Alcohol as a risk factor for global burden of disease. *European Addiction Research* **9**, 157-164.

Rehm J et al. (2004). Alcohol use. In: Comparative quantification of health risks. Global and regional burden of disease attributable to selected major risk factors. Vol. 1. Ezzati M et al., eds, pp. 959-1108. World Health Organization, Geneva.

Rehm J, Taylor B, Room R (2006). Global burden of disease from alcohol, illicit drugs and tobacco. *Drug and Alcohol Review* **25**(5), 503-13.

Rothman KJ, Greenland S (1998). Causation and causal inference. In *Modern epidemiology*, second edition, Rothman KJ, Greenland S, eds, pp. 7-28. Lippincott-Raven Publishers, Philadelphia, PA.

Statistics Canada (2003) *Canada e-book*. Catalogue no. 11-404-XIE. Statistics Canada, Ottawa. Retrieved August 24, 2006, from http://www.statcan.ca/bsolc/english/bsolc?catno=11-404-XIE

World Health Organization (2000). *World health report – Health systems: improving performance.* World Health Organization, Geneva.

World Health Organization (2002). *World health report: Reducing risks, promoting healthy life.* World Health Organization, Geneva.

SECTION II: ISSUES RELATED TO EMERGENCY DEPARTMENT STUDIES
INTRODUCTION

Robin Room - Turning Point Alcohol and Drug Centre and
University of Melbourne | Melbourne, AUSTRALIA
Cheryl J. Cherpitel - Alcohol Research Group | Emeryville, CA US

This section of the book addresses a number of areas of special importance to the study of alcohol and injuries in studies of emergency department (ED) patients. The first chapter in this section, by Roizen, provides a critical overview of the epidemiological studies of alcohol and injury to date. She has been intimately involved in the critique of such ED studies since the late 1970's, and provides an in-depth perspective involving both developed and developing countries. She puts a particular emphasis on issues related to the methodological problems and limitations of existing studies, as well as issues regarding the future development of studies in the ED, especially in relation to the role of alcohol and its effects on injury.

The second chapter in this section, by Rehm and Room, addresses the important issue of estimating alcohol-attributable fractions for injury. Estimates obtained from aggregate-level data are compared to those obtained from individual-level data. While the two estimates are reasonably comparable, both approaches show considerable weaknesses, as well as strengths. Alternative methods for determining alcohol-attributable fraction are presented, and the potential of ED data for improving these estimates is discussed.

The third chapter, by Cherpitel, presents a pragmatic approach to conducting epidemiologic studies of alcohol and injuries in the ED, following the "Cherpitel Model" outlined in the first chapter in this section (see Chapter 5), with the intent of guiding those wishing to conduct similar studies. This chapter lays out practical issues and biases, and caveats in the development of an ED study design, including sample selection and procedures for obtaining patient interviews and estimates of blood alcohol concentration.

The final chapter in this section, by Gmel and Daeppen, presents an in-depth look at conceptual issues in studies of alcohol and injury in the ED. Their view is that we are well beyond substantiating that alcohol consumption is, indeed, associated with injury, and must now move forward with a focus on explaining the actual magnitude of this association in terms of relative risk and associated alcohol attributable fractions. This will include a closer examination of the diversity of findings across ED studies in relation to the interaction of frequency and usual volume of drinking with drinking in the injury event, and new thinking on the choice of control subjects, including applications of the case-crossover design.

Taken together, the chapters in this section provide a comprehensive overview of the important issues relevant in conducting sound epidemiological studies of alcohol and injury in the ED, limitations inherent in such studies, and the resulting quality and usefulness of the results obtained.

CHAPTER 5 :
AN OVERVIEW OF EPIDEMIOLOGICAL EMERGENCY ROOM STUDIES OF INJURY AND ALCOHOL

Judith Roizen - Consultant in Medical Demography | Denver, CO US

Scope and Types of Studies

Introduction

An alcohol presence in casualties of all types has been the subject of scientific investigation for over a century. However, before that an alcohol presence in injury causing accidents was of little social or even medical concern in the U.S. and in most countries. The colonial period in North America saw frequent notices of death from drunkenness and although the dangers of drinking were acknowledged, little attention, even through the temperance period, was concentrated on preventing accidents by urging the control of drinking (Levine, 1983). Injuries, whether fatal or non-fatal, were not seen as the major social, economic and medical problem they have now become in both developed and developing countries.

Thirty years ago, the then Social Research Group (now the Alcohol Research Group in Emeryville, CA) undertook a large-scale review of the literature of alcohol involvement in casualties and crime for the National Institute of Alcohol Abuse and Alcoholism (NIAAA). The purpose of this well-funded, year-long review and re-analysis was to identify the literature (which proved to extend back more than a century) and to reanalyse data (where available), but, as important, to begin to develop a conceptual framework within which to understand alcohol's role in the events which lead to injury, injury-related death and criminal behaviour (Aarens et al., 1978).

The perspective the team took was largely questioning of the assumed causal role of alcohol found in much of the contemporary research. The emphasis in the literature was on assessing the level and kind of an alcohol "presence" rather than questioning whether and how causality had been demonstrated. As discussed below, this issue remains salient in evaluating the research on alcohol and injuries as seen from the "window" of emergency rooms (ERs), including urgent care departments of hospitals, in the past twenty years.

In this paper, the following questions are asked of evidence from ER studies: How is the ER "window" described? What types of injury events are included? Which aspects of alcohol, e.g. "alcohol use prior to injury" in contrast to patterns of drinking or "alcohol abuse", are thought to be most important? What effects of alcohol are assumed to have contributed to the injury event? Is alcohol shown or assumed to be a contributing factor (i.e. "involved")? Is the focus of the study largely on the person or on characteristics of the event? How does evidence come together to define the **role** of alcohol?"

Research using ERs as a "window" has developed rapidly as emergency medicine has developed over the last three decades, but primarily in developed countries. The use of ERs in many countries has also increased, sometimes reflecting the only possible medical point of entry into the health system (as for the uninsured in the U.S.). In the original Casualty study, we reported on only two all-injury ER studies from the U.S. and 15 studies from abroad, three of which were carried out by Honkanen in Finland. Ten years later, 20 years ago, at the Toronto conference on drinking and casualties in 1985, the paper that I then gave included some 40 studies of which half were U.S. studies (Roizen, 1989). By then the research literature on alcohol and injuries had grown in many developed countries.

When I undertook the WHO literature review on alcohol and injuries in 2003 for the WHO Collaborative Study on Alcohol and Injuries, on which this paper is based, I still believed a review such as the previous two was a manageable task. The task was to prepare a review of the "evidence of the role of alcohol in fatal and non-fatal injuries/accidents and violence, including domestic violence". A global review on these harmful events is now a massive undertaking where partial success is the only reasonable goal. As an example, limiting a search simply to PUBMED gives 10 040 entries for "(Injur* or Trauma) and Alcohol" between 1990 and 2004. "(Injur* or Trauma)" and "(Alcoholism or Alcohol Abuse or Drinking Problem)" gives 1,594 for the same time period. Adding "emergency" can help for some searches on developed countries but not for many developing countries, where emergency care is limited. One of the major problems for an international review is that, by and large, the country in which the study occurs is not indexed by its region. So for sub-Saharan Africa a search has to be carried out for each country, similarly for other parts of the world. Searches cannot be limited by qualifying the search with, for example, "U.S." because "U.S." is not a usual text indexing term. This, in many cases, means state by state searches. Although many studies located by the multiple searches were not relevant to the WHO remit, about half would have been.

The task was not simply to identify the best studies, methodologically and theoretically, but those which would inform us of the state of research on alcohol and injuries in all countries where studies were accessible. Even a relatively poor published study or one with poor measurement of alcohol use can indicate the direction research is likely to take in a particular country, the perspective on alcohol problems researchers and institutions have, and whether or not researchers are connected to the larger alcohol research community.

Another measure of how interest in alcohol and injury has grown is the very large number of journals containing relevant articles. Some 200 different journals were consulted, mostly volumes published in the last 15 years. (Further details of the search and review and list of journals can be obtained from the author.) Only a small number of the studies located for the WHO review are used in this discussion paper including studies of ERs.

The Cherpitel Model

In the early 1980s, I argued that if there ever was a field of research that was "pre-paradigmatic", research on alcohol and casualties, including accidents, was it. I meant "paradigm" in one of the ways used by Thomas Kuhn (1962):

as a recognized scientific achievement that for a time provides model problems or questions and solutions to a community of practitioners and which provides the conceptual and instrumental tools that are needed to answer these questions.

Twenty years ago such a "paradigm" began to emerge with the first ER study of Cherpitel (Cherpitel, 1989). At the time of this first Cherpitel study there had been only one U.S. emergency room study which examined an alcohol presence in all injuries. There had been a number of other emergency room studies, largely Scandinavian, that could be built upon.

The key elements of the Cherpitel model and those studies that have followed over the last decade and a half are these:

KEY ELEMENTS IN THE DESIGN, EXECUTION AND REPORTING OF ER STUDIES CONSISTENT WITH THE CHERPITEL MODEL

1 A probability sample of those attending the ER to assure that all times of day and days of week are equally represented to control for temporal variation in alcohol use and known variation in alcohol presence in different types of injury events.

2 Breath or blood testing for alcohol within 6 hours of the presenting injury. Reporting of time from injury to alcohol test and drinking after injury. Measures of patterns of alcohol use, harmful drinking, dependence and consequences related to drinking, as well as alcohol use within 6 hours prior to the event and the amount consumed.

3 Detailed presentation of numbers and reasons for patients not participating. Reasons may include refusals, severity of injury, loss to follow-up, type of injury or missed cases.

4 Criteria spelled out for inclusion/exclusion from sample, including alcohol and drugs taken after the injury, age ranges and first language.

5 Specification of whether informed consent was needed or not for participation in study.

6 Detailed demographic description of sample.

7 Where patients are interviewed, estimate of time taken for interview; description of interview contents and completion rates.

8 Use of drugs other than alcohol in the 6 hours before the injury, and types and patterns of drug use.

9 Place of injury occurrence, places of drinking within 6 hours prior to injury and drinking companions for last drink prior to the event.

10 Feeling drunk at the time of injury and causal attribution of the event to drinking.

11 Ideally, case-control or case cross-over design to provide risk measurements.

12 Causes/mechanisms and types of injuries reported.

13 Study generalizability and limitations reported.

14 Description of the ER/hospital receiving unit, in terms of possible sample selection bias and other possible measurement errors.

15 Contextual variables: Organizational/administrative (trauma center, public or pirvate ER); Socio-cultural (per capita consumption, legal drinking age, legal level of intoxication for driving, stigmatization of alcohol use, detrimental drinking pattern level, homicide rate, gross national product)

By and large this defines the "gold standard" which – apart from the work of a few individuals and research groups – is not often met. Most of the studies of Cherpitel and her colleagues meet most of these design elements, although there has been some evolution over time (e.g. inclusion of drug use is a very recent development). A selected number of these studies are shown in Table 1. Despite their relatively high levels of sample attrition, these epidemiologically sound studies using a common methodology enhance the scope for comparative analysis. Wide differences in research design of often non-probablility samples of injured ER patients has been a major confounder in interpreting results from reviews of studies in the past.

Table 1 shows, for each study using this common methodology, the percentage of alcohol-positive cases for those presenting to the ER, as well as the percentage of those who were intoxicated. Several of these studies have a non-injured hospital/ER comparison group. Although these comparison groups are now generally agreed to be less than ideal, as both groups tend to show heavier drinking than the general population (see chapter 8 in this volume for a further discussion), at worst they provide a conservative assessment of alcohol presence. In almost all studies with these comparison groups, the non-injured have lower levels of alcohol presence and intoxication than the injured.

Looking first at the U.S. studies of injured only, we see that both the percentages of alcohol-present cases and the percentages intoxicated vary: the percentage of alcohol-positive cases in the Mississippi study is 15%; in suburban Contra Costa County, CA (1985 sample) 11%; in urban San Francisco 23% and in the Contra Costa Health Maintenance Organizations (HMOs) 6%. Percentages intoxicated are 6%, 4%, 15% and 1% respectively. These differences largely reflect the different catchments of the ERs. For example, the San Francisco ER serves an inner city clientele, including many homeless, while the HMO sample is representative of a suburban general population. The Mississippi study shows higher proportions of alcohol than might be expected in a "dry" part of the country. The two Canadian samples also show differences in alcohol presence and intoxication in two regions of the same country. The differences are surprisingly large. Levels of intoxication, which may be considered presumptive of a contributory effect of alcohol, are relatively low except for the San Francisco and Alberta samples, but the latter is a relatively small sample. These are lower levels of alcohol presence than are often found in non-probability samples or samples which include only a single cause of injury, e.g. assault or head injuries.

The Mexico samples show relatively high percentages of alcohol-positive cases and intoxication cases – in Acapulco 22% and 8% and in Mexico City 21% and 11%, respectively. These differences again may reflect ER catchments as well as differences in patterns of alcohol use in the different populations, but distinguishing these is difficult. General population studies suggest that, in comparison to the U.S., Mexican drinkers show less frequent regular drinking but more frequent heavy drinking leading to drunkenness. The one Spanish study shows percentages of alcohol presence and intoxication similar to the Contra Costa County studies. This may reflect the daily but relatively light drinking in Spain and the suburban drinking patterns in Contra Costa, at least amongst those in Contra Costa who attend an HMO, but these would be generalisations based on few ER sites.

Differences in alcohol presence are shown in Table 1 within the same state (California), in two different states in the U.S., in two Canadian provinces and across countries. What can be seen from these studies is a fairly high percentage of alcohol present cases in most ERs. What cannot be assumed is that these differences reflect drinking cultures in the areas that they are found. Except for studies in California (and these are from a small area of Northern California) and Mexico, there are few ER studies from any given geographic area – state, province or country. The single most significant difference shown here is the difference in alcohol use in the two early Cherpitel studies – reflecting the difference in two ER populations in the same geographic area – raising the serious question of generalisability addressed below.

Table 1. Selected ER studies measuring BAC* among probability samples of ER patients

Reference	Locale	Year	Length of collection	Completion rate	Age	Alcohol Measure	Injured % positive BAC (n)	Non-injured % positive BAC (n)
Cherpitel (1988)	San Francisco, CA, USA,	1984-85	2 months	75	≥ 18	Breath within 6 hours	25 (502) 15 ≥ 0.10	10† (1192) 6† ≥ 0.10
Cherpitel (1988)	Contra Costa Co., CA USA (4 Ers)	1985	3 months	73	≥ 18	Breath within 6 hours	11 (1026) 4 ≥ 0.10	5† (1306) 3 ≥ 0.10
Rosovsky & Garcia (1988)	Acapulco, Mexico (3 ERs)	1987	1 month	82	≥ 18	Breath within 6 hours	22 (376) 8 ≥ 0.10	7† (308) 1† ≥ 0.10
Cherpitel & Rosovsky (1990)	Mexico City, Mexico (8 ERs)	1986	1 week	91	≥ 18	Breath within 6 hours	21 (1644) 11 ≥ 0.10	3† (462) 2† ≥ 0.10
Cherpitel et al. (1991)	Barcelona, Spain	1987-88	1 year	80	≥ 18	Breath within 6 hours	10 (1640) 5 ≥ 0.10	4† (657) 1† ≥0.10
Cherpitel (1992)	Contra Costa Co., CA, USA	1986-87	1 year	72	≥ 18	Breath within 6 hours	13 (1004) 6 ≥ 0.10	–
Cherpitel (1993)	Contra Costa Co., CA., USA (3 HMO ERs)	1989	6 weeks	76	≥ 18	Breath within 6 hours	6 (452) 1 ≥ 0.10	3† (614) 1 ≥ 0.10
Cherpitel (1996)	Jackson, Mississippi, USA	1992	6 months	**	≥ 18	Breath within 6 hours	15 (172) 6 ≥ 0.10	–
Cherpitel et al. (1999)	Alberta, Canada	1989	1 week, Feb	48	≥ 18	Urine	21 (196) 13 ≥ 0.10	6 (348) 2 ≥ 0.10
Cherpitel et al. (1999)	Quebec, Canada	1989	1 week, July	80	≥ 18	Urine	27 (256) 2 ≥ 0.10	3 (377) 0.3 ≥ 0.10

* BAC is recorded as mg %: positive ≥ 0.01 (10 mg of alcohol per 100 ml of blood).
† p < 0.5 comparison of positive breathalyzer readings between injured and non-injured in the same sample.
** Not reported for injury-only sample
(Adapted from Cherpitel, 1993c)

The study of Borges et al. (1998), described in Table 2, is noteworthy because it is one of very few ER studies that can claim to be a population-based sample of injuries (at least of all moderate to serious injuries) in that it includes all three ERs in Pachuca, Mexico, and also includes a general population control group. (See also Watt et al., 2004.) However, the Borges study does not report the contribution of each of the ERs to the demographic make-up and alcohol measurement in the overall study sample.

Table 2 looks at three studies using this same methodology to look at alcohol patterns and problems in three countries. These again show considerable variation in alcohol use in the injury event and in patterns of use. What is noteworthy about these studies compared to studies of injury and alcohol use reviewed 20 years ago is the reporting of drinking in the event as well as drinking patterns and problems. The use of multiple measures of alcohol use in ER studies on injury in many sites and countries is one of the major developments in ER research in the past 20 years. This is an important achievement.

Noteworthy are the considerably higher levels of drinking problems and higher quantity and frequency of drinking than would be predicted by the levels of drinking in the event. The high levels of drinking problems and harmful drinking clearly show the importance of the ER as a venue for education and intervention both among the injured and medical cases.

Exemplary ER studies from developed countries

Table 3 reports studies of patterns of drinking using differing designs, several of which report drinking in the event and drinking problems. Taken together they offer some interesting findings. In the single Cherpitel study reported in Table 3, the differences in blood alcohol concentrations (BACs) between injured and non-injured are not reflected in the levels of harmful and dependent drinking between the two groups, with high levels of harmful drinking in both groups. For the non-injured there is little relationship between drinking in the previous six hours and drinking problems. Vingilis et al. (1994) in this case-series report no significant difference in alcohol dependency and use between BAC positives and negatives. The Australian study (Holubowycz and McLean, 1995), however, suggests a strong relationship between BAC level and drinking problems. The relationship between BAC and drinking patterns and problems varies considerably among studies, suggesting the need for both types of measures in future studies. This has major implications, however, for routine and unburdensome testing and questioning of ER patients. These studies and others like them raise the question of whether the main focus of ER research on injuries is or should be ascertaining the alcohol/drug contribution to the injury event or whether the ER is simply an important venue to identify problem drinkers and offer interventions. Trying to maximize both agendas puts a considerable burden on the ER, especially when it is still unclear what the relationship is between drinking before the injury event and harmful drinking patterns. What these few selected studies show (and they are selected from many others) is that interest and research effort into these problems is reflected in many countries, albeit mostly developed countries, and that a wide range of drinking patterns and problems are now included in ER studies. This is a considerable gain over research carried out over two decades ago. Again, however, these studies are non-comparable and results are not generalisable, since they are based on different designs and single sites.

Research on alcohol and injury from developing countries.

Although many studies of injury and alcohol from developing countries were located for the WHO review, the great majority come from only a few countries. In the great majority of countries, apart from a few with a long tradition of the study of alcohol problems including Mexico and South Africa, epidemiological research on alcohol and injuries is in its infancy – if it exists at all. Emergency room studies are even rarer – reflecting the recency of the development of emergency medicine in these countries. Table 4 describes selected studies from developing countries, although Taiwan is included here only because of the fairly recent development of emergency medicine there. These are some of the best studies and in many cases reflect collaboration with or knowledge of the work of alcohol researchers in developed countries. More commonly a study from a developing country will comment on alcohol use or "drunkenness" with no serious attempt at alcohol measurement. The great majority of studies carried out in developing countries are case-series, reflecting both lack of resources for more ambitious studies, the lack of a research tradition on these problems, and very likely lack of access to the reports, books and journals which can connect these researchers to a wider community.

The feasibility of accurate alcohol measurement prior to injury in most health centres and hospitals in developing countries is limited. There are many studies which report the difficulties. These include the expense of breathalysers, the expense and possible cross-contamination of blood tests, time constraints on staff, and reluctance of patients to admit alcohol use. However, many of the case-series studies suggest possible ways forward, including, for example, the "building of a training curriculum for Africa" for trauma life support (Andrews et al. 1999). Substance abuse could be built into this both for prevention and treatment but also for clinical assessment within hospitals. However, prevention of injuries which occur after drinking may be addressed without the full picture which research in many developed countries is able to give. Case-series studies of injury in developing countries are at their strongest in assessing the probable immediate causes of an injury: lack of pedestrian crossings, lack of restraints in motor vehicles, failure to use helmets when cycling. These can be triangulated with developed country research, and local research studies can be initiated where, with forethought, they can be generalised to other health facilities and, in suggesting prevention strategies, to similar geographic settings.

Table 2.   Three selected studies of alcohol use in the injury event and drinking patterns

	Investigator/locale/date of publication	Years of study	Study population and design	Measures of alcohol use	Results: patterns of alcohol use (including acute)	Study limitations
A	Borges et al. Mexico (Pachuca) (1998)	1996 October to Nov.

1997 June to July | Prospective. Case / control.

Probability. All patients

≥ 18 years who arrived at one of 3 ERs because of medical emergencies or trauma injuries.

Of 1624 potential participants, 1511 agreed to interview (93%). Of these 67% of "injured" were male, compared to 62% of "medical" patients.

(General population data not reported here) | Breath samples (not reported); alcohol quantity/frequency*; frequency of drunkenness; CAGE; TWEAK; previous alcohol-related accident.

* defined as:
Moderate
(moderate or high frequency/low quantity)
High
(moderate or high frequency/ mod. quantity or mod. frequency / high quantity) and
Heavy
high frequency/ high quantity) | | Injured % (N=756) | Medical % (N=755) |
|---|---|---|
| Q/F moderate | 4 | 2 |
| Q/F high | 8 | 4 |
| Q/F heavy | 2 | 1 |
| Usual # drinks 5–11 | 21 | 8 |
| >12 | 17 | 8 |
| Alcohol consumed within 6 hrs: yes | 16 | 3 |
| CAGE positive (≥2 of 4 items) among drinkers | 25 | 21 |
| TWEAK positive (≥3 of 5 items) among drinkers | 37 | 28 |
| Previous alcohol-related accident | 5 | 5 | | Study time periods not clear.

Breath tests not reported.

Majority of non=participants were refusals. |
| **B** | Cherpitel US (California) (1993) | 1985 3 months (County/ community sample A.)

1989 6 weeks (HMO sample B) | Prospective. Probability.

Patients sampled every other day for all study weeks.

In each of the four hospitals in Sample A, every second patient was selected. In Sample B, every fourth patient was selected.

N for Sample A = 3609, of whom 73% completed.

N for Sample B = 1458, of whom 76% completed.

49% were male for both samples.
Of injured in Sample A, 60% were males.
Of injured in Sample B, 56% were males. | BAC (breath) levels; alcohol quantity/frequency*

Self-reported consumption prior to injury; alcohol problems (social or health); indicators of alc. dependence, including reports of blackouts, relief drinking, binge drinking, hands shaking after morning drinking.

*defined as:
Moderate
(moderate or high frequency/ moderate quantity or moderate frequency/high quantity)
Heavy
(high frequency/high quantity) | Sample A: County/community. (In percents)

	Males injured	Males non-inj.	Females injured	Females non-inj.
BAC pos.	13	9	7	3
BAC ≥.10	5	5	3	2
Self-report	22	22	14	10

Sample B: HMOs (In percents)

	Males injured	Males non-inj.	Females injured	Females non-inj.
BAC pos.	8	6	5	3
BAC ≥.10	2	0.4	1	1
Self-report	11a	8a	8a	7a

Drinkers (male and female) (In percents)

	Sample A		Sample B	
	Injured	Non-inj.	Injured	Non-inj.
Moderate Q/F	40	30	29	25
Heavy Q/F	19	13	13	4a
> 3 social/health consequences	6	5	2a	1a

a. p<.05 comparison of injured and non-injured between Samples A and B. | Considerably longer time of study for Sample A.

Some possible seasonal variation. |

C			
Cherpitel, Moskalewicz and Swiatkiewicz Poland (Warsaw) (2004)	2002 7 months	Prospective. Case/control. 1000-bed public hospital. Probability sample of injured and non-injured patients ≥18 years. 1089 sampled, 734 interviewed (68%). Of these, 80% breathalysed within 6-hour limit. Of those interviewed, 55% male (61% of injured, 47% non-injured were males.)	BAC (breath) levels; Alcohol quantity/frequency * Self-reported consumption prior to injury; alcohol problems (social or health); harmful drinking/ alcohol abuse (ICD-10; DSM-IV) *defined as: **Moderate** (moderate or high frequency/ moderate quantity or moderate frequency/high quantity) **Heavy** (high frequency/high quantity)

BAC positive: 5.5% injured males. BAC ≥0.10 = 2.6%
No BAC positive non-injured males or any females.

	Males injured	Males non-inj.	Females injured	Females non-inj.
Moderate Q/F	52	32	16	12
Heavy Q/F	16	2	2.5	1.7
Drunkenness (Among drinkers) ≥monthly	20	6	2	3
Harmful drinking/Alc. abuse	16	6	6	3
Alc.related accidents	9.6	4.9	2	2

* $p<0.05$ between injured, non-injured males.
Self-reports: among injured males, 35% reported 7+ drinks during 6 hours prior to injury.

Large N not interviewed.
Complicated sample design.
Specific hours, days, months sampled not reported.
Variation in time periods in availability of ER services may have affected results.
Self-reports for injured males widely discrepant from reported BAC.

Table 3. Exemplary ER studies from developed countries

	Investigator/ locale/ date of publication	Years of study	Study population and design	Measures of alcohol and drug use	Results: alcohol and drug use			Study limitations

A

Cherpitel
US (Missisipi)
(1995)

Years of study: –
6 months

Study population and design: Prospective, probability (sampled shifts). Every third consecutive patient ≥18 years admitted to Level 1 trauma center, excluding psychiatric patients.

N=1330 interviewed and breath-tested.

N=168 not interviewed, including refusals, non-English speakers, those in police custody.

Measures of alcohol and drug use: BAC(breath within 6 hours); CAGE (positive = 2+/4); MAST (positive = 6+/10); AUDIT (weighted score of ≥ 8 based on 10 items); TWEAK (weighted score of 3 based on 5 items).

Questions from Composite International Diagnostic Interview. (CIDI adapted to measure ICD-10 criteria for alcohol dependence and harmful drinking.)

Results: alcohol and drug use:

	** Injured	** Non-injured
AUDIT	30%	22%
TWEAK	26%	32%
BAC	17%	4%
Self-report (6 hours)	25%	9%
Harmful drinking	16%	17%
Dependence	19%	19%

* *p* <.05% for injured vs. non-injured

** Based on 58% of sample who were current drinkers (i.e. "drinking during preceding year")

"Breath alcohol analyzer and self-reported drinking (in previous 6 hours) were found to have very poor sensitivity in identifying harmful drinking or alcohol dependence. TWEAK and AUDIT were the most sensitive identifiers of subjects who met ICD-10 criteria for harmful drinking and alcohol dependence."

Study limitations: Order of questions may have affected performance of instruments.

Performance may have also been affected by quantity/frequency questions preceding administration of screening instrument.

B

Holubowycz & McLean
Australia (Adelaide)
(1995)

Years of study: June 1985–June 1987

Study population and design: Prospective. Sample included 16 study periods of 2-3 weeks followed by 16 "catch-up" periods in which outstanding interviews were completed. Eligibles were patients in hospital on second day after admission who had a blood sample taken. Males only. 235 male drivers admitted, 131 participated. 218 male motor-cyclists admitted, 171 (78%) participated. Only two refusals.

Measures of alcohol and drug use: BAC (blood) levels; drinking behaviour (e.g. quantity/frequency, reasons for drinking); perceived drinking problem; alcohol dependence scale(ADS); GGT.

Results: alcohol and drug use:

		BAC mg/dl			
	zero%	1-79%	80-149%	≥150%	
Intoxicated ≥1 week	31	54	51	82	
Binge drinking ≥1 year	12	32	29	56	
(N=)	(153)	(26)	(35)	(41)	
Drinking problem	2	4	3	20	
(N=)	(197)	(25)	(36)	(41)	
ADS score ≥ 14	3	8	–	22	
(N=)	(183)	(24)	(32)	(37)	
Drivers with illegal BAC ≥1week	10	20	16	62	
Drink-driving Suspension	8	19	29	38	
(N=)	(151)	(25)	(35)	(40)	

Study limitations: Substantial % of missed BACs.

Some interviews long after event.

Complex sample design.

GGT levels not reported.

	Study	Time period	Sample	Measures	Results	Comments
C	Lejoyeux et al. France (Paris) (2000)	– 1 month.	Consecutive patients to psychiatric emergency ward during "day-time" hours. (Exclusions not stated) N=104 (41% male) (All patients agreed to participate.)	"Diagnosis of alcohol dependence was determined by clinical evaluation and DSM-IV criteria." Assessment of acute alcohol intoxication determined by using "DSM-IV criteria". MAST. Quantity/frequency of alcohol use. Mini International Neuropsychiatric Interview	Overall prevalence of alcohol dependence 38%. Alcohol dependent (AD+)/non-dependent (AD-) |	AD+ (N=39)	AD- (N=65)
---	---	---
Unemployed	51%	23%
Suicide atts.	23%	29%
Alc. intox.	62%	12%	 AD+ patients consume a higher number of drinks per day and drink more days per week. Patients were not sent for treatment, so the observed rate "reflects the prevalence of alcohol disorders among 'typical' patients examined in the emergency ward."	Small N. No BAC. Reporting of MAST scores not clear.
D	Maio et al. US (Ann Arbor, MI.) (1997)	1992-4 29 mos. at site 1 1993-4 15 mos. at site 2	Prospective study of motor vehicle crash (automobiles or small trucks) patients, ≥18 years, presenting to 2 EDs directly from crash scene. within 6 hours. One site Level I Trauma Center, second site university affiliate. Pregnant women, institutionalized and transfer patients excluded. Stratified, random sample. Full sampling from 3.30- 11.30pm and time sampling for other shifts. "Representative" samples created by weighting. Data sources included hospital records, crash reports. 1833 eligible, of whom 318 had no alcohol test (185 refused); 354 no interview (161 refused). 1161 (63%) comprised main analysis, of which 46% male.	Diagnosis of alcohol abuse or dependence (AA/AD) based on the alcohol section of the Diagnostic Interview Survey and met criteria for "mild, moderate or severe" AA/AD. BAC (blood). GGT.	20% of MVC patients had current AA/AD (N=222); BAC was negative for almost half. Of those positive, 41% had BAC ≥100mg/dL. Of those without current AA/AD (N=936), 6% had positive BAC, 3% with BAC ≥100mg/dL. Analysis by demographic variables, restraint use, culpability, ISS, admitted. $GGT > 85 IU/L$ AA/AD+ 9.9% AA/AD- 2.7%	Alcohol abuse and dependence combined. Suburban setting.
E	McLeod et al. Australia (Perth) (1999)	1997 4 seasonal data collection periods (2-4 weekends)	Prospective. Controls. Non-injured matched by area of residence, randomly selected, interviewed at home within 8 days of cases injury, 15+ years. Cases. Presented to ED within 24 hours with an injury or poisoning; between 14 & 50 years. 797 cases and controls. (67.3% of potentially eligible cases and 62% of controls),13% of eligibles refused. Male cases = 67%; controls 42%	Interview, questionnaires and medical records. Cases: drug and alcohol consumption patterns in 6 and 24 hours prior to injury and over past three months. Predicted BAC estimated from quantity of alcohol consumed in previous 6 hours. BAC (breath) for cases (N not reported); measures of harmful drinking.	Cases drank more, on average, in all three time periods, consumed more than the mean number of drinks in the 6 and 24 hours before injury, and drank more frequently at harmful levels. Controlling for confounders, the risk of injury for cases increased as quantity of alcohol consumed increased. "The ORs were not significant until greater than 60g. of alcohol when the risk increased ...by threefold...." No significant differences between male and female cases in mean number of standard drinks in the 6 hours prior to injury. Risk of injury for women is higher at higher levels of drinking.	High non-response. Demographic differences in cases and controls. Presence of another person during interview. Possible recall bias.

	Investigator/ locale/ date of publication	Years of study	Study population and design	Measures of alcohol use	Results: alcohol and drug use	Limitations
F	Ponzer et al. Sweden (Stockholm) (1999)	– 6 months	Prospective. Patients attending ED in major hospital treating largely moderate injuries; 15–65 years requiring inpatient care. 345(70% male) recruited to study. 55(14%) refused.	Interviews included psychiatric, alcohol and drug assessment including CAGE. Repeated Trauma group defined as patients reporting at least 2 trauma episodes. Single trauma group reported one or none. "Alcohol intoxication" noted at admission.	Patients divided into Repeated Trauma (RT) and Single Trauma (ST) groups. RT N= 120, ST N=225. 11% of RT and 2.2% of ST positive on all four CAGE items. RT patients had more psychosocial problems, higher weekly alcohol consumption, and were more likely to be intoxicated on admission. External causes "similar".	Sampling not specified. No BAC
G	Soderstrom et al. US (Baltimore, MD) (1997)	1994-6 15 mos.	See Table 6 for design 1909 eligible 1118(72%) consented to interview: others refused or not approached. Of these 709(63%) interviewed and had alcohol/drug screens. 70% of these were male.	CAGE; CAGE modified for drugs; brief MAST; AUDIT; SCID. BAC (blood) levels. (PSUD = Psychoactive Substance Use Disorders) Drug (urine) screen	54% lifetime abuse or dependence ≥1 PSUD 24% current alcohol dependent at time of injury. Patients with current dependence diagnosis (%) Alc. dependent with or without other drug BAC neg 12, pos. 54 Other drug dependent with or without alc. BAC neg 17, pos 19 Alc. and other drug dependence BAC neg. 6, pos 15 48% of patients with lifetime alcohol dependence and 34% with current alcohol dependence were BAC negative.	See Table 6
H	Vingilis et al Canada (Toronto) (1994)	1986-9 37 mos.	Prospective. Motor vehicle driver admissions to Regional Trauma Unit. ≥18 years; competent to drive and able to consent; comprehended English. 596 admitted, of whom 306 (51%) were approached for consent. Of 306, 102(33%) refused, others lost to contact.	BAC (blood) positive. Alcohol use. Deviant/illicit drug use.	Frequency of alcohol use in general and alcohol dependency scores not significantly different for BAC positive/BAC negative. BAC positives: "In month before accident, significantly greater frequency of alcohol use; greater frequency of intoxication..." "A greater proportion of BAC positive drivers [reported] driving within 2 hours of drinking or driving with a self-perceived BAC greater than .08% a month before the accident". 42% of BAC positive drivers said they were not drinking prior to the event.	Small N. High refusal rate. Time/day of sampling for interview not reported.

The value of case-series studies

The greatest number of studies of alcohol and injury, including ER studies, are case-series. These have been reviewed numbers of times (Roizen, 1989, Romelsjö, 1995, Hingson and Howland, 1993; more recently on suicide see Cherpitel et al., 2004). Case-series studies have been widely criticised for their non-comparability, lack of generalisability because of different designs, and lack of control/comparison groups (preventing evaluation of the risk of injury associated with either acute or chronic alcohol use). Some studies use retrospective designs based on hospital records or other local data bases, giving the investigators little control over the variables of interest. Many have incomplete data on study participants' acute alcohol use and/or drinking patterns, a problem shared with the studies reviewed above. There are hundreds of such studies worldwide (including studies of fatalities) and a review is outside the scope of this paper. However, case-series studies, when well carried out, are valuable in their own right. Their value is enhanced by what can now be seen as some of the problems encountered by the large numbers of more epidemiologically sound but far more expensive case-control studies.

What, then, are some gains from case-series studies to knowledge about the relationship of alcohol and injury, and what are some of the uses to which these studies can be put? Only a few selected studies are included in the discussion below.

– Treatment decisions within the ER for different groups of patients

Jurkovich et al. (1993) were able to show in this now dated, but important, U.S. study that chronic rather than acute alcohol use affects <u>outcome</u> from trauma. In this series patients with both biochemical and behavioural markers of chronic alcohol abuse were at significantly greater risk of complications, including pneumonia, during treatment for trauma. These investigators argued for greater focus on chronic alcohol use and brief alcohol interventions. They also argued that their results could be generalised to other similar trauma centers. (See also Jurkovich et al., 1992)

Fabbri and colleagues (2001), in an Italian ER study of road crash patients, show the higher trauma severity of those who are alcohol positive and the need for differential management of them. They also show that these alcohol positive patients have higher numbers of unsuspected injuries.

– Locally based knowledge of the specifics of ER provision, as a basis for estimating generalisability or for contrasts

For instance, Blake et al. (1997), in a study of alcohol and drug use with musculoskeletal injuries (U.S.), compared length of hospital stay and total hospital charges, and pay or class (insured or not) for alcohol and drug users compared to non-users. They found that alcohol users had longer stays, greater hospital charges and were more likely to be uninsured. They emphasise the burden that these patients create for their hospital and for similar hospitals, adding to the rationale for alcohol/drug ER and community interventions.

In another example, a comparative study of maxillo-facial trauma in Bristol and Bordeaux carried out by Timoney et al. (1990) contrasts differences in drinking cultures, types and causes of injuries, as well as the differences in the structure of health care and treatment and care for this single type of injury.

Table 4. Selected studies from developing countries

	Investigator/ locale/date of publication	Years of study	Study population and design	Measures of alcohol and drug use	Results: alcohol and drug use	Limitations
A	Adeloye & Ssembatya-Lule Malawi (Blantyre) (1997)	1995 6 months	Prospective. Consecutive acute head injury cases admitted to surgical wards. N=104 (85% male)	Self-report. Alcohol use/abuse, including "Drank habitually".	10% reported "alcohol use" 45% assaults "drank habitually"	Small N; inconsistent alcohol reporting; children not excluded; alcohol use variable not defined
B	Aptel et al La Reunion (1999)	1993 1 year July–June	Prospective. RTA – Injuries resulting in hospital admission collected retrospectively every other week and all weeks in last month of year in all wards involved in management of injured patients. Matched to police records. N=1000 (84% male)	BAC (blood) levels.	25% BAC positive (12% ≥0.8 g/l) 75% "unknown"	Base includes unknown N of children; only 37% cases matched with police records; higher BACs less likely to have police match as more likely to leave crash scene; no exclusions presented; no time of injury to BAC test.
C	Borges & Rosovsky Mexico (Mexico City) (1996)	–	Prospective/case control. In six of eight Mexico City hospitals all ER patients ≥18 years were sampled. In two hospitals, every other patient was sampled. Data collected in each hospital 24 hours a day for a week. 88% case completion and interviewed (N=2197). 40 suicide attempters from 8 hospitals (52% male). 372 (86% male) non-suicide, trauma controls.	Breath alcohol. Self-reported alcohol use. Self-reported drug use.	Cases: 16/36 (44%) "consumed alcohol prior" to event compared to 32/367 (9%) controls. Cases: 10/36 (28%) BAC positive compared to 22/347 (6%) BAC positive controls.	Very small Ns from numbers of hospitals. Week/month not reported. Characteristics of study described in earlier article. Articles not consistent on age limits.
D	Chen et al. Taiwan (Taipei) (1999)	1996 4 mos.	Prospective, but trauma registry used. Traffic injuries to ER. Exclusions: fatalities (8), patients <15 and ≥65 (26). N=381 (64% male)	BAC (blood) positive.	21% BAC positive.	Alcohol use probably underestimated – "sensitivity of TD analyzer limited to 10mg/dL." No analysis of BAC levels; no time from injury to hospital/BAC.
E	Gururaj India (Bangalore) (2004)	–	Prospective. Traumatic brain injury (TBI) patients in 7 city hospitals. Males >16 included. Of 2900 TBIs, 1553 met inclusion criteria.	Self-reported alcohol use; Person smelling of alcohol at registration; "documented medical or police evidence".	Patients grouped in alcohol-users (N=243), non-users (N=1310). 27% of "alcohol users" had consumed alcohol prior to injury; three quarters had consumed 3 hours prior to injury. 62% of TBIs in "user" group occurred between 7 & 12 p.m. in contrast to 34% in non-user group.	Time of injury to admission not reported. 60% injuries were RTAs but not reported by type of vehicle. Pedestrians not analyzed separately from vehicle occupants. Length of study not reported nor % of patients not seen.
F	Lapham et al. Thailand (3 Regions) (1998)	–	Prospective/case control. 3 regional ERs, consecutive admissions, ≥14, excluding transfers, women in labour, Buddhist monks, respiratory difficulties. Of 1351 potential cases, 993 (74%) completed interviews; 89% of trauma patients completed (N=531). % male unclear.	AUDIT + Study questionnaire. Drinkers, 6pm to 2am. (AUDIT positive = score of 8 or more)	36% trauma patients AUDIT positive (score of 8 or above). Controls: non-trauma cases including those with possible alcohol-related illnesses. (Male = 62%) Cases: 43% males and 13% females had positive AUDIT scores compared with 35% male non-trauma cases and 6% female.	No analysis of full range of AUDIT scores.

	Author/location (year)	Date	Study description	Factors that could reduce risk of crashes and injury.	Results	Limitations
G	Mock et al. Ghana (Kumasi) (1999)	1995 May–Oct.	30 commercial drivers, principally buses and taxis. Focus groups of 5–7 drivers, all male.		"Most voiced opinion that drinking alcohol in any amount could lead to crashes and all denied drinking before driving." Previous roadside survey showed that 10% of taxi drivers and 4% of bus drivers had BACs above 80mg/dL.	Small N. Insufficient qualitative analyses of focus group responses.
H	Obembe & Fagbayi Nigeria (Kaduna) (1988)	1986–7 Dec.–Feb.	Prospective. RTAs attending A & E departments of four participating hospitals. Of 25448 patients in the study period, 209 were "road accidents related".	Self-reported alcohol/drug use; "smell".	Of drivers (N=50), 7 (14%) were considered "drunk", 15 (30% had abused drugs. Of non-drivers (N=159), 18% were "drunk", 16% had abused drugs. Of 22 drivers with alcohol or drug problems, only 6 agreed to seek medical help. Consensus was: "RTA was a result of providence." Alcohol users mainly drank local brew.	Small N. No report of time of injury to admission. Children not excluded
I	Odero Kenya (Eldoret) (1998)	1994–5 6 mos. Dec.–May	Prospective/case control. Consecutive RTA injuries to 4 hospitals; ≥age 16; arrived within 10 hrs. of crash. Of 387 eligible patients, 51% not breathalysed. N=188(77% male). Roadside survey = controls.	BAC (breath) levels.	23% BAC positive/ clinical evaluation 15% 12% BAC>50 mg % Night-time 21% BAC>50mg % 40% cases vs. 8% controls BAC>50mg.%	Exclusions/refusals not clear. Time limit for exclusion >10hrs. Of eligibles, only 49% evaluated for alcohol.
J	Odero & Zwi Kenya (Eldoret) (1999)	1994–5 6 mos. Dec.–May	Prospective. 2073 injured persons ≥16years arrived at casualty department in study period (all types of injury). 850 arrived 10 hours after injury; 52% failed to provide breath tests. 179 had both blood and breath screens.	Alcohol (breath and blood).	74% breath test positive (58% ≥50 mg,%) 67% blood analysis positive (60% ≥ 50mg,%) Close agreement between "BACs in excess of 50mg,% as determined by breathalyser and blood analysis in casualty department."	Small Ns. No report of cause of injury x BAC.
K	Peden et al. South Africa (Cape Town) (2000)	1997 4 weeks Random shifts per day	Prospective. "All patients presenting to trauma unit with injuries less than 6 hrs.old". Excluded: re-attenders; <age 18. 9% refusals. N=254(79% male).	BAC (levels). Self-report of alcohol use. Drug screens	BAC positive 59%. "One third" >.08mg/100ml. Over half drank in 6 hrs. before injury. High % possible problem drinkers. (See Table 5 for details)	Possible bias from use of urine samples. Accuracy of drug wipe "disappointingly poor".
L	Swaddiwudpong et al. Thailand (Mae Sot district, Tak Province) (1994)	1991	Prospective. Seriously injured drivers admitted to hospital or died. N=431 crashes, 581 victims (app.90% drivers (app.90% drivers used). 36 deaths/ 75 vehicles in multi-vehicle crashes left scene and were not included. 341 crashes (79%) involved motor-cycles. Police accident reports and post mortem medical exams included.	History of alcohol and drug use; self-report of alcohol consumption prior to collision.	56% motor-cyclists, 46% bicyclists, 24% other motorists "had consumed alcohol just before or during the driving". 75% motor-cyclists had no driving licence. 94% motor-cyclists did not use helmets.	Refusals and non-participants not reported. Poor description of alcohol measure. No reported analysis of drug/alcohol history from interview.

– Focus on a single type or cause of trauma

Both the studies cited just above show detail of mechanism and type of injuries. In beginning to understand the "role" of alcohol in injury this type of detail is essential. The Blake study of musculoskeletal injury shows mechanism (broadly defined) of injury by BAC level and drug use and detailed types of injuries by alcohol presence. This is an important start in assessing how alcohol and/or drugs may have been involved in an injury and how prevention efforts can be directed toward patients with particular injuries rather than interventions across all injured patients.

– Planning for short and longer-term interventions

Case-series studies can be useful in setting a base-line for both short (e.g. hospital based) interventions and community or longer term interventions. Recent research on "surrogate" measures of alcohol presence in trauma cases uses the "case" side of case-control studies in the ERCAAP research programme to look for minimal data that is "likely to be recorded in even the busiest ER" in order to predict acute alcohol involvement in injury (Young et al., 2004). This is considered an inexpensive alternative to ER studies based on a probability sample covering all causes of injury. However, this approach uses models with many assumptions and a methodology still in its infancy.

Andrews et al. (1999), in their case-series of road traffic accidents (RTAs) in Uganda, show the high percentage of all trauma due to RTAs and suggest interventions that would limit the number and/or severity of injuries. These include more "zebra crossings", since pedestrian injuries are the largest cause of injury, and greater use of safety restraints for vehicle drivers and helmets for cyclists. Although the alcohol measurement is weak in this study, alcohol is believed to be an important contributor to these causes of injury. Research in developed countries can inform these prevention proposals, based on case-series data, even in the absence of the expensive and difficult-to-obtain alcohol measurements found elsewhere.

– Special populations

Case-series research can be especially useful in looking, in more detail than is possible in all-injury probability samples, at discrete populations of ER/ hospital patients – e.g. women victims of domestic violence, adolescents, military personnel, women drivers. (See, for example, Turner and Shu (2004), U.S.; Moraes and Reichenheim (2002), Brazil; Wright and Kariya (1997), Scotland.) The survey instrument can focus on key questions which might not be of interest in relation to a total ER cohort and may use methods which vary depending on the population in question.

– Research-poor countries

Countries with little research on substance use and abuse can benefit from well-designed but simple case-series ER studies, especially in attempting to assess whether alcohol is likely to be a major contributor to injury. These could focus on one type of cause of injury (e.g. adult pedestrians) and reduce the burden on medical staff by including only a few demographic and assessment questions. Keeping in mind the limitations of studies of this type, they could be the basis for further research if resources permit.

– Discussion

Many case-series studies clearly reflect the clinicians' interest in alcohol and a type of injury, the burden of alcohol and drugs on the ER and the management of these patients. Given the difficulty of encouraging hospitals routinely to test for alcohol and drug use and to set up brief interventions (see Gentilello, 2005 and Soderstrom et al., 2001b), many of these case-series studies show that the interest is there to be encouraged. However, these clinician-led studies often show their distance from the alcohol epidemiological research by their methods and measures, modes of analysis and lack of familiarity with the research literature. These suggest a need for more intensive collaboration.

There is a need for continuing investment in case-series research, especially where studies include detailed data on type and mechanism of injury, detailed demographic analysis and analysis of the context of the injury event. These studies will, hopefully, be informed by the best epidemiological research and will benefit from collaboration between substance abuse researchers and practising clinicians. There is also a need for less ambitious studies in areas which are research- and resource-poor in order to contribute to building up a picture of alcohol and injury and alcohol abuse.

Methodological Problems and Limitations in ER Research on Alcohol and Injuries

Despite the gains that have been made in emergency room research in the past two decades, and in particular the wide use of the Cherpitel model, there are many limitations in using the ER "window" to study injuries and alcohol. Some have proven, so far, to be largely intractable. Several are outlined in this next section.

Sample selection in emergency room/hospital samples of the injured

Perhaps the most serious and most intractable methodological problem faced in ER research, as in other studies of alcohol and injury, is sample selection and the biases associated with it. Possible reasons for these selection biases have been analysed by many researchers in different countries and settings over a long period of time. (See, for example, Aarens et al., 1978; Cherpitel, 1993; Treno et al., 1998.) Detailed analyses are hampered by the very limited data on non-participants in most settings, but also by the limited reported analysis of data which investigators do have. Sample selection biases occur where those who are included in a study differ from the wider population they are thought to represent (e.g. the catchment-area population of the ER), from the injured within a geographic area or from the full injured or control population at ER point of entry. These differences may be demographic, in type/cause of injury, in injury severity and in levels and patterns of alcohol/drug use. Sample selection biases will vary by community, drinking culture, and research resources for an individual ER study, and by type of ER (e.g. major trauma centres will have more seriously injured, but they will not come only from the usual catchment area of that ER). Sample selection biases are likely to be more serious in studies in developing countries. Reasons for selectivity in ER samples are briefly described below:

– Selection before or shortly after presentation to an ER

No injury treatment sought. General population surveys in developed countries show that many of those who are injured but not severely injured will not seek treatment (Treno et al., 1997). This is more likely to occur where there is a cost at point of entry to the health service or where injured do not have medical insurance.

Injured seek alternatives to the ER. The injured may seek alternative sources of care from their primary care physicians, health workers (in developing countries) and/or pharmacists. In developed countries, this may be influenced by greater treatment options among the more affluent. In developing countries and in some rural areas of developed countries, there may be no ER alternatives. The ER may only be used when other sources of health care have failed, perhaps days after the initial injury.

Fatalities are not included in ER samples of injured. Few studies of alcohol and injury include the fatalities which occur in the same injury events. These injured may include those who are dead on arrival to the ER, who die shortly after admission or who died in the same incident that brings non-fatal injuries into the ER but who are not taken to the ER. Thus, the most severely injured (i.e. those who die of their injury) are unlikely to appear in the samples of injured. Few studies have tried to link the non-fatally injured and the fatally injured in the same community at the same time. (See below.)

Specialist trauma centres may receive the severely injured or those with specific types of injuries. The severely injured may be sent to trauma centres which take only or mainly the severely injured; some specialist units, e.g. burns units, may take only one type of injury. These re-directions may take place prior to or shortly after admission to a general ER, or patients may be sent to other hospital departments after triage but before acceptance into an ER study. This can especially affect the assessment of the relationship between severity of injury and alcohol use and also create a distorted picture of the presence of alcohol in injury events of particular types. Burn injuries are one example: the least serious burns may be admitted to the ER; the more serious may go directly to a burns unit. Patients with mental health problems provide another example. Injured patients with mental health problems may be quickly redirected to another department or facility once immediate treatment is given.

– Selection within the ER or by study criteria

Eligibility/inclusion criteria vary by study. Minimum age for inclusion varies by study. (In the WHO Collaborative Study the lower age limits for inclusion in studies ranged from 15 to 18. Studies vary in whether they ask for "informed consent" for inclusion; possibly encouraging more "opt-outs" from the study because of confusion about meanings and terms. Patients who have had medication or were drinking after the injury, or who have had previous treatment by a "first medical responder" or at another hospital, are often excluded.

The severely injured may not participate. Severely injured patients may not choose to participate, those accompanying them may not consent, the ER staff may not allow them to participate, or interviewers may not choose to disturb their treatment.

Refusal to participate. Refusals may occur for reasons other than severity of injury: e.g. those who may be concerned with legal liability (e.g. injuries as a result of traffic crashes or assault); those who arrive under police protection; victims of domestic assault who arrive with a family member.

Interviewer discretion. Where there is discretion, interviewers may avoid some patients: those obviously drunk and/or aggressive or those perceived to be uninjured but thought to be visiting the ER for other services e.g. meal vouchers, but who may also have an injury or other medical problem.

Loss to follow-up. Patients may leave the ER to attend other departments for tests or they may be sent for immediate surgery or go home to seek other alternatives for care. Interviewers' shifts may end. There may be language barriers and lack of competence on the part of patients for consent and interview (including intoxication).

– Selection biases in alcohol measurement

BAC and other drug testing, as well as self-reports about alcohol/drug use before injury, are the most important measures for determining the contributory role of alcohol/drugs in the injury event. However, BAC and drug testing may be carried out in only a subset of cases and may not cover all those who consent to participate in the study. Factors which may affect this include:

Arrival in the ER within 6 hours of injury. Although most recent studies measure BAC on those arriving at the ER within 6 hours following the injury or illness event, this can vary by study. Some patients may be eligible for an interview which may assess self-reported drinking before the injury or illness or drinking patterns, but not be eligible for a BAC test. Thus, the cross-correlation of these different measures used in analysis may be based on different sub-samples in a given study. Most studies do not estimate the BAC back to the time of injury/ onset of illness.

Refusal to have a BAC test. Patients may refuse BAC testing because of concern with legal liability for an injury, embarrassment/stigma and/or because they have had repeated alcohol related injuries.

BAC testing at discretion of medical staff. In most ERs alcohol testing is not routine and may be at the discretion of the admitting or other staff or interviewers. Some evidence suggests that the young and males are more likely to be tested (see Waller, 1988). Discretionary testing is likely to vary by city, region and country, biasing cross-community and cross-national comparative studies (see, for example, Fabbri et al., 2001, Italy). In many developing countries, where BAC testing is not or cannot be carried out, alcohol use in the event may be entirely discretionary depending on a patient's smell of alcohol or behavior.

– Discussion

The possible biases outlined here are under-analysed in most ER research. (Exceptionally see McLeod et al.,1999). More could be done to address types and magnitude of bias by refining and enhancing the data gathered at ER point of entry or by a "first responder" to the patient's injury. Whether the "search for surrogate" measures is a solution remains to be seen and will only be possible where there is substantial data on alcohol and drug use from local or regional communities for building appropriate models. This is not now possible in most countries nor in the less well-researched parts even of developed countries.

Estimating blood alcohol at the time of injury

Estimating blood alcohol at the time of injury has proven difficult and is often estimated as the BAC at the time of arrival at the ER. Working back to the time of injury depends on a very accurate self-reported estimate of the time of injury and an estimate of alcohol metabolism between that time and arrival at the ER (assuming no drinks post-injury). Alcohol metabolism can vary by a number of factors related to the individual e.g. amount of food eaten in the intervening period, body weight, health of pancreas and liver and other factors affecting tolerance of alcohol. More work is needed to factor in, in aggregate analyses, the short-term decay in reported alcohol consumption at the time of the injury event as well as increasing the accuracy of reporting of time of injury. While estimates of alcohol at the time of injury are accurately made for individual cases, e.g. drunk driving defenses, this has not translated to analyses at the aggregate level.

Discrepancies between BAC, self-reported alcohol use and clinical assessment

Although breathalyzer measurement of BAC is the measure of choice, this is expensive and unavailable in many countries. Whether use of breathalyzers even in developed countries will become routine in the ER is still an open question. Also, as noted above, BAC measurement may not be obtained in all ER cases. Self-reported alcohol use prior to injury is one way forward, but the agreement between self-reports and BAC readings varies by country and ER (WHO, 2005). It also appears that self-reports are often obtained on a greater number of participants than are BAC readings for reasons that are often not reported. (See, for example, Cherpitel et al., 2003b). In most of the 13 Emergency Room Collaborative Alcohol Analysis Project (ERCAAP) studies reported in this paper a BAC was reported for only between a half and two thirds of participants who offerred self-reports. In considering the agreement of these measures, the differential numbers on which the correlations are based must be considered as well as the levels of and reasons for refusals for one measure or another and loss to follow-up. Treno et al. (1998) argue that self-reported drinking is "the more inclusive indicator" and suggest that this is a satisfactory measure. However, the willingness of study participants to provide a self-report and its accuracy will depend on a number of cultural factors including insurance barriers and types of injury. The WHO Collaborative Study found only moderate agreement between clinical assessment of alcohol use, BAC and participant self-assessment. (See Chapter 9 in this volume; also Cherpitel et al., 2005).

A Taiwanese study of the feasibility of ER nurses' identification of alcohol use shows that at higher BACs their assessment of alcohol use based on alcohol odor was quite good. For BACs above 50mg/dL detection sensitivity was 75%, specificity was 90% and accuracy was 85% (Li, 2003). This type of clinical assessment may, however, depend on specific characteristics of the ER, e.g. crowding and other ambient smells.

For the management of alcohol-present cases in the ER and for identifying at least some of patients for whom an alcohol/drug intervention may be appropriate, even moderate agreement on these different alcohol assessments may be sufficient. Estimating alcohol costs and burdens at the aggregate level will need either other types of measures or modeling and weighting from the achieved ER samples. The current evidence suggests that reporting systems based on clinical evaluation are likely to be weak. (See Chapter 9 in this volume).

Selection of control populations in case-control designs

Difficulties in finding appropriate controls in social research of all types has been discussed in many reviews. (See, for example, Aarens et al, 1978; Schlesselman, 1982, and Maclure and Mittleman, 2000). Until a little over a decade ago few ER studies had used a control group of any kind. In the intervening period many studies have case-control/case-comparison designs using either non-injured ER patients or a general population sample. In the past, several studies have used a site-based case-control design but these are not common and have their own limitations. (See, for example, Honkanen et al., 1983; and criticisms of site-based samples in Roizen 1989). Despite recognizing the limitations of using non-injured ER patients, a number of studies in different countries have used medical patients as controls. (Some of these studies are reviewed by Cherpitel, 2005; Cherpitel and Driggers, 2005, and these are used in the ERCAAP studies.) The major limitations of non-injured patients as a comparison group are that they have not been found to be representative of the non-injured in what is thought to be the general population of the ER catchment under study (see Chapter 8 in this volume) and that they differ on key alcohol use variables. General population (GP) samples, on the other hand, are expensive and also may not cover the actual catchment of the ER. GP controls are also subject to variable participation rates, in some cases quite high. In one recent Australian study, in only 40% of households selected was contact actually made with any person and/or one who met the inclusion criteria for controls.

While case-crossover designs (see Chapter 8 in this volume) offer one solution to the problem of finding appropriate controls, they are likely to have their own limitations, especially in areas and countries with limited data on alcohol use and problems. In these situations there will be nothing with which to compare the ER findings on alcohol use. Thus, although there will be a risk estimate, the representativeness of the injured sample on which it is based will be unknown. Despite the limitations of non-injured controls they offer at least one other population for comparison.

Generalisability of ER findings

Science is by and large about the replication and generalisability of research studies and findings, hopefully leading to new theories. Tables 1 and 2 show studies which use a common design and several are by the same researcher (Cherpitel); other studies are in collaboration. The question which needs to be answered is how generalisable do the relevant research communities and their funders perceive these findings to be, and in what circumstances? The answer to this has important consequences for future funding of research, surveillance and prevention. It also has consequences for the use of these studies in higher order analyses, i.e. for other than descriptive analyses.

Most of these studies end with the warning that the results cannot be generalised. As examples, Cherpitel and Rosovsky (1990) note, "Findings reported here cannot be generalised beyond the emergency rooms sampled" Cherpitel (1996b) notes, … data from this study, as well as from other similar studies of other regions, cannot necessarily be considered to be representative of the larger area from which the injury cases come." (See also WHO Collaborative Study Final Report, 2005.) Individual ER studies are thus rightly deemed as not generalisable without regional and national analysis of what the ER population is likely to represent. Nevertheless, these studies are pooled by other researchers for meta-analyses, for the calculation of the economic costs of alcohol abuse, and for calculation of the global burden of alcohol.

The lack of generalisability comes in large part from the differences in the relationship of alcohol and injury in different geographic areas, different types of ERs and hospitals. But it is also due to the still largely unanalysed effects of the differential quality of studies and measurement error or bias.

It has been argued in relation to the Cherpitel and related studies that the value of these similar studies stems from their use of the same methodology and their similar "quality" (Cherpitel et al., 2003a). But while the many studies of Cherpitel and collaborators largely use the same methodology, they are unlikely to be all of the same quality. The quality will vary with the setting for the research, interviewers' interest and training, study director's experience, respondents' honesty in reporting alcohol problems and sample selection biases. Numbers of non-participants/respondents are typically considered one aspect of quality in social research. One of the major problems with the studies reviewed here is the variable (but relatively large) number of non-participants (especially refusals) and the variable number of those injured on which no BAC is taken. These studies also differ in the time period allocated for each study, varying from two weeks to intervals over an entire year.

The implications of "lack of generalisability" have not been spelled out by researchers involved in ER studies. Is it suggested that a finite but very large number of studies need to be undertaken to measure risk? Does every prevention effort need to be linked to a local ER study or can it be assumed that enough is known from the studies that exist? It seems unlikely that funding agencies will continue funding particularist efforts with no prospect of generalisation.

Confounding of alcohol effects and the effects of other drugs

A development of considerable importance in recent years for research on alcohol use and injuries is the increasing interest in analysing drug use as well as alcohol use in ER studies. It has been argued by Soderstrom et al. (2001a) that the increases found in cocaine and opiate use in the U.S. by trauma center patients have reached "epidemic" proportions, especially in violence-related injuries. While most emergency room studies do not include drug screening, some recent studies do and these investigators argue for routine drug screening (see, for example, Soderstrom et al., 2001b; Borges et al., 2005).

The greater focus in recent ER studies on drugs may, in part, reflect the interest in determinants of injury by the medical community generally rather than, as in the past, reflecting the interests of those particularly concerned with substance abuse, where funding may be, or has been, based on a remit to study particular drugs. The reluctance to routinely screen for illicit drugs will also reflect costs of testing, concern with law enforcement and private insurance issues, and concerns about patient cooperation. One U.S. study suggests that fewer than 40% of Level I and II centers rountinely test for drugs other than alcohol (Soderstrom et al., 1994). Even research studies which report that the trauma center "routinely" screens for alcohol and drugs often show a high proportion of cases not screened for either.

However, if ER studies which show high alcohol presence have failed to investigate drug use or the interaction of alcohol and other drugs, drug effects may be attributed (and very likely are) to alcohol alone. This renders these studies less than useful in calculating an etiologial fraction and attributable risk for alcohol and for evaluating alcohol's **role** in injury in contrast to its **presence** in injury events. Unmeasured drug use may also bias clinical assessments of alcohol use and intoxication within the ER where drug testing is not routine and where it may be assumed that a case involves only alcohol.

Table 5 shows the percentages of those injured having tested positive for drugs and alcohol from selected studies. Apart from the Norwegian study (which included only fatalities), drugs alone or combined with alcohol show a greater presence in the events than alcohol alone. Table 6 shows alcohol and drug use by mechanism and intent of injury in one U.S. ER study. The consderable varia-tion in alcohol/drug positives by mechanism (e.g. knife injury in contrast to falls) suggests that overall percentages of alcohol/drug positives are likely to vary by type of ER/hospital, by geographic area, and by ER catchment (especially with regard to socio-economic/ethnic variables). Table 7 shows a selected sample of studies from several countries which analyse both drug and alcohol. Here, where possible, I have tried to estimate the proportion of alcohol positive cases that are also drug positive. It is unfortunately not possible to estimate this for many studies because the analysis is not based on the individual, so the same person may enter tables several times with no cross-correlation of alcohol and drugs. While there is considerable **variation** in the proportion of drug positives nested within alcohol positive, several studies show very high proportions of alcohol positive cases also drug positive. (It should be noted that one study in Table 5 and four in Table 7 refer to alcohol and drugs among fatalities. ER studies usually refer to non-fatally injured patients.)

One recent ER study using a general population control showed alcohol to be more important than drugs as a risk factor for injury, but this was based on drug and alcohol use disorders in the 12 months prior to the interview rather than at the time of an injury event (Borges et al. 2005).

As long as the "spotlight" stays for many researchers unquestioningly on alcohol, alcohol will be perceived as a major, if not the major, cause of injury – fatal and non-fatal.

Table 5. Selected studies: alcohol and other drug screening of injured patients

Study	N	Negative for all %	Positive for alcohol alone%	Positive for other drug alone %	Positive for alcohol & for other drug %
United States	709	39	17	29	16
Australia	2500	77	9	10	4
South Africa	254	19*	38*	24*	19*
Norway	159	63	21	9	8
Jamaica	111	38	15	31	16

United States: (Soderstrom et al.,1997) All seriously injured patients in trauma centre in Maryland.
Australia: (Longo et al., 2000) All injured drivers in South Australia over two 3-month periods.
South Africa: (Peden et al., 2000) All injured patients at Level I trauma centre in Cape Town. (*Approx. estimates from graph.)
Norway: (Gjerde et al.,1993) Fatally injured drivers in Norway over 2 years.
Jamaica: (McDonald et al.,1999) All injured patients at A&E Unit over 3 months.

Table 6. Blood alcohol concentration and other drug status in one U.S. ER
by mechanism of injury and intention

	Testing negative %	Testing positive for alcohol or other drug %		
		Alcohol alone	Other drug alone	Other drug and alcohol
Mechanism of injury				
Unintentional Vehicle occupant (273)	48.0	16.9	20.2	15.0
Motorcycle (28)	35.7	14.3	25.0	25.0
Pedestrian (75)	36.0	26.7	14.7	22.7
Falls (99)	52.5	13.1	25.3	9.1
Other (23)	34.8	17.4	43.5	4.4
Intentional Gunshot (128)	30.5	8.6	49.2	11.7
Knife (47)	14.9	12.8	36.2	36.2
Assault (36)	11.1	22.2	44.4	22.2
Summary Unintentional (498)	45.8	17.5	21.7	15.1
Intentional (211)	23.7	11.9	45.5	19.0
TOTAL (709)	39.2	15.8	28.8	16.2

(Soderstrom et al., 1997)

Cooperation of emergency medical professionals

Especially in developed countries, as injury prevention strategies become more effective (e.g. greater penalties for drinking driving or speeding), the medical community may be increasingly reluctant to choose to routinely test for alcohol and drugs, fearing insurance problems, cost factors and invasion of patients' privacy (Gentilello, 2005b). However, the limited evidence available suggests that many health professionals are unware or unclear about insurance consequences of alcohol use in their legal jurisdiction. Cooperation may also be hampered by the low priority given to education on alcohol problems. Even in the U.S. medical students receive about only 4 hours of education related to alcohol treatment, putting alcohol problems on the periphery of medical education. Thus, building a research capability into routine ER treatment faces many difficulties. This is despite the fact that most trauma surgeons (89% in a recent survey) believe alcohol is a major burden on their trauma center and that, if tested, 30% of patients admitted to their hospital would test alcohol positive (Gentilello et al, 2005a).

In developing countries a large proportion of accidents which present to the ER are from road crashes or vehicular injuries of pedestrians. The majority of these come from commercial vehicles. In these cases there may be a strong incentive to bribe or otherwise affect the accurate reporting of alcohol in order to keep one's job and driving license.

Differences in the delivery of emergency medicine worldwide

Little analysis of individual ER studies of alcohol and injury has considered the type of emergency care delivered in relation to selection biases and accurate and complete alcohol/drug measurement. Most of the best of the recent research has been carried out in Level 1 trauma centers or equivalent; however, even where type of ER or hospital is reported little or no analysis of the consequences for data gathering is given. Thus a table which reviews findings from numerous countries and settings, as in Table 1 here (or Cherpitel et al., 2004, and WHO, 2005) conveys no context of the research setting. This is also most often the case in the original studies.

It has been suggested, for example, that there are two different models for emergency care in developed countries – the Anglo-American model which "brings the patient to the hospital" and the Franco-German model "which brings the hospital to the patient" (Arnold, 1999). Even under the Anglo-American model, emergency care often begins outside the hospital and is given by emergency medical technicians and paramedics. These cases may be triaged in a different way from "walk-in" patients. There may be no single central point for initial assessment, including BACs or equivalent. In the Franco-German model, emergency units are "rudimentary"; patients are triaged in the field and then, where necessary, admitted as inpatients to specialist departments. Emergency medicine is or has been more likely to be a recognized medical specialty in countries with the Anglo-American model, although emergency medicine is a relatively new specialty even where this model applies (Arnold, 1999). By and large the development of emergency medicine follows the level of social and economic development of the society as a whole.

Russia is an example of the Franco-German model, with doctors working from ambulances to evaluate and give initial treatment. Emergency medicine is not yet a medical specialty. EDs are "little more than triage areas". The nearest equivalent to an ED is the "resuscitation room" where the critically ill would be taken and attending doctors are trained as anaesthesiologists. There is considerable disparity in health care between urban and rural centres and no co-ordinated system to bring trauma patients from rural to urban areas (Townes, 1998). This model of emergency health care would appear an unpromising environment to carry out research on alcohol and injury, even prospectively, without substantial resources. Care is dispersed within a medical institution; the triage system means that the patient may be seen by different people, in different locations prior to treatment.

Table 7. Selected studies of both alcohol and drug use

	Investigator/locale/ date of publication	Years of study	Study population and design	Measures of alcohol and drug use	Results: alcohol and drug use	Study limitations
A	Carmen del Rio et al. Spain (Valladolid) (2002)	1991-2000	Retrospective. Data from National Toxicological Centre. Fatally injured drivers. N=5745 on whom samples were taken. (59,091 died in RTAs in the study period.)	BAC levels/screening for illicit and medicinal drugs (reported as positive).	Alcohol positive 44%. BAC> 0.8 g/l 32%. Of alcohol pos, 5% pos. for illicit drugs. Of cocaine users (N=300), 54% were alcohol positive.	Selection of fatalities to be tested unclear; exclusions (inc. age) not reported. Until 1996 toxicology testing not formally required.
B	Demetriades et al. US (Los Angeles, California) (2004)	Jan 2000–May 2003	Retrospective. Traumatic deaths in Level I trauma center. Trauma registry and autopsy records occurring in the period were examined. Of 931 trauma deaths, 600 victims were tested for alcohol and illicit drugs.	Blood and urine screens for alcohol and drugs (reported as positive).	Alcohol positive 27%. 51% of these positive for illicit drugs.	12 cases were < 14 years. Cases > 60 were included, although low positive screens.
C	Deutch et al. Denmark (Aarhus) (2004)	1999–2000 24 months	Prospective. All patients admitted to regional trauma centre as "victims of major trauma."; ≥15 yrs; with blood/urine samples (92%). Exclusions: Patients admitted from other hospitals; those with post-injury drug administration. N=495 admitted, 41 excluded, 37 samples missing or damaged. Sample N=417.	BAC (blood) levels/Blood and urine screens for drugs (reported as positive).	BAC positive 26% (above legal limit 50 mg/dl 25%) Of BAC positive, 20% drug positive. (Based on 200 drug screens)	Full drug screen for only half patients. Small sample size for some sub-groups e.g. assaults N=18. Time from injury to test not specified.
D	Drummer Australia (3 states: Victoria, NSW, Western Australia) (2004)	1991-9 Years vary by state	Retrospective. Driver fatalities, forensic records (>90% autopsy; others screened in hospital); 85% of all driver fatalities in period. Exclusions: suicide; time from crash to death >4 hrs; those with post-injury drug administration.	Most analyses (>90%) were conducted on blood taken at autopsy, others blood in hospital. BAC levels/Screening for number of "impairing and non-impairing" drugs (reported as positive).	Reported here for 1997-99 (N=1490) 31% alcohol positive (88% ≥.05%) 27% impairing drug positive (including: cannabis 16%, opiates 7%, stimulants, benzo.) Both alc and drug pos.= 10%. Of alc positive, 31% also drug positive.	Numbers and reasons for exclusion from study not reported; lower ages of drivers not shown; alc and drug positive not disaggregated by drug type.
E	Gjerde et al. Norway (Oslo) (1993)	1989 & 1990	Retrospective. Fatally injured drivers who died within 30 days after an RTA. Cases with ante mortem blood sampling within 6 hours or post mortem sample if driver died at scene or on same day. Blood samples from National Institute of Forensic Toxicology. Of 277 driver fatalities, 195(70%) had blood samples sent to NIFT. In 159 cases there was "sufficient" blood to satisfy inclusion criteria.	Blood samples. BAC levels/Drug screens various (reported as positive).	Alcohol positive 28% (median BAC was 0.18%). Drug positive only 9%. Alcohol & drugs positive 8%. Of cases alc.pos., 27% were drugs pos. Analysis of plausible effects of BAC level and drug concentrations showed considerably higher % of alcohol and drug positives in single vehicle accidents, 42% and 22% respectively.	Relatively small Ns. Not known whether "cases where blood sampling was performed" were randomly selected.

	Study	Dates	Design/Methods	Measurement	Results	Comments
F	Kurzthaler et al. Austria (Innsbruck) (2003)	1995 12 months	Prospective. All consecutive RTAs to ER subsequently admitted as in-patients; ≥18 years. Exclusions: patients receiving benzo, post-accident. Overall sample size = 269 (71% male)	BAC (blood) levels/Blood screening for benzodiazepines only (reported as positive)	Of all RTAs: 29% BAC positive. Of all drivers: 37% BAC pos, 33% BAC>0.8 g/l. Of all alcohol pos, 3% benzo. positive (all were male).	Focus on benzodiazepines only.
G	Lowenstein & Koziol-McLain US (Denver) (2001)	1995-6 Until 400 enrolled.	Prospective. Sample of injured drivers presenting to Level III trauma center. Vehicle drivers, arriving at the ED within one hour of crash, ≥ 18 years. 652 eligible; 414 entered into study, 238 "missed" or no adequate urine sample. Of 414 participants, 371 crash records were retrieved.	Urine samples for alcohol and drugs (reported as positive).	Based on N=371. 6% alcohol alone 9% marijuana alone 18% drugs other than alcohol (inc. marijuana) 6% alcohol and drugs other than alcohol.	Limited reporting of reasons for missed cases; no time/day accident reporting.
H	Madan et al US (New Orleans, Louisiana) (1999)	— 6 months	Trauma Registry used to identify those with "life-threatening" injuries presenting to Level I trauma center over 6 months who had both serum ethanol and urine toxicology studies. (N=450/550)	BAC (blood) positive/Urine screening for illicit and a few medicinal drugs (reported as positive).	Of 319 positive screens for alcohol and/or drugs, alcohol alone = 177(55%); alcohol + 1 or more drugs = 123(39%); drugs alone = 109(34%). (90 cases presented with more than one drug other than alcohol.)	Presentation in graphs hard to interpret. Six-month length of study does not cover possible seasonal variation. "A major limitation of our study is the retrospective nature…"
I	Mura et al. France (Multi-centre) (2003)	June 2000-Sep 2001	Prospective, case-control. Six emergency care units and their toxicological laboratories. Sampled for month, day of week. Cases were car-drivers only. Controls attended for non-traumatic reasons and had driver's licences. Exclusions: patients who received one of the standard drugs or those admitting voluntary or accidental intoxication. Case-controls matched for age(≥18) and sex. 933 cases and controls recruited; 33 of each group excluded for insufficient samples. N=900 cases; 900 controls. (74% males).	Blood samples and urine (in a few cases sweat samples.) BAC levels. Screening for cannabis, amphetamines, opiates, cocaine, other psychoactive drugs.	Of all drivers (N=900), 26% alcohol positive. Of all controls (N=900), 9% alc.pos. In both groups, alc pos. defined as exceeding 0.5 g/l. Of drivers <27 years, 26% alc. positive; 15.3% THC positive. Of controls <27, 7% alc. pos, 6.7% THC pos. Of alc. positive drivers <27, 36% THC pos. Of alc. pos. controls <27, 32% THC pos.	Age limits not given in detail. Controls' reasons for admission not reported. THC reported for < 27 years only.
J	Peden et al. South Africa (Cape Town) (2000)	1997 4 weeks Random shifts	Prospective. All patients presenting to Groote Schuur Hospital trauma unit, a Level I facility. Exclusions: injuries more than 6 hours old; re-attenders;<18 years. 494 attended trauma unit in study period; 216(43%) did not meet inclusion criteria, and 24(9%) refused consent. Study N=254.	Alcohol: Breath BAC (N=250) levels; self-reports including CAGE (N=213)/ illicit drugs: self-reports (N=213); urine samples (N=196); Drugswipe strip for cannabis.	Alcohol positive 59%, alcohol levels were "on average" 0.12 g/100ml. 42% had used at least one illicit drug; 29% cannabis positive. 13% Mandrax (methaqualone) pos. 12% "White Pipe" (methaqualone and cannabis) pos. Of patients alcohol positive, 32% were positive for at least one illicit drug.	Toxicology results for drugs inconsistent with self-reports. Opiates, some of which were for pre-treatment, were excluded.

	Investigator/locale/ date of publication	Years of study	Study population and design	Measures of alcohol and drug use	Results: alcohol and drug use	Study limitations
K	Smink et al. The Netherlands (2005)	1998-9 12 months	Retrospective. All drivers with severe injuries admitted to hospital or dead who could be linked to Transport Registry. N=993 (74%) of 1347 drivers.	BAC (blood) levels/Blood screening for illicit and psychoactive drugs.	962 cases (97%) were analyzed for alcohol and drugs. Of these, 83% exceeded the legal limit of 0.5mg/ml. for alcohol. In 34% of the BAC positive samples (0.5 mg/mL), at least one drug was present.	Blood samples small % of all crashes - taken only if breath test not possible, or if there was suspicion of alcohol use.
L	Soderstrom et al. US (Baltimore,MD) (1997)	Sep 1994- Nov 1996	Prospective. Seriously injured patients at level I Trauma Center:≥18years, admission from scene of injury, length of stay ≥2 days, "intact cognition", ability to speak and understand English. 1909 eligible patients admitted,: 1220 (64%) approached, 1118 consented to interview. N=709 interviewed and had blood and drug screens.	BAC (blood) levels/urine drug screens/ self-reported substance abuse. "Drugs" refer to "substances of abuse other than alcohol and nicotine". Drugs reported as positive. Substance use disorders.	32% alcohol positive; alcohol alone 15.8%; alcohol and drugs 16.2%; other drug alone 29%; of alc positive,51% for at least one drug.	N=689 not approached, 29% of these discharged on first or second day of eligibility. Interviewers part-time, did not work weekends. Not apparently a probability sample.
M	Walsh et al. US (Baltimore, MD.) (2004)	2001 6 mos.	Prospective. Motor Vehicle Crash Victims (MVC) admitted to Level I Trauma Center. Inclusion: those for whom a BAC test result and a urine specimen (vol. >45mL) were obtained. Possible treatment drugs coded as negative. N= 322 (71% male)	BAC levels/Drug screens reported as positive.	26% alcohol positive, of whom 39% positive for one or more drugs, primarily marijuana. 16% alc.alone (mean BAC 207 mg/dL) 34% drug pos. alone. 10% pos. for alcohol and drugs.	Sampling details for study period not reported. % of patients admitted who had useable alc and drug screens not reported. Time of accident to alc/drug screen not reported.

South Africa offers another example. Here the picture is very mixed with dramatic contrasts between the metropolitan areas (formerly for whites only) and the provinces. The Johannesburg General Hospital has an Accident and Emergency ("A and E") unit divided into medical and trauma sections, with "24-hour access to all diagnostic, laboratory and therapeutic support services of a modern teaching hospital...The trauma unit [is] equivalent to a Level I trauma centre" (Clarke, 1998). It is supported by a range of first-response vehicles with trained paramedics and occasionally doctors, including a 24 hour helicopter service. On the other hand, Umtata General Hospital in the Eastern Cape province is the only tertiary care hospital in a region serving a population of 4 million, with 30 minor district hospitals. Facilities for A and E are small in relation to patient numbers and the nurse-led triage system from an A and E "area" sends patients to appropriate departments and clinics – the largest number are sent to the Medical Outpatient Department. The unit is not integrated physically into the rest of the hospital. There are some transportation services in the area but they are in need of upgrading. There is considerable need for inter-hospital transfer (Clarke, 1998). Poor or delayed transportation services, dispersed district hospitals with the attendant need for hospital transfers, and understaffed facilities without a single focal point for patient entry do not offer a promising environment for research on alcohol and injuries or for surveillance of alcohol and drug presence in injury.

These examples illustrate the challenges in providing an accurate picture of the role of alcohol and injuries using the emergency room as a research "window". For many people living in rural areas and many living in the slums of urban areas in less developed countries access to hospital care of any kind, especially hospitals with timely emergency services, will be limited if not rare. The great majority of injuries will not be attended to by any health professional, and where they are so treated they are likely to be initially attended in village or local community health stations. Injuries seen in the hospital are more likely to be urban based and/or to have considerable delays between injury and treatment and to involve payment for services. They will also be more serious injuries and are more likely to be related to motor vehicles, including those to pedestrians. Research in developed countries suggests that these are the injuries most likely to show the presence of alcohol. These factors and the demands this type of research place on already stretched hospital staff in poorer and rural areas will inevitably bias the association between alcohol and injury and offer little chance of gaining further knowledge on the **role** of alcohol, rather than its presence in the injury event.

Several other studies illustrate the potential for bias in hospital-based studies of alcohol and injury in developing countries. Mock et al. (1993) compare trauma medical provision in one trauma centre in the US with a hospital in rural Ghana. In this study 59% of injured patients in Ghana were seen more than 24 hours after injury, in contrast to 4% of the US patients. While most of the Ghanaian injured did not receive pre-hospital care (only 25% did), these care providers would be a source of more immediate evidence on the causes of injury including alcohol use than hospital-based health workers. Hijar et al. (1998), in a study of road traffic injuries on a Mexican highway, show a strong relationship between injury severity and reported alcohol consumption 6 hours prior to the accident. In this series, 4.5% of those with no injury, 20.3% of those with minor injury and 28.6 of those with severe injury reported alcohol intake. However, only 38.4% of the injured were transported to hospital, undoubtedly the most severely injured survivors. Reliance on private transport to the hospital is significantly greater in developing countries and is likely to delay assessment of alcohol or drug use prior to injury as well as initial injury severity. Fabbri et al. (2001) in their Italian ER study report that most of the injured were transported "by a basic trauma life support-staffed ambulance" and that fewer than 1% of the severely injured were privately transported to hospital. This is in contrast to 80% of injured in a Ugandan study (Andrews et. al., 1999).

The extent that these factors lead to selection biases is not known because in most research the type of ER care in the country or area in question is not analysed as part of the context of the study. However, the limited evidence that exists suggests that for many, perhaps most, developing countries the ER "window" is of limited value for research on alcohol and injuries.

The Future Development of ER Research

There has been notable progress in ER research in the past two decades, especially in the use of a common research model in a large number of countries. At the same time, gaps in this research and its limitations have also been shown. This section briefly discusses several areas where new work is needed. These include the need for sustainable models for research on injury and alcohol in developing countries where much more work is required; and progress in establishing the relationships between alcohol use and severity of injury and dose and response. There is also a compelling need to move away from the almost econometric approaches of much current research in favor of more qualitative, ethnographic research which will look at the context of injuries.

Lack of resources in developing countries to replicate the developed countries epidemiological models

In most developing countries, progress in research will be largely related to resources – financial, infrastructure and medical personnel. Whether or not the Western epidemiological model of alcohol use can be repeated in most countries in the foreseeable future is doubtful. In trying to assess the contribution of alcohol to injuries in, for example, North America and Scandinavia, a range of types of studies has been carried out – including national general population surveys of drinking practices, local community surveys, ER studies of different types of service providers in different areas of the same country, medical examiner reports, comparisons of ER and mortality data. This process of gathering evidence is expensive and slow in producing results, and it consumes research and medical resources that could perhaps better be used for even more pressing health problems.

It can be argued that at the same time as increasing numbers of emergency room studies are carried out in developing countries, and the appropriate methodologies for these are refined, other approaches which are less expensive and labour intensive are needed to complement or substitute for them. These might include the Rapid Assessment Process (RAP) (See Beebe, 2001) and/or small ethnographic studies with a survey component, or interviews with law enforcement officers and local health professionals. These studies would be especially useful in rural areas. However, the simple use of "informants" to measure patterns of alcohol use and the corresponding burden of alcohol problems belies the complexity and variability in research findings found in developed countries. Qualitative interviews are unlikely to be of much value for model building, including assessing the global burden of alcohol related injuries. They may, however, be useful for research planning and localised attempts at prevention.

Further research is needed on how to obtain self-reports of alcohol use, for example refining the problem of what is a "standard drink" in cultures/sub-cultures with non-standard units of measurement. Temporal patterns of drinking present another challenge in cultures which are not "clock " oriented. This will be an important test of the value of case-crossover methodologies in less developed countries and in differing sub-cultures within countries. (See for example Levine (1997) on U.S. "clock time" vs. Brazil "event time".)

Severity of injury and alcohol use prior to injury

Whether injury severity is related to alcohol use and/or amount of alcohol prior to injury is still unresolved. In part this is because not all of those who are injured attend the same ER and not all ER patients are tested for alcohol use. More severe injuries in many settings are sent to specialist units, may be triaged before alcohol testing, may be intubated at the place of injury or may be otherwise incapacitated. Assessing this relationship is also affected by the fact that many investigators report only a measure of alcohol positive rather than reporting different levels of alcohol use prior to injury: very low levels of alcohol use in the six hours prior to injury may not contribute to serious injury or to the injury at all. Li et al. (1997) report a series of epidemiological studies which show that alcohol is associated with injury severity and higher case fatality rates, but also a series of studies which do not. These studies differ in design, types of ER from which cases are drawn and types of injuries.

There are major confounders in assessing the alcohol and injury severity association includ-ing factors contributing to the injury event such as, for example, speeding in relation to RTAs. However, severity of RTA injuries, for example, can be the result of failure to use a helmet or seatbelts, age of motor vehicle, road lighting and condition and time of day. Thus detailed data on context of injury are essential. Other confounders may include the mistaken assessment that the intoxicated or those with very high BACs are more severely injured because of alcohol-related behaviours, and the greater number of post-injury complications for these patients which may result from both acute and chronic drinking (Jurkovich et al., 1992 and Jurkovich et al., 1993).

A surprising result from one study of injury severity and alcohol intoxication was found by Millham et al. (2004) based on a very large U.S. trauma data base: intoxicated patients were almost 3 times more likely to survive after penetrating trauma and 2 times more likely to survive after blunt trauma, controlling for severity and other confounders.

Few studies compare fatal and non-fatal injuries in the same jurisdiction or geographic area. Two studies which do (one California and one Mississippi study) suggest that a larger propor-tion of fatalities (coroners' samples) were alcohol positive and intoxicated than those non-fatally injured (ER samples). Alcohol levels varied by cause of injury (Cherpitel 1994 and 1996). Large scale regional studies are needed which have sufficient numbers of cases of different causes and types of injury, fatal and non-fatal, which control for demographic and contextual confounders and which cover all major injury care providers.

Dose-Response

Most recent ER studies contribute relatively little to our understanding of a dose-response relationship in relation to alcohol consumption and injury. There are at least three related questions, rarely separated: does a greater "dose" of alcohol increase the risk of injury, what sort of curve describes this relationship, and is an increasing dose correlated with increasing severity of injury? While there is considerable evidence that alcohol consumption prior to injury increases the risk of some injuries, the nature of this relationship is not established for most injuries. It is useful to compare what is known about alcohol and injury to the established relationship between cigarette smoking and mortality from lung cancer. A roughly linear death rate from lung cancer exists in relation to average number of cigarettes smoked daily (Doll and Hill, 1964) and this relationship has been known for decades. Similarly, there is a monotonic relationship between BAC and risk of traffic crashes (Borkenstein et al., 1964 and Hurst et al., 1994). Whether this relationship holds for other types of injury has not been established. Indeed, ER studies have made little contribution to understanding this relationship, in part because many ER injury studies do not report the full range of BACs or self-reported number of drinks prior to time of injury or arrival at an ER. Thus dose-response cannot be assessed even for all-injury studies where numbers are, on average, larger than for most studies of single injuries. An exception is the Australian case-control study of McLeod et al. (1999), which shows that the risk of injury increases as the quantity of alcohol consumed increases. This study also shows the importance of a close analysis of consumption cut-off points and categories by gender. These data show no significant increase in risk until more than 60 grams of alcohol are consumed within six hours prior to injury – at this consumption level there was a three-fold increase in risk. Women were at significantly greater risk then men at these levels of consumption. (See also Watt et al., 2004.)

Corrao et al. (1999), in their meta-analysis of dose-response between alcohol consumption and a number of alcohol-related conditions, included only 18 studies of injury, none using blood alcohol. Alcohol consumption was limited to estimated grams of alcohol per day. Studies were excluded where only two levels of consumption were reported. The included studies covered only suicides (N=3), road traffic accidents (N=9) and fractures (N=7). Risks varied for differ-ent medical conditions and different levels of alcohol consumption, but significant risks were found for injuries. No methodological analysis of the included studies of injuries is given and the included studies are not cited. The heterogeneity across all studies was acknowledged, explained in part by differential quality of the studies. Studies of injuries were not reported separately with regard to heterogeneity and/or quality.

Whether increasing amounts of alcohol consumed prior to injury result in more severe injuries within and across types of injury awaits further investigation, as does the dose/response relation-ship more generally for separate causes and types of injury.

Movements in ER research toward "Thin Description"

Vingilis argued a decade ago (Vingilis et al., 1994): "the exact **role** [emphasis added] that alcohol plays in crashes is still clouded in controversy.… Numerous researchers have suggested that although alcohol plays a role in increasing crash risk, continued emphasis on its effects as the primary contributory factor takes the focus away from the role of other important human factors that also contribute to [in this case] driving risk". ER research has largely ignored the role alcohol might play in association with other human factors in injury by severely limiting both contextual variables, alcohol *effects* in relation to the activity causing an injury, and human factors linked to particular injury events. Contextual effects are often at the country level rather than the subculture in which the actor lives or the circumstances of the injury as an event. Alcohol effects are typically dealt with in a paragraph listing possible effects based on laboratory research. (See, for example, Gmel, 2005.) The person who falls after slipping on a banana peel after one pint in the pub is analytically indistinguishable from a person who falls because he is too intoxicated to keep to the sidewalk, loses his balance, and falls off a curb into the path of an oncoming car.

Rather than the "thick description" that an anthropologist such as Geertz uses, current ER research has become thinner and thinner. "Thick description" involves the actor and putting the actor in a larger social and personal context. It involves "setting down the meaning particular social actions have for the actors whose actions they are, and stating as explicitly as we can manage, what the knowledge thus attained demonstrates about the society in which it is found…" (Geertz, 1973). The actor is all but lost in current ER research. As an example, in many studies no age/gender analysis is reported, in part because of small numbers of cases. As for the context in which injuries occur, most recent ER studies report little or nothing, although the ERCAAP studies have examined places of injury occurrence, drinking places and companions prior to the event, and causal attribution of alcohol to injury on the part of the patient (Cherpitel, 1993c). Additionally, analysis of the ERCAAP and WHO datasets across 16 countries has taken into account both organizational/administrative variables describing the ER, and socio-cultural contextual variables aggregated to the region/country where the ER is located, (see Chapters 1 and 3 in this volume) to explain observed variations found in the association of alcohol and injury (as seen in Tables 1 and 2, for example).

Young et al. (2004) in their search for a "reliable surrogate measure" exemplify this "thinning" in a study which attempts to dispense with all-injury, probability, ER samples, by reducing the (already small) number of variables in ER studies to four or five quantifiable variables. In this proposal "alcohol related/involved" incidents are based on at least one drink (self-reported) in the six hours prior to injury, a weak measure by any reckoning and one, unsurprisingly, that showed a "low probability of injuries being related to alcohol". (See also Treno et al., 1994 and Soderstrom, 1997d.) In a further example, the attributable risk of injury associated with alcohol use in a recent multi-country analysis is reported as only 2% based on BAC (Cherpitel et al., 2005). This is simply not consistent with the risk shown in most individual studies and is likely to be the result of analysing all injuries together, not controlling for severity of injury or level of alcohol use. Contextual analysis is slight, often too global and not centred on the injury event. Some recent analyses are proving to be counter to what even the most conservative researchers on alcohol and injury would suggest the probable contribution of alcohol is likely to be.

"Thicker" analyses of injury events would include multidisciplinary investigations which closely describe the sequence of behaviors (including drinking) before the injury, but would also include a rich description of the context of the injury event, especially safety-related contextual variables (or the lack thereof) and close specification of the alcohol "effects" which could have contributed to the occurrence of the injury. Central to these inquiries is the question "Why did the injury occur on this drinking occasion but not on others, given the same level and type of drinking and the same or similar actions on the part of the injured person?"

Conclusion

As with review papers generally, this chapter is a mix of review, discussion and conclusions. A few discussion points and conclusions stand out as important for future work in research and on prevention. The work of Cherpitel and colleagues is a rare achievement in cross-national research. To carry this on, and extend it into other countries and geographic areas will require a commitment of considerable financial and professional resources. What has been shown is that the model works in many settings. However, one important area for future work is to look at alcohol and injuries in ERs on a regional or state-wide basis – to cover all ERs and hospitals which take emergency cases. This is a necessary but expensive undertaking. Effort is needed to accomplish this without "thinning" research on these problems to a level that is unlikely to help prevent them.

Developing countries are unlikely to have the financial resources in the near future to commit to the problems outlined above on any major scale. There are, however, efforts at injury prevention that can be begun which are closely linked to research on alcohol and injury in developed countries. Forjuoh and Li (1996) suggest a number of possible interventions for developing countries. These include government policies which allow importing only those cars which have functioning safety restraints and banning the sale of alcohol at lorry parks. Although education programmes may not be entirely successful in developed countries (perhaps due to so many of them in all forms of media), they may be successful in developing countries in creating an enhanced consciousness of the effects of alcohol especially, for drivers and pedestrians. Women's health centers could be an important outlet for such information, as these centers, in many developing countries, are a focus for health care and information.

The cooperation of health professionals and insurance companies is needed to make alcohol and drug testing in emergency rooms and with "first responders" to an injury a part of routine medical testing in order to promote the best patient care. This will require close collaboration between alcohol and drug researchers and health workers to influence local and government mandated policy.

Continuing attempts to establish the "global burden" of alcohol problems using ER and other research may be less than useful in contributing to the future development of ER research and the prevention of alcohol-related injuries. (See Chapter 4 in this volume). The research on "global burden" is weak in most, if not all, of the largest developing countries, and it is difficult to establish and quantify the harm to other than the injured parties from alcohol-related injuries – a major part of the burden. What is needed now is a focus on the role and effects of alcohol on injuries; this is harder to come to terms with but it is an area where ERs can play an important part in the future.

References

Aarens M et al. (1978). Alcohol, *Casualties and Crime*. Social Research Group, University of California, Berkeley, CA.

Adeloye A, Ssembatya-Lule GC (1997). Aetiological and epidemiological aspects of acute head injury in Malawi. *East African Medical Journal* **74**, 822-828.

Andrews CN, Kobusingye OC, Lett R (1999). Road Traffic Injuries in Kampala. *East African Medical Journal* **76**, 189-194.

Aptel I et al.(1999). Road accident statistics: discrepancies between police and hospital data in a French island. *Accident Analysis and Prevention* **31**,101-108.

Arnold JL (1999). International emergency medicine and the recent development of emergency medicine worldwide. *Annals of Emergency Medicine* **33**, 97-103.

Beebe J (2001). *Rapid Assessment Process: an Introduction*. Alta Mira Press, Lanham, MD.

Blake RB et al. (1997). Alcohol and drug use in adult patients with musculoskeletal injuries. *American Journal of Orthopedics*, **26**, 709-710.

Borges G et al. Case-control Study of Alcohol and Substance Use Disorders as Risk Factors for Non-fatal Injury (2005). *Alcohol and Alcoholism* **40**, 257-262.

Borges G et al. (1998). Alcohol consumption in emergency room patients and the general population: a population-based study. *Alcoholism, Clinical and Experimental Research* **22**, 1986-91.

Borges G, Rosovsky H (1996). Suicide attempts and alcohol consumption in an emergency room sample. *Journal of Studies on Alcohol* **57**, 543-548.

Borkenstein RF et al. (1964). *The Role of the Drinking Driver in Traffic Accidents*. Indiana University, Indiana.

Chen SC, Lin FY, Chang KJ (1999). Body region prevalence of injury in alcohol- and non-alcohol-related traffic injuries. *Journal of Trauma* **47**, 881-4.

Cherpitel CJ (in press). Alcohol and injuries: A review of international emergency room studies since 1995. *Drug and Alcohol Review*.

Cherpitel CJ et al. (2005). Clinical assessment compared to breathalyzer readings in the ER: concordance of ICD-10 Y90 and Y91 codes. *Emergency Medicine Journal* **22**:689-695.

Cherpitel CJ, Driggers P (2005). *Alcohol and Injuries: Review of Emergency Room Studies since 1995*. World Health Organization, Geneva.

Cherpitel CJ, Ye Y, Bond J (2005). Attributable risk of injury associated with alcohol use. *American Journal of Public Health* **95**, 266-272.

Cherpitel CJ, Moskalewicz J, Swiatkiewicz G (2004). Drinking patterns and problems in emergency services in Poland. *Alcoholism and Alcohol* **39**, 256-61.

Cherpitel CJ, Borges GLG, Wilcox HC (2004). Acute Alcohol Use and Suicidal Behavior: a Review of the Literature. *Alcoholism, Clinical and Experimental Research.* **28**, 18S-28S.

Cherpitel CJ, Giesbrecht N, Macdonald S (1999). Alcohol and injury: a comparison of emergency room populations in two Canadian provinces. *American Journal of Drug and Alcohol Abuse* **25**, 743-59.

Cherpitel CJ (1996a). Regional differences in alcohol and fatal injury: a comparison of data from two county coroners. *Journal of Studies on Alcohol* **57**, 244-8.

Cherpitel CJ (1996b). Alcohol in fatal and nonfatal injuries: a comparison of coroner and emergency room data from the same county. *Alcoholism, Clinical and Experimental Research* **20**, 338-42.

Cherpitel CJ (1995a). Alcohol use among HMO patients in the emergency room, primary care and the general population. *Journal of Studies on Alcohol* **56**, 272-6.

Cherpitel CJ (1995b). Alcohol and casualties: comparison of county-wide emergency room data with the county general population. *Addiction* **90**, 343-50.

Cherpitel CJ (1995c). Alcohol and injury in the general population: data from two household samples. *Journal of Studies on Alcohol* **56**, 83-9.

Cherpitel CJ (1994). Alcohol and Casualties: a Comparison of Emergency Room and Coroner Data. *Alcohol and Alcoholism* **29**, 211-218.

Cherpitel CJ (1993a). Alcohol consumption among emergency room patients: comparison of county/community hospitals and an HMO. *Journal of Studies on Alcohol* **54**, 432-40.

Cherpitel CJ (1993b). Alcohol and violence-related injuries: an emergency room study. *Addiction* **88**, 79-88.

Cherpitel CJ (1993c). Alcohol and injuries: a review of international emergency room studies.

Cherpitel CJ (1992). Drinking patterns and problems: a comparison of ER patients in an HMO and in the general population. *Alcoholism, Clinical and Experimental Research* **16**, 1104-9.

Cherpitel CJ, Pares A, Rodes J (1991). Drinking patterns and problems: a comparison of emergency room populations in the United States and Spain. *Drug and Alcohol Dependence* **29**, 5-15.

Cherpitel CJ, Rosovsky H (1990). Alcohol consumption and casualties: a comparison of emergency room populations in the United States and Mexico. *Journal of Studies on Alcohol* **51**, 319-26.

Cherpitel CJ (1989). A Study of Alcohol Use and Injuries among Emergency Room Patients. *Drinking and Casualties.* Giesbrecht N et al., eds, pp. 288-99. Routledge, London.

Cherpitel CJ (1988a). Alcohol consumption and casualties: a comparison of two emergency room populations. *British Journal of Addiction* **83**, 1299-307.

Cherpitel CJ (1988b). Drinking patterns and problems associated with injury status in emergency room admissions. *Alcoholism, Clinical and Experimental Research* **12**, 105-10.

Clarke ME (1998). Emergency Medicine in the new South Africa. *Annals of Emergency Medicine* **32**, 367-72.

Corrao G et al. (1999). Exploring the dose-response relationship between alcohol consumption and the risk of several alcohol-related conditions: a meta-analysis. *Addiction* **94**,1551-73.

Doll R, Hill AB (1964). Mortality in relation to smoking. *British Medical Journal* 5396, 1460- 67.

Fabbri A et al. (2001). Blood alcohol Concentration and Management of Road Trauma Patients in the Emergency Department. *Journal of Trauma* **50**, 521-8.

Forjuoh SN, Li G (1996). A Review of Successful Transport and Home Injury Interventions to Guide Developing Countries. *Social Science and Medicine* **43**, 1551-1560.

Geertz C (1973). *The Interpretation of Cultures*. Basic Books, New York.

Gentilello LM et al. (2005a). Effect of the Uniform Accident and Sickness Policy Provision Law on Alcohol Screening and Intervention in Trauma Centers. *Journal of Trauma* **59**, 624-31.

Gentilello LM (2005b). Confronting the Obstacles to Screening and Interventions for Alcohol Problems in Trauma Centers. *Journal of Trauma* **59**, S137-S143.

Gjerde H, Beylich KM, Morland J (1993). Incidence of alcohol and drugs in fatally injured car drivers in Norway. *Accident Analysis and Prevention* **25**, 479-83.

Gmel G et al. (2005). Drinking Patterns and Traffic Casualties in Switzerland. *Public Health* **119**, 426-436.

Gururaj G (2004). Alcohol and road traffic injuries in South Asia: challenges for prevention. *Journal of the College of Physicians and Surgeons Pakistan* **14**,713-8.

Hijar M et al. (1998). Alcohol Intake and Severity of Injuries on Highways in Mexico. *Addiction* **93**, 1543-51.

Hingson R, Howland J (1993). Alcohol and non-traffic unintended injuries. *Addiction* **88**, 877-83.

Holubowycz OT, McLean AJ (1995). Demographic characteristics, drinking patterns and drink-driving behavior of injured male drivers and motorcycle riders. *Journal of Studies on Alcohol* **56**,513-21.

Honkanen R et al. (1983). The role of alcohol in accidental falls. *Journal of Studies on Alcohol* **44**, 231-245.

Hurst PM, Harte D, Frith WJ (1994). The Grand Rapids dip revisited. *Accident Analysis and Prevention* **26**, 647-54.

Jurkovich GJ et al. (1992). Effects of alcohol intoxication on the initial assessment of trauma patients. *Annals of Emergency Medicine* **21**, 704-8.

Jurkovich GJ et al. (1993). The effect of acute alcohol intoxication and chronic alcohol abuse on outcome from trauma. *Journal of the American Medical Association* **270**, 51-56.

Kuhn TS (1962). *The Structure of Scientific Revolutions*. University of Chicago Press, Chicago.

Lapham SC et al. (1998). Prevalence of alcohol problems among emergency room patients in Thailand. *Addiction* **93**, 1231-9.

Lejoyeux M et al. (2000). Alcohol dependence among patients admitted to psychiatric emergency services. *General Hospital Psychiatry* **22**, 206-12.

Levine H (1983). In: Room R, Collins G, eds. *Alcohol and Disinhibition*. National Institute on Alcohol Abuse and Alcoholism, Rockville, MD.

Levine RV (1997). *A Geography of Time*. Basic Books, New York.

Li G et al. (1997). Alcohol and injury severity: reappraisal of the continuing controversy. *Journal of Trauma*, **42**(3), 562-9.

Li Y-M (2003). Feasibility of Identification of Alcohol Intoxication by Nurses in Emergency Departments. *Kaohsiung Journal of Medical Science* **19**, 391-396.

Longo MC et al. (2000). The prevalence of alcohol, cannabinoids, benzodiazepines and stimulants amongst injured drivers and their role in driver culpability: part ii: the relationship between drug prevalence and drug concentration, and driver culpability. *Accident Analysis and Prevention* **32**, 623-32.

Maio RF et al. (1997). Alcohol abuse/dependence in motor vehicle crash victims presenting to the emergency department. *Academic Emergency Medicine* **4**, 256-62.

McDonald A, Duncan ND, Mitchell DI (1999). Alcohol, cannabis and cocaine usage in patients with trauma injuries. *West Indian Medical Journal* **48**, 200-2.

McLeod R et al. (1999). The relationship between alcohol consumption patterns and injury. *Addiction* **94**, 1719-34.

Maclure M, Mittleman MA (2000). Should we use a case-crossover design? *Annual Review of Public Health*, **21**, 93-221.

Millham FH, LaMorte WW (2004). Factors Associated with Mortality: Re-Evaluation of the TRISS Method Using the National Trauma Data Bank. *Journal of Trauma* **56**,1090-1096.

Mock CN et al. (1993). Trauma Outcomes in the Rural Developing World: Comparison with an Urban Level I Trauma Center. *Journal of Trauma* **35**, 518-23.

Mock C, Amegashie J, Darteh K (1999). Role of commercial drivers in motor vehicle related injuries in Ghana. *Injury Prevention* **5**, 268-71.

Moraes CL, Reichenheim ME (2002). *International Journal of Gynecology and Obstetrics*. **79**, 269-277.

Obembe A, Fagbayi A (1988). Road traffic accidents in Kaduna metropolis: a-3 month survey.*East African Medical Journal* **65**(9), 572-7.

Odero W, Zwi AB (1999). An evaluation of sensitivity and specificity of blood alcohol concentrations obtained by a breathalyser survey in a casualty department in Kenya. *Accident Analysis and Prevention* **31**,341-5.

Odero W (1998). Alcohol-related road traffic injuries in Eldoret, Kenya. *East African Medical Journal* **75**, 708-11.

Peden M et al. (2000). Substance abuse and trauma in Cape Town. *South African Medical Journal* **90**, 251-5.

Ponzer S et al. (1999). Patients with recurrent injuries – psychosocial characteristics and injury panorama. *European Journal of Emergency Medicine* **6**, 9-14.

Roizen J (1989). Alcohol and Trauma. *In: Drinking and casualties*. Giesbrecht N et al., eds, pp. 21-66. Tavistock/Routledge, London.

Romelsjö A (1995). Alcohol Consumption and Unintentional Injury, Suicide, Violence, Work Performance and Inter-generational Effects. *In: Alcohol and Public Policy*. Holder H, Edwards G, eds, pp. 115-142. Oxford University Press, New York.

Rosovsky H, Garcia G (1988). Alcohol and related casualties in Mexico: a comparison between populations. Presented at the Kettil Bruun Society for Social and Epidemiological Studies in Alcohol, Berkeley, CA.

Schlesselman JJ (1982). *Case-Control Studies*. Oxford University Press, New York.

Soderstrom CA, Dailey JT, Kerns TJ (1994). Alcohol and Other Drugs: an Assessment of Testing and Clinical Practices in US Trauma Centers. *Journal of Trauma* **36**, 68-73.

Soderstrom CA et al. (1997a). The accuracy of the CAGE, the Brief Michigan Alcoholism Screening Test, and the Alcohol Use Disorders Identification Test in screening trauma center patients for alcoholism. *Journal of Trauma* **43** 962-9.

Soderstrom CA et al. (1997b). Alcoholism at the time of injury among trauma center patients: vehicular crash victims compared with other patients. *Accident Analysis and Prevention* **29**, 715-21.

Soderstrom CA et al. (1997c). Psychoactive substance use disorders among seriously injured trauma center patients. *Journal of the American Medical Association* **277**, 1769-74.

Soderstrom CA et al. (1997d). Predictive Model to Identify Trauma Patients with Blood Alcohol Concentrations ≥50 mg/dl. *Journal of Trauma* **42**, 67-73.

Soderstrom CA, Birschbach JM, Dischinger PC (2001a). Alcohol/Drug Abuse, Driving Convictions and Risk-taking Dispositions among Trauma Center Patients. *Accident Analysis and Prevention* **33**, 771-82.

Soderstrom CA, Cole FJ, Porter JM (2001b). Injury in America: the role of Alcohol and Other Drugs. (2001b) *Journal of Trauma* **50**, 1-12.

Stuhlmiller DF et al. (2005). Adequacy of Online Medical Communication and Emergency Medical Services Documentation of Informed Refusals. *Academic Emergency Medicine* **12**, 970-7.

Swaddiwudhipong W et al. (1994). Epidemiologic characteristics of drivers, vehicles, pedestrians and road environments involved in road traffic injuries in rural Thailand. *Southeast Asian Journal of Tropical Medicine and Public Health* **25**, 37-44.

Townes CA (1998). Emergency Medicine in Russia. *Annals of Emergency Medicine* **32**, 239-42.

Treno AJ, Cooper K, Roeper P (1994). Estimating Alcohol Involvement in Trauma Patients: Search for a Surrogate. *Alcoholism, Clinical and Experimental Research* **18**, 1306-1311.

Treno AJ, Holder HD (1997). Measurement of Alcohol-involved Injury in Community Prevention: Search for a Surrogate III. *Alcoholism, Clinical and Experimental Research* **21**, 1695-1703.

Treno AJ, Gruenewald PJ, Ponicki WR (1997). The contribution of drinking patterns to the relative risk of injury in six communities: a self-report based probability approach. *Journal of Studies on Alcohol* **58**, 372-81.

Treno AJ, Gruenewald PJ, Johnson FW (1998). Sample Selection Bias in the Emergency Room; an Examination of the Role of Alcohol in Injury. *Addiction* **93**, 113-129.

Turner JC, Shu J (2004). Serious Health Consequences Associated with Alcohol Use among College Students. *Journal of Studies on Alcohol* **65**,179-183

Vingilis E et al. (1994). Psychosocial characteristics of alcohol-involved and non-alcohol-involved seriously injured drivers. *Accident Analysis and Prevention* **26**, 195-206.

Waller JA (1988). Methodologic Issues in Hospital-based Injury Research. *Journal of Trauma* **28**, 1632-6.

Watt K et al. (2004). Risk of Injury from Acute Alcohol Consumption and the Influence of Confounders. *Addiction* **99**, 1262-71.

WHO (2005). WHO Collaborative *Study on Alcohol and Injuries–Final Report*. World Health Organization, Geneva.

Wright J, Kariya A (1997). Aetiology of assault with respect to alcohol, unemployment and social deprivation: a Scottish accident and emergency department case-control study. *Injury* **28**, 369-72.

Young DJ et al. (2004). Emergency Room Injury Presentations as an Indicator of Alcohol-related Problems in the Community: a Multilevel Analysis of an International Study. *Journal of Studies on Alcohol* **65**, 605-612

CHAPTER 6 :
AGGREGATE VERSUS INDIVIDUAL DATA AS BASES FOR MODELING THE IMPACT OF ALCOHOL ON INJURY

Jürgen Rehm - Centre for Addiction and Mental Health | Toronto, ON CANADA
Robin Room - University of Melbourne | Melbourne, AUSTRALIA

Global Extent and Trends of Alcohol-Attributable Injury

Alcohol is a major risk factor for burden of mortality and disease (Rehm et al., 2003a; 2004). In total, 3.2% of deaths and 4.0% of the burden of disease as measured in disability adjusted life-years (DALYs; see Lopez et al., 2006) in the year 2000 was attributable to alcohol (see also WHO, 2002). Alcohol exposure affects both acute and chronic disease. Relatively, the impact of alcohol on acute conditions is more pronounced than on chronic disease. Thus, in 2000, 16.2% of deaths and years of life lost from injuries, and 13.2% of injury DALYs, were attributable to alcohol (see Table 1). In the estimations for the year 2005, the proportion of all injuries attributable to alcohol increased not only in absolute terms, but also as a proportion of all injuries. The increase is mainly due to shifts in rates of different types of injuries, with an increased proportion of types of injuries more often related to alcohol (Rehm et al., 2006a).

This contribution will not try to analyze these numbers further from a content perspective (see Chapter 4 in this volume), but will illustrate the methodology to derive them, and compare it to alternative methodologies. First, the traditional individual-level based estimation method is introduced and potential shortcomings are outlined. Then, an alternative based on aggregated data will be presented in the same way. After an empirical comparison of both methods in estimating alcohol-attributable injury in eight countries, alternative estimation methods are introduced. A general discussion follows.

How to Estimate the Proportion of Injury which is due to Alcohol?

Once a risk factor has been established as causally relevant by the usual criteria (Hill, 1965; Rothman & Greenland, 1998), the following elements are necessary to derive population estimates of attributable risk (see also formula below):

- Estimates of exposure

- Estimates of outcome

- Estimates of the relationship between exposure and outcome, often derived from cohort or case-control studies (e.g., Walter, 1976; 1980).

There are exceptions to this rule, which are relevant for injuries. For instance, if there are direct estimates from available statistics on the proportion of alcohol-attributable injuries from traffic accidents or fires, as is the case in many countries, the proportion can be taken directly from the statistic. Thus, in many studies on alcohol-related mortality and morbidity in traffic injuries (e.g., Rehm et al., 2006b), estimates on the relationship between exposure and outcome are based essentially on the fact that the injured person or the person judged responsible for the injury had more than a specified level (usually 0.10%) of alcohol in their blood. Since it is clear that some injuries are caused by drinking at levels below 0.10%, the working assumption has been that the underestimation of the relationship resulting from omitting them more or less balances the overestimation from attributing cause whenever there is a blood alcohol level of 0.10% or greater. It is clear that making this assumption is a weakness, which needs to be addressed in future work.

The customary practice in epidemiology is to derive estimates of the relationship between the exposure and the outcome from individual-level studies of those who are injured or have died of trauma. Very often, these studies are based on measurements of aspects of the drinking (in the event or more generally) of those coming to hospital emergency rooms (ERs), or of those who are examined by a coroner after death from an injury. This raises the question of what to use for appropriate controls (see Chapter 8 in this volume), and how to control for confounders (Aarens et al., 1977). The epidemiological counterfactual scenario underlying the derivation of alcohol-attributable fractions stipulates that the injury would not have occurred in the absence of factor alcohol, but it is clear in some instances the injury would have occurred to a person with a substantial level of alcohol in their blood even if the person had not been drinking. For instance, Reed (1981) estimated that this would be true for about half of alcohol-involved driving casualties in the U.S.

A further deficiency in current estimates of alcohol's contribution to injuries is that they measure only injuries to the drinker him/herself. It is clear that many traffic injuries, for instance, happen because of someone else's drinking, and there are also many who suffer violence from someone who is intoxicated. Studies have shown that this risk can be substantial (e.g., Rossow and Hauge, 2004) For example, in the US, omitting BAC negative victims results in a 7% understatement of alcohol-involved serious and fatal injury, including an 11% understatement of highway crashes and a 14% understatement for serious and fatal intentional injury (Levy et al., 2002). But the epidemiological studies, irrespective of methodology (i.e. whether based on cohort, case-control, ER or coroners' samples), typically do not have a good measure of aspects of the drinking of someone who is not there on the gurney or coroner's slab, or is not part of the cohort. Lacking both estimates of exposure and of the relationship between exposure and outcome for alcohol's role in injuries to others, current epidemiological estimates of alcohol's contribution to the burden of injuries thus do not cover a substantial and politically important part of the burden.

Table 1: Global extent and trends of alcohol-attributable injury 2000-2005 (in '000)

	2000			2005			2000-2005
	Alcohol attributable	All cause	% of all injury	Alcohol attributable	All cause	% of all injury	Percent change
All injury							
Deaths							
Women	134	1,728	7.8%	141	1,771	8.0%	2.9%
Men	690	3,360	20.5%	752	3,632	20.7%	0.9%
Both	824	5,088	16.2%	894	5,403	16.5%	2.2%
YLL							
Women	2,814	39,527	7.1%	2,977	38,420	7.7%	8.8%
Men	16,505	79,647	20.7%	18,154	84,179	21.6%	4.1%
Both	19,319	119,174	16.2%	21,130	122,599	17.2%	6.3%
DALYs							
Women	3,604	61,414	5.9%	3,791	62,212	6.1%	3.8%
Men	19,953	116,974	17.1%	21,675	122,926	17.6%	3.4%
Both	23,557	178,387	13.2%	25,466	185,139	13.8%	4.2%
Unintentional injury							
Deaths							
Women	92	1,260	7.3%	99	1,291	7.7%	5.3%
Men	484	2,274	21.3%	519	2,409	21.5%	1.2%
Both	576	3,534	16.3%	618	3,700	16.7%	2.5%
YLL							
Women	1,839	28,716	6.4%	2,009	27,753	7.2%	13.0%
Men	11,408	53,808	21.2%	12,295	55,334	22.2%	4.8%
Both	13,246	82,524	16.1%	14,304	83,087	17.2%	7.3%
DALYs							
Women	2,487	47,596	5.2%	2,703	48,929	5.5%	5.7%
Men	14,008	83,040	16.9%	14,940	85,669	17.4%	3.4%
Both	16,495	130,636	12.6%	17,642	134,599	13.1%	3.8%
Intentional injury							
Deaths							
Women	42	468	9.0%	42	480	8.8%	-2.2%
Men	206	1,086	19.0%	233	1,223	19.1%	0.6%
Both	248	1,554	16.0%	275	1,703	16.2%	1.4%
YLL							
Women	975	10,811	9.0%	968	10,667	9.1%	0.6%
Men	5,097	25,839	19.7%	5,858	28,845	20.3%	3.0%
Both	6,073	36,650	16.6%	6,826	39,512	17.3%	4.3%
DALYs							
Women	1,117	13,818	8.1%	1,088	13,283	8.2%	1.4%
Men	5,945	33,934	17.5%	6,735	37,257	18.1%	3.2%
Both	7,062	47,751	14.8%	7,823	50,540	15.5%	4.7%

Sources: Rehm et al. (2004) for the 2000 estimates, and own calculations based on WHO data for the 2005 estimates.

Dimensions of alcohol relevant for risk of disease and injury

In general, risk factor epidemiology considers one dimension per risk factor, e.g. level of choles-terol or blood pressure, amount of cigarettes or cigars and pipes smoked per day, level of lead exposure etc. (Murray et al., 2003). Alcohol was the only risk factor in the Comparative Risk Assessment (CRA) of the World Health Organization (WHO) with two dimensions examined: average volume of consumption and patterns of drinking (Ezzati et al., 2004; Rehm et al., 2004; WHO, 2002). Both factors have been shown to impact independently on some disease catego-ries and injury (Rehm et al., 2003b; 2004). In injury, the main factor causally relevant is blood alcohol concentration (Eckardt et al., 1998), but average volume consumed has also been shown to impact the risk of injury (Corrao et al., 1999), and there may be interactions between both (Hurst et al., 1994). Causality of this relationship has been discussed in the past, but overall, most reviews agree on a causal impact of alcohol on both unintentional and intentional injury (Rehm et al., 2003b; 2003c; but see Gmel and Rehm, 2003).

Deriving attributable fractions from exposure and relative risks

In the CRA and other studies based on this epidemiological framework (Murray et al., 2003), alcohol-attributable fractions (AAFs) of disease were derived from combining prevalence of exposure and relative risk estimates based on meta-analyses (e.g., English et al., 1995; Gutjahr et al., 2001), using the following formula (Walter 1976; Walter 1980):

$$AF = [\sum_{i=1}^{k} P_i(RR_i - 1)]/[\sum_{i=0}^{k} P_i(RR_i - 1) + 1]$$

Where

i: exposure category with baseline exposure or no exposure i=0

RR(i): relative risk at exposure level i compared to no consumption

P(i): prevalence of the ith category of exposure

AAFs, as derived from the formula above, can be interpreted as reflecting the proportion of disease that would disappear if there had been no alcohol consumption. The application of this formula to estimating AAFs for injury assumes that alcohol impacts injuries the same way in all cultures. This is a strong assumption, which can be doubted, as we know that injuries depend on a variety of factors specific to the culture (e.g. road conditions for traffic injury, law enforcement for homicide, to give just two examples). The same assumption has to be made for the relation-ships between alcohol and other diseases as well, but seems less crucial. For example, the same biological pathways for alcohol's effect on liver cancer can be assumed to be acting in all cultures, and are presumably less impacted by cultural factors (Taylor and Rehm, 2005).

The most comprehensive meta-analysis on alcohol and injury is from Corrao et al. (1999), and the relative risks (RRs) of this analysis allow calculating alcohol-attributable fractions based on individual-level studies. It should be noted that Corrao et al. (1999) only included one dimension of alcohol, average volume of alcohol consumption.

Using an aggregate level analysis to estimate alcohol-attributable fractions

Rehm et al. (2004) used aggregate level statistics to derive alcohol-attributable fractions for injury which take into account pattern as well as volume of drinking in a population. The methodology is described in detail elsewhere (Gmel et al., 2001; Rehm et al., 2004). For this purpose, it suffices to outline the essential statistical methodology. Available time series data on alcohol consumption and injury mortality were used together with country-specific drinking pattern scores to estimate the impact of alcohol consumption on injury, using multi-level regression techniques (Gmel et al., 2001, Raudenbush and Bryk, 2002). Yearly data on level of economic development and calendar year were used as control variables. The resulting regression coefficients were then applied to AAFs derived from individual-level studies for Australia to derive AAFs in societies with other drinking pattern scores.

Please note that neither this method nor any other aggregate-level analysis allows gender-specific exposure for average volume of consumption, as the times series on adult alcohol per capita consumption are only available as a global measure. For the drinking pattern score, in principle a gender-specific score could be derived, but in the analyses reported, the pattern score was only country-specific and assumed to be constant over time.

An empirical example on comparing individual vs. aggregate data to estimate alcohol-attributable fractions

Both approaches were compared empirically in a sensitivity analysis using data from the European Commission Public Health Project:* Table 2 gives an overview.

The two approaches correlate very highly (Pearson r >.99 for both intentional and unintentional injuries), but there are differences in the estimated absolute number of deaths. In general, the individual-level approach yielded higher estimates, with the exception of Russia. This may be explained by the fact that in Russia the estimated pattern value was the most detrimental possible, indicating a culture of irregular binge drinking outside of meals (Rehm et al., 2004). This drinking pattern has been linked specifically to injuries. Conversely, France, with the relatively least detrimental drinking pattern, showed a much higher estimate when based on average volume of drinking alone.

In sum, both approaches yield comparable estimates with respect to overall level of alcohol-attributable injury, and differences between countries. However, the CRA approach, not surprisingly, captured better the impact of drinking patterns.

---

* *HEM – Closing the Gap – Reducing Premature Mortality: Baseline for Monitoring Health Evolution Following Enlargement* (grant agreement no. 2003121) to the Cancer Center and Institute of Oncology, Cancer Epidemiology and Prevention Division, Warsaw (Principal Investigator: Witold Zatoński).

Table 2.   Sensitivity analysis of using aggregate versus individual RR-approaches for alcohol-attributable mortality due to unintentional and intentional injuries in eight European countries, age 20 - 64, for the year 2002

Condition and Approaches[#]	Czech Republic		France		Hungary		Lithuania		Poland		Russia		Sweden		UK		Total	
	M	W	M	W	M	W	M	W	M	W	M	W	M	W	M	W		without Russia
Unintentional injuries (aggregate mixed)	881	170	2,704	466	1,157	208	1,122	232	4,077	533	77,418	15,522	293	55	1,608	337	**101,197**	8,257
Intentional injuries (aggregate mixed)	259	51	1,112	303	626	142	485	59	1,280	198	30,188	6,040	153	48	805	172	**41,921**	5,693
Unintentional injuries (RR)	998	132	3,436	503	966	163	991	177	4,107	410	71,271	10,085	292	49	1,577	313	**95,470**	14,114
Intentional injuries (RR)	490	56	2,683	523	777	141	576	46	2,043	192	31,466	3,651	242	56	1,251	208	**44,401**	9,284
Differences between approaches																		
Unintentional injuries	-117	38	-732	-37	191	45	131	55	-30	123	6,147	5,437	1	6	31	24	**11,313**	-271
Intentional injuries	-231	-5	-1,571	-220	-151	1	-91	13	-763	6	-1,278	2,389	-89	-8	-446	-36	**-2,480**	-3,591

Sources: and own calculations based on WHO data.

For the aggregate approach the regional AAFs from the WHO CRA were used (Rehm et al., 2004), derived from the pooled cross-sectional time series analysis approach described in the text. For the individual-level RR approach, the RRs from Corrao et al. (1999) were combined with country specific prevalence rates of the following exposure categories: current abstainer (drinking up to 0.25 gram pure alcohol per day), 0.25 - < 20 g/day, 20 - < 40 g/day, 40 - < 60 g/day, and 60 and more g/day

Alternative methods to determine AAFs

Variations within both general approaches are possible:

- The CRA approach could be refined to determine AAFs for subcategories of injury, and to introduce gender and time-specific values for pattern of drinking.

- There is also the possibility to use conventional time-series analyses on country-level data, and then derive country-specific attributable fractions (e.g., Norström and Skog, 2001). This approach is discussed further below.

- The approach based on relative risks could be refined by including RRs from emergency room studies, and by conducting a meta-analysis including patterns of drinking from studies where this was measured.

Refining the CRA approach by using different outcomes is relatively straightforward. However, the smaller the disease subcategory, the more problematic the potential differences between countries in coding. Thus, one still has to choose relatively large subcategories, such as unintentional vs. intentional injury, or maybe the sub-categories of the Global Burden of Disease Study related to alcohol; for unintentional injuries: road traffic accidents, poisonings, falls, drownings, "other unintentional injuries", as well as self-inflicted injuries, violence, and "other intentional injuries" for intentional injuries.

With respect to country-specific time series, the same problems as with the pooled time-series method persist: adult alcohol per capita data cannot be disaggregated into consumption by gender and age. On the other hand, in principle the aggregate analyses on alcohol and fatal injury based on country-specific time series account for effects of alcohol occurring to someone other than the drinker – effects of alcohol on the perpetrator in the assault, and also alcohol-related contextual factors which may play a causal role, such as drunken bystanders egging on the fight (Wells and Graham, 1999). The estimates are also specific to a country as a whole – neither based on studies done elsewhere, nor based on studies which are not necessarily representative of a country (like one or two ERs in a country). These advantages have led some writers to prefer aggregate analysis as the basis for estimation of the AAF for homicide and other violence (Room and Rossow, 2001).

On the other hand, alcohol-attributable fractions derived from country time series tend to show lots of variability, which may not be plausible. To give one example: such analyses showed no impact of alcohol on suicide in Switzerland, whereas the vast majority of suicides in Hungary were attributable to alcohol (Gmel et al., 1998; Skog and Elekes, 1993).

With respect to improving on individual level-derived AAFs, the methodologies to include ER studies are less clear. While there exist RRs from ER studies (e.g., Borges et al., 2006), they are usually based on drinking in the situation, and we lack population distributions of drinking in the event for most countries. On a more general level, for most ER studies we lack the possibility to generalize to populations. What does it mean if 22% of injuries in one emergency room in a city in India are related to prior consumption of alcohol (data from WHO study; see Borges et al., 2006)? India is a country, where injury is a major cause of death (e.g. World Health Reports), with about

80% lifetime abstention in men and more than 98% in women (based on World Health Survey; cf. Rehm et al., 2006; see also WHO, 2004). Even though the consumption among drinkers reaches Western levels, it is unlikely that such a high percentage of injury can be attributable to alcohol consumption in the country as a whole. To make ER data useful for population estimates, we need to restructure such studies to allow generalizations. In the meantime, one could work with different assumptions and additional data from other sources to help estimate AAFs. In the current ERCAAP/WHO project (Cherpitel et al, 2006), there are plans to explore this route. The basic idea is to triangulate the existing ER datasets with known information about the regional populations to estimate population RRs. Some additional corrections for the apparent overestimation of RRs may be possible using the case-crossover methodology compared to case-control studies (see Chapter 8 in this volume).

Finally, conventional epidemiological meta-analyses such as the one by Corrao and colleagues (1999) could be enlarged to have two dimensions analysed: average volume of alcohol consumption and patterns of drinking. However, there may not be enough studies with comparable pattern variables.

What to Do in Future Calculations of the Alcohol-Attributable Burden of Injury?

General Conclusions for Use of Aggregate- vs. Individual-Level Studies to Derive Alcohol-Attributable Fractions

Overall, both aggregate and individual-level studies used so far to derive alcohol-attributable fractions for injury show considerable weaknesses. Aggregate-level techniques cannot disaggregate adult per capita consumption and there are no standard ways to derive attributable fractions consistent with the epidemiological theory. In addition, aggregate-level studies, no matter how complex the statistical methods applied, are ecological in nature and cannot be used to establish causality, i.e. causality has to be established by other means (Gmel et al., 2004; Morgenstern, 1995; 1998). On the other hand, traditional comparative risk analyses assume a similarity of effect size (i.e., RR) across cultures, when they apply the same RRs in different sub-regions. Until recently these studies also neglected the effect of patterns of drinking, and usually did not deal with the effect of drinking on others. While all of these problems could in principle be overcome (i.e. region- or country-specific estimates based on suitable statistical techniques; inclusion of patterns; inclusion of the effect on others), it would require a lot of new and better designed studies to see such results.

In an empirical analysis, despite the different assumptions and sources of bias, both types of analyses yielded similar results, i.e. the different methodologies converged to a large degree (see above). This allows some optimism with respect to the validity of the results of these methods. However, we believe that both methods should be improved. In terms of individual-level analyses, we have to include measures of patterns, especially heavy drinking occasions and drinking in the event. Also, we need studies to better integrate the effect of drinking on others than the drinker. For assaults, homicides and traffic casualties, the tradition pioneered by Marvin Wolfgang of painstaking analysis of police reports (Aarens et al., 1977) could be brought into play, although such studies have been done in only a few countries, and most are from some time ago. There is, unfortunately, no such tradition of analyses in the context of emergency health services, and studies should be undertaken to initiate such a tradition. Finally, we need to identify the factors impacting on the alcohol-injury relationship. The above requires a better planning and analyses of future studies. At the present time, re-analyses of existing ER studies as planned in the ERCAAP/ WHO project in combination with other information may be able to improve the estimates.

For aggregate-level analyses, better outcome measures and trying triangulation of different methodologies may improve the estimates. However, as long as we have no way of disaggregating adult per capita consumption, the results of aggregate-level analyses for deriving attributable fractions will remain unsatisfactory, even though such methods may be the best we can currently achieve. For some regions, there may be solutions. The Swedish monthly monitoring data have now been collected since June 2000, which means there more than 60 data points. In principle, data like this could be used for time-series analyses of mortality, hospitalizations or police statistics with the month as the unit, allowing disaggregation of adult per-capita consumption.

References

Aarens M et al. (1977). *Alcohol, casualties and crime*. Social Research Group, Report No. C-18, Berkeley, California.

Borges G et al. (2006). Multicenter study of acute alcohol use and non-fatal injuries: data from the WHO Collaborative Study on Alcohol and Injuries. *Bulletin of the World Health Organization* **84**(6), 453-460.

Cherpitel CJ et al. (2006). Multi-level analysis of causal attribution of injury to alcohol and modifying effects: data from two international emergency room projects. A research report from the Emergency Room Collaborative Alcohol Analysis Project (ERCAAP) and the WHO Collaborative Study on Alcohol and Injuries. *Drug Alcohol Depend*. **82**(3): 258-268.

Corrao G et al. (1999). Exploring the dose-response relationship between alcohol consumption and the risk of several alcohol-related conditions: a meta-analysis. *Addiction* **94**, 1551-1573.

Eckardt MJ et al. (1998). Effects of moderate alcohol consumption on the central nervous system. *Alcoholism, Clinical and Experimental Research* **22**, 998-1040.

English D et al. (1995). *The quantification of drug caused morbidity and mortality in Australia 1995*. Commonwealth Department of Human Services and Health, Canberra, Australia.

Ezzati M et al. (2004). Comparative quantification of health risks. Global and regional burden of disease attributable to selected major risk factors, WHO. Geneva, Switzerland.

Gmel G, Rehm J. (2003). Harmful alcohol use. Alcohol Research and Health, 27(1), 52-62.

Gmel G, Rehm J, Frick U (2001). Methodological approaches to conducting pooled cross-sectional time series analysis: the example of the association between all-cause mortality and per capita alcohol consumption for men in 15 European states. *European Addiction Research* **7**(3), 128-137.

Gmel G, Rehm J, Ghazinouri A (1998). Alcohol and suicide in Switzerland – an aggregate-level analysis. *Drug and Alcohol Review* **17**, 27-37.

Gmel G, Rehm J, Room R (2004). Contrasting individual level and aggregate level studies in alcohol research? Combining them is the answer. *Addiction Research and Theory* **12**(1), 1-10.

Gutjahr E, Gmel G, Rehm J (2001). Relation between average alcohol consumption and disease: an overview. *European Addiction Research* **7**, 117-127.

Hill AB (1965). The environment and disease: association or causation? *Proc R Soc Med*. **58**, 295-300.

Hurst P, Harte W, Frith W (1994). The Grand Rapids dip revisited. Accident Analysis and Prevention 26, 647-654.

Levy, DT et al. (2002). Blood alcohol content (BAC)-negative victims in alcohol-involved injury incidents. *Addiction*, **97**, 909-914.

Lopez, AD et al. (2006). *Global burden of disease and risk factors*. Oxford University Press and the World Bank, New York and Washington.

Morgenstern H (1995). Ecologic studies in epidemiology: concepts, principles, and methods. *Annual Review of Public Health*, **16**, 61-81.

Morgenstern H (1998). Ecologic studies. In *Modern Epidemiology*, Rothman KJ, Greenland, eds, pp. 459-480. Lippincott-Raven Publishers, Philadelphia, PA:

Murray CJL et al. (2003). Comparative quantification of health risks: conceptual framework and methodological issues. *Popular Health Metrics* **1**, 1-20.

Norström T, Skog OJ (2001). Alcohol and mortality: methodological and analytical issues in aggregate analyses. *Addiction* **96**(Suppl. 1), S5-17.

Raudenbush S, Bryk A (2002). *Hierarchical linear models: applications and data analysis methods*, 2nd ed. Sage Publications, Newbury Park, CA.

Reed DS (1981). Reducing the costs of drinking-driving. *In Alcohol and Public Policy: Beyond the Shadow of Prohibition*, Moore MH, Gerstein DR, eds, pp. 336-387. National Academy Press, Washington, DC.

Rehm J et al. (2003c). Alcohol-related mortality and morbidity. *Alcohol Research and Health* **27**(1), 39-51.

Rehm J et al. (2006a). *Alcohol consumption and global burden of disease 2002*. Final report to World Health Organization.

Rehm J, Patra J, Popova S (2006b). Alcohol-attributable mortality and potential years of life lost in Canada 2001: implications for prevention and policy. *Addiction* **101**(3), 373-384.

Rehm J et al. (2003b). The relationship of average volume of alcohol consumption and patterns of drinking to burden of disease – an overview. *Addiction* **98**, 1209-1228.

Rehm J et al. (2003a). Alcohol as a risk factor for global burden of disease. *European Addiction Research* **9**, 157-164.

Rehm J et al. (2004). Alcohol Use. In *Comparative Quantification of Health Risks. Global and Regional Burden of Disease Attributable to Selected Major Risk Factors*, vol. 1. Ezzati M et al., eds, pp. 959-1108. WHO, Geneva, Switzerland.

Room R, Rossow I (2001). The share of violence attributable to drinking, *Journal of Substance Use* **6**, 218-228.

Rossow I, Hauge R (2004). Who pays for the drinking? Characteristics of the extent and distribution of social harms from others' drinking. *Addiction* **99**(9), 1094-1102.

Rothman KJ, Greenland S (1998). Causation and causal inference. In *Modern Epidemiology, 2nd ed.* Rothman KJ, Greenland S, eds, pp. 7-28. Lippincott-Raven Publishers, Philadelphia, PA.

Skog OJ, Elekes Z (1993). Alcohol and the 1950-90 Hungarian suicide trend: Is there a causal connection? *Acta Sociologica* **36**, 33-46.

Taylor B, Rehm J (2005). Moderate alcohol consumption and diseases of the gastrointestinal system: a review of pathophysiological processes. *Digestive Diseases* **23**(3-4), 177-180.

Walter SD (1976). The estimation and interpretation of attributable risk in health research. *Biometrics* **32**, 829-849.

Walter SD (1980). Prevention of multifactorial disease. *American Journal of Epidemiology* **112**, 409-416.

Wells S, Graham K (1999). The frequency of third-party involvement in incidents of barroom aggression. *Contemporary Drug Problems* **26**, 457-480.

World Health Organization (2002). *World health report: Reducing risks, promoting health life.* WHO, Geneva, Switzerland.

World Health Organization (2004). *Global status report on alcohol 2004.* WHO, Geneva, Switzerland.

CHAPTER 7 :
METHODS OF EPIDEMIOLOGICAL STUDIES IN THE EMERGENCY DEPARTMENT

Cheryl J. Cherpitel - Alcohol Research Group | Emeryville, CA USA

Introduction

Numerous studies to date of alcohol and injury have been conducted in hospital-based emergency departments (EDs), internationally. These studies, which are cross-sectional in nature, have generally obtained an objective measure of blood alcohol concentration (BAC) at the time of admission to the ED and administered a face-to-face interview with the patient, eliciting data from a range of questions to establish the association of alcohol and injury, including the type and cause of injury and place of injury occurrence, alcohol use prior to the injury event (typically six hours), usual drinking patterns and higher consumption times, alcohol-related problems, and dependence symptoms (see the Cherpitel Model, Chapter 5). For those reporting drinking prior to the event, data have also been obtained about the number and types of drinks consumed, time lapsed between the last drink and injury, drinking companions and places of drinking prior to injury, whether the patient was feeling drunk at the time of injury and whether he believed the injury would have happened even if he had not been drinking (the patient's causal attribution of injury to alcohol).

Sampling and Interviewing Procedures

Since it usually has not been possible to include all patients coming to the ED for treatment for their injuries in patient samples, a variety of sampling schemes have been used to obtain probability samples of patients, in which each patient has an equal chance of being selected for the study (by selecting every nth patient), and in which each ED staffing-shift over a 24 hour-period is sampled an equal number of times across all days of the week during the study period. One example of such a sampling design would be to select each 4th injured patient (with a pre-determined age cut off of, say, 18) admitted to the ED during the day shift for the first week, the evening shift for the second week and the night shift for the third week, which would assure equal representation of all shifts for all days of the week for every three-week block that data are collected. Some studies have restricted the patient sample to those arriving at the ED within a specified time period following the injury event, and a six-hour time frame commonly has been used. This circumvents the situation in which patients may use the ED for follow-up of an injury which previously had been treated (either in the ED or elsewhere) and that may have occurred a substantial amount of time prior to the present ED visit. One ED study found that only 36% of the patients (which included non-injured as well as injured patients) arrived to the ED within six hours following the event (Cherpitel, 1989). (The "event" for the non-injured patient in ED studies has been defined as first awareness of the symptom for which the patient was now admitted to the ED).

Patient samples are most often obtained from ED admission forms which generally reach a central location shortly after the patient is admitted. To prevent sampling bias related to severity of the injury it is important to identify a patient sampling point which includes not only patients who walk in but also those who arrive by ambulance. Every nth patient meeting study criteria (for example, patients sustaining an injury no more than six hours prior to arrival at the ED and who are 18 years or older) is then approached as soon as possible with an informed consent to enter the study. This commonly occurs in the ED waiting room, especially for those patients less severely injured and for whom a relatively long period of waiting is often the case.

This sampling procedure normally requires that one person remains in charge of identifying patients who fall into the sample, while one or more individuals (depending on the number of patients admitted to the ED) are available to approach patients to participate in the study and obtain study data. It is important at this point that patients falling into the sample are not "missed" because an interviewer was not available at the time the patient is selected for the sample. It has been unrealistic to expect ED staff to obtain a BAC estimate from the patient or to administer an interview in addition to their usual duties in the ED. It has therefore been important that special staff is recruited for the study, whose primary responsibility is to the study itself. The number of interviewers in the ED setting at any one time must be sufficient to obtain all of the study data for each patient selected into the sample, and includes allotting time for locating the patient and obtaining informed consent as well as obtaining a BAC estimate and administering the questionnaire. The number of study staff must not be so large, however, as to encroach on the space required by the ED staff to perform both their routine and emergency functions. EDs do not see a continuous flow of patients around-the-clock, and this must also be taken into account in determining both the number of interviewers needed for each sampled shift and the sampling fraction to be used. For example, it may be possible to sample every 3rd patient during most days and times, but on weekend evenings when a larger number of patients are admitted to the ED, it may only be possible to sample every 6th patient. If the sampling fraction changes during the period of data collection, it is important to document the times when this occurs and the sampling fraction used, and to make adjustments for this difference in the data analysis by weighting the data accordingly. In this example, patients sampled during the period when every 6th patient was sampled would be given twice the weight as those sampled during the period when every 3rd patient was sampled.

If the patient refuses to participate in the study, or is unable to participate due to other reasons such as the present condition, another patient is not selected in their place, but rather the next nth patient, as usual, is selected. This sampling design, without patient replacement, ensures the integrity of the probability selection of all patients in the sample. Additionally, there is always the possibility of "refusal conversion" or "non-participation conversion" in which the patient later agrees to participate in the study. Any patient who is too severely injured to participate in the study at the time, and is admitted to the hospital, should be approached at a later time in the hospital after the condition has stabilized. Obtaining the interview while the patient is waiting for treatment usually has been found to be an opportune time, and patients, especially if they are anticipating a long wait, generally are more than enthusiastic to engage in a discussion about themselves and their problems with an interested individual. The waiting room, itself, may afford a good location for obtaining the interview when patient privacy can be maintained, but when this is not possible, another location in close proximity to the waiting room, so the patient can be found for treatment, also provides a good location for obtaining the interview. If interviews are not completed prior to patient treatment, interviewers have often followed patients into the treatment area (with permission of the clinician) to continue the interview. At times ED clinicians have found that engagement of the patient in an interview during treatment is actually beneficial in diverting attention away from a procedure, for example, while the patient is being sutured. Interviewers have also followed patients who were required to visit other areas of the hospital (for example, lab, x-ray or pharmacy), to complete the interview.

BAC Measurement

A number of different methods exist for estimating blood alcohol concentration at the time of the ED visit. These include blood draws, which normally are not part of special ED studies, but may be stand-ard ED procedure for particular types of injury victims (for example, those admitted to level I trauma centers in the U.S.). Breath analysis has been the most commonly used procedure for estimating BAC in ED studies, but saliva and urine testing have also been used successfully. Breath analyzers have been found to provide estimates of BAC that are highly correlated with chemical analysis of blood among cooperative patient (Gibb et al., 1984). A device is also available that can be attached to the breath analyzer to obtain ambient exhaled air from patients who are less cooperative or unable to exhale sufficiently to provide a breath sample. While obtaining a BAC estimate as soon as possible after a person has been admitted to the ED is important, this may not always be possible, especially if the patient is undergoing intensive treatment and/or is not able to be approached to participate in the study for some other reason. In such circumstances it has been possible for the interviewer to request that the attending clinician obtain a breath sample and record the results, which will then only be obtained from the clinician at a later time after the patient has provided consent. This proce-dure circumvents the problem of having biased (conservative) BAC estimates for those patients who may be more severely injured than those who do not require immediate treatment and/or are able to provide immediate consent to participate in the study.

BAC estimates have been obtained most often prior to the patient interview and rates of those registering positive for BAC who deny drinking, when the time of injury and time of arrival in the ED have been taken into account, have been found to be small, ranging from .5% to 3.6%. However, larger proportions have been found for those who reported drinking in the six hours prior to injury but who are negative for BAC, with rates ranging from 18% to 41% (Cherpitel, 1989; Cherpitel et al., 1992), and this has been attributed to the time lag between the last drink and arrival in the ED, which may be a number of hours in some cases. Since patient self-reports of drinking have been obtained after the BAC estimate was obtained in these and other studies, it is possible that patients may have been less likely to deny drinking, knowing that the inter-viewer already had the BAC results in hand. In a separate study, timing of the self-report relative to estimated BAC was examined, and no difference was found in whether the self-report was obtained before or after the BAC was estimated (Cherpitel, 1993). However, interestingly, some patients on whom the BAC estimate was obtained following the interview felt providing a BAC constituted a breach of trust that had been established with the interviewer during the interview, although the patient had previously consented to both procedures (BAC and interview) prior to initiation of the study. Refusal rates for a BAC estimate were no higher for those on whom the estimate was obtained after the interview compared to before; however, other reasons for not obtaining a BAC were more prevalent for those on whom the estimate was obtained following the interview, and it is possible that some of these patients may have terminated the interview early because their BAC would have contradicted their self-reports. For these and other reasons having to do with the importance of the timeliness of the BAC reading in relation to arrival in the ED, it seems prudent to obtain the BAC estimate at the outset, prior to interviewing the patient. In this regard, some studies also have restricted obtaining data on BAC estimates to those arriving within six hours following the event.

Methodological Issues

Types of studies and control groups

The overall purpose of the ED study and planned uses for the data dictate, in part, the best methods to be used to collect the data. If the chief aim of the study is to establish the prevalence of drinking and/or intoxication at the time of the ED visit for an injury, or the prevalence of alcohol-related injury, based on a measure of estimated BAC at the time of arrival and/or patient self-report of drinking (typically within six hours) prior to the injury event, representativeness of the data in terms of times of day and days of week is especially importance. One would expect a higher prevalence of alcohol associated with injury on a weekend evening, when drinking may be more prevalent, than on a weekday morning, for example. Although data from any one ED will likely not be generalizable beyond the particular ED where the data were collected, the data collected in that ED should be representative of all injured patients treated in the ED, and strict adherence to the sampling procedures described above is crucial in this respect.

While prevalence studies are important in determining the burden that alcohol and alcohol-related injuries place on the ED system, studies in EDs also have been interested in examining the risk at which alcohol places the individual for injury. Determining the risk of injury due to alcohol requires a control group of those without injuries for comparison to those with injuries. It is also important that these studies are representative of all times of day and days of the week, since there is reason to believe that injury risk due to alcohol may differ across day and time. For example, not only is it more likely that one may be drinking during evening hours compared to day hours, but it also might be expected that drinking and driving at night would be more likely to result in a motor vehicle accident due to poorer visibility than drinking and driving during day-light hours. Therefore, the best studies, again, include samples in which all times of day and days of week are equally represented for both case (injured) and control (non-injured) patients, so findings are representative of the actual risk of injury from alcohol.

Three types of control patients have been used in ED studies which have examined the risk of injury. The most common control patient in these studies has been the non-injured patient seeking care in the same ED at the same time as the injured patient. Studies using non-injured controls have sampled them simultaneously with injured patients, so that every nth patient, regardless of injury status but meeting other selection criteria, is included in the sample. The advantage of using non-injured as control subjects is that they are similar to the injured in many respects (both known and un-known) having to do with seeking care in a particular ED. On the other hand, a disadvantage of using non-injured patients as controls is the fact that they may be seeking care for non-injured conditions related to their alcohol use, including conditions related both to the acute effects of alcohol such as alcohol intoxication, overdose or withdrawal, as well as to the chronic effects such as liver cirrhosis. Studies using non-injured patients as control subjects have found them more likely to report frequent, heavy and problem-related drinking compared to their counterparts in the general population (Cherpitel, 1992). Non-injured patients as control subjects, then, likely provide a conservative estimate of the association of alcohol with injury.

A second type of control patient that has been used is the general population or community control. The advantage of using community controls is, like the non-injured patient controls, they come from the same catchment area as the injured patients, but, unlike the non-injured controls, they are not seeking treatment for a condition possibly related to their use of alcohol. Community controls have been selected by matching to injured patients on age and gender, and, during the interview, have been asked about drinking during the time period corresponding to the time prior to injury occurrence for the injured patient (Vinson et al., 2003 ; Watt et al., 2004).

Finally, a relatively new methodology for determining the risk of injury from acute consumption of alcohol prior to the injury event is called the case-crossover study, in which each patient acts as his own control subject (Maclure, 1991; Mittleman et al., 1993). This method has the advantage of theoretically reducing the influence on the risk of alcohol-related injury of stable risk factors such as demographic characteristics and usual drinking patterns, but has only relatively recently been applied to the study of alcohol and injury in ED populations (Borges et al., 2004; Vinson et al., 1995). In the case-crossover study, drinking prior to the injury event is compared to the same patient's drinking at the same time of day the day before injury or the week before injury as a control period (and is called the self-matched approach). If these data are not available, then the usual frequency approach is used, in which drinking prior to the event is compared to the same patient's usual frequency of drinking in a more general time frame, such as the last 30 days or last 12 months, which provides an estimate of the expected likelihood of drinking prior to the injury event based on usual drinking patterns. The case-crossover method for analyzing the risk of injury from alcohol consumption, as well as those methods using non-injured patients or community control subjects, have generally not taken into account other important risk factors which could potentially influence whether or not drinking may result in an injury, for example, the use of other drugs in combination with alcohol prior to the event and the activity in which the patient was engaged at the time of injury. It will be important to consider such potential risk factors in future ED studies of the risk of injury related to alcohol use.

Biases in ED samples

Even given sound methodological principles in conducting ED studies, a number of limiting factors must be considered when the emergency department is used as the 'window' for viewing associations of alcohol and injuries. A major limitation to ED studies is the potentially biased selection of individuals who use the ED. Studies carried out in EDs only provide information on alcohol's association with injuries for patients who are treated in the ED, and who may well differ on socio-demographic and drinking characteristics from those who go elsewhere for treatment or those who go untreated. Additionally, by the very nature of ED studies, those patients falling into ED samples are those who are most frequent users of the ED, and these individuals also may vary in socio-demographic and drinking characteristics from those who are less frequent users of the ED. One ED study found 47% of the injured patients reported at least one previous ED visit during the preceding year compared to only 29% of those in a general population sample who belonged to the same health maintenance organization and who also reported an ED visit to the same facility for an injury during the previous year (Cherpitel, 1992). ED studies also only provide information about alcohol involvement on the part of the patient seeking treatment, and not on others who may have been instrumental in precipitating the event, for example, the 'other driver' who may have been drinking and caused the accident, but was not injured and/or not treated in the ED, or the perpetrator of an assault who was likewise uninjured and/or not treated.

– Severity of injury

Findings from ED studies may also likely be biased with respect to severity of injury, even when sound study methods are used. Those patients arriving by ambulance and/or those hospitalized following admission to the ED (who are likely to have sustained more severe injuries than those arriving on their own and/or those not hospitalized), are more likely to not be included in the study although falling into the sample, than those with less severe injuries and on whom an interview can be completed in the ED waiting room. The relationship between alcohol consumption and severity of injury is not completely clear, however, and has been an issue of on-going debate (Li et al., 1997) with mixed findings (Brickley and Shepherd, 1995; Fabbri et al., 2001; Honkanen, 1993). Data indicating no association of alcohol and injury severity have come primarily from trauma center-based studies (those studies which include the most severely injured patients), while data from studies based in emergency departments have generally found a significant association, although findings from these studies suggest the association may be due largely to other risk factors associated with alcohol use, such as speeding, not wearing seat belts or helmets, or other risk-taking behaviors. It has been argued that those more severely injured are likely to reach the ED sooner, and consequently are more likely to have a positive (and higher) BAC than those less severely injured who arrive later. Alcohol intoxication itself also can bias injury severity scores upward, resulting in apparently more severe injury for those who are alcohol impaired at the time of ED admission (Waller, 1988). Estimates of alcohol's association with injury, based on ED studies, may then be either inflated or conservative in relation to injury severity, depending on a number of factors including the probability of obtaining comparable data on those more severely injured and those less severely injured.

– Types of emergency services delivery systems

This chapter has largely focused on methods to conduct studies in emergency departments which can be considered as belonging to the "Anglo-American" model of emergency services delivery, in which the patient is brought to the hospital-based ED facility for treatment (Arnold, 1999). In this model, designated hospital-based emergency departments are available 24 hours a day during which time patients arrive on their own initiative or by ambulance, with few physician-referred patients. An alternate system of emergency services delivery, called the "Franco-German" model, is predominant in many countries, however, and, in contrast to the "Anglo-American" model, brings the hospital to the patient (Arnold, 1999). In this model ambulance services provided by physicians (primarily anesthesiologists) travel to the patient to provide emergency treatment and are limited to life-sustaining measures in the home or at the scene of an accident, as well as during transportation to an appropriate in-patient or out-patient facility, including facilities run by the ambulance station itself. In this model many patients may only be treated by the ambulance team and not subsequently admitted to a hospital for emergency care. In this system emergency treatment provided in the hospital is offered on a rotating schedule across a number of hospitals to assure access to such services and to distribute the emergency service burden among all existing facilities in an area. Each hospital in a given region provides a certain number of emergency days per month, and offers general emergency treatment as well as emergency treatment in a specific number of specialty areas. Patients obtain hospital-based emergency services by either ambulance transport or by referral from a doctor, and only a relatively small number of patients arrive for emergency services on their own. This model, with the lack of on-going availability of a specific emergency department to serve a given catchment area, precludes a sampling scheme where patients are identified at a single central processing point, which is possible in studies of EDs that follow the "Anglo-American" model, and necessitates a sampling scheme at possibly several points in the emergency services delivery system. It would be expected that this model of care would provide lower estimates of the prevalence of alcohol involvement in injury cases than the "Anglo-American" model, given lack of proximity of the injury event to treatment in an emergency care facility. The model of emergency services delivery which is followed, then, will not only affect the quality, representativeness and generalizability of the patient sample and data obtained, and the best methodologic strategy to use in obtaining probability samples of patients, but will also affect the prevalence of alcohol involvement in the emergency services caseload and the risk of injury at which alcohol places the individual.

Finally, as noted earlier, regardless of the rigor of methods used in obtaining samples of ED patients and their representativity of all patients using the respective ED facility, the data collected in an individual ED cannot be generalized beyond the facility in which the data were collected. With this and other limitations of ED studies in mind, such studies have, nevertheless, provided, and can continue to provide, valuable data on the prevalence and burden alcohol places on ED caseloads and the risk at which drinking places the individual for an alcohol-related injury.

References

Arnold JL (1999). International emergency medicine and the recent development of emergency medicine worldwide. *Annals of Emergency Medicine* **33**(1), 97-103.

Borges G, Cherpitel CJ, Mittleman M (2004). The risk of injury after alcohol consumption: a case-crossover study in the emergency room. *Social Science & Medicine* **58**(6), 1191-1200.

Brickley MR, Shepherd JP (1995). The relationship between alcohol intoxication, injury severity and Glasgow Coma Score in assault patients. *Injury* **26**(5), 311-314.

Cherpitel CJ (1989). Breath analysis and self-reports as measures of alcohol-related emergency room admissions. *Journal of Studies on Alcohol* **50**, 155-161.

Cherpitel CJ (1992). Drinking patterns and problems: a comparison of ER patients in an HMO and in the general population. *Alcoholism: Clinical and Experimental Research* **16**(6), 1104-1109.

Cherpitel CJ 1993). Timing of the breath analyzer: Does it make a difference? *Journal of Studies on Alcohol* **54**, 517-519.

Cherpitel CJ et al. (1992). Validity of self-reported alcohol consumption in the emergency room: data from the U.S., Mexico and Spain. *Journal of Studies on Alcohol* **53**(3), 203-207.

Fabbri et al. (2001). Blood alcohol concentration and management of road trauma patients in the emergency department. *Journal of Trauma* **50**(3), 521-528.

Gibb K et al. (1984). Accuracy and usefulness of a breath alcohol analyzer. *Annals of Emergency Medicine* **13**, 516-520.

Honkanen R (1993). Alcohol in home and leisure injuries. *Addiction* **88**, 939-944.

Li G et al. (1997). Alcohol and injury severity: reappraisal of the continuing controversy. *Journal of Trauma* **42**, 562-9.

Maclure M (1991). The case-crossover design: a method for studying transient effect on the risk of acute events. *American Journal of Epidemiology* **133**(2), 144-53.

Mittleman MA et al., for the Determinants of Myocardial Infarction Onset Study Investigators (1993). Triggering of acute myocardial infarction by heavy physical exertion. Protection against triggering by regular excertion. *New England Journal of Medicine* **329**, 1677-1683.

Vinson DC (1995). Alcohol and injury. A case-crossover study. *Archives of Family Medicine* **4**, 505-511.

Vinson DC et al. (2003). A population-based case-crossover and case-control study of alcohol and the risk of injury. *Journal of Studies on Acohol* **64**(3), 358-366.

Waller J (1988). Methodological issues in hospital-based injury research. *Journal of Trauma* **28**, 1632-1636.

Watt K et al. (2004). Risk of injury from acute alcohol consumption and the influence of confounders. *Addiction* **99**, 1262-1273.

CHAPTER 8 :
CONCEPTUAL ISSUES IN EMERGENCY ROOM STUDIES AND PATHS FORWARD

Gerhard Gmel, Jean-Bernard Daeppen - Swiss Institute for the Prevention of Alcohol and Drug Problems | Lausanne, SWITZERLAND

Introduction

Just to mention only two of the many examples, the Emergency Room Collaborative Alcohol Analysis Project (ERCAAP, e.g., Cherpitel et al., 2003a; Cherpitel et al., 2003b) and the WHO-Collaborative Study on Alcohol and Injury (e.g. Borges et al., 2004b; Cherpitel et al., 2005b) demonstrate the internationality of research on alcohol and injury in emergency rooms (ERs). ER studies generally show that alcohol consumption is associated with injury and its cognitive and psychomotor effects are well established (Rehm et al., 2003).

Thus, there is no need to further substantiate **that** alcohol consumption is associated with injury. However, there is need to explain what the real **magnitude** of the association is in terms of relative risks and of alcohol attributable fractions (percentages of injuries attributable to alcohol consumption), e.g. to measure the burden of disease stemming from alcohol consumption (Rehm et al., 2004). Typically a large variation across ER studies has been found in the percentage of alcohol positive cases (e.g. Hingson and Howland, 1993; Treno et al., 1998; Cherpitel et al., 2004a); for example with respect to both fatal and nonfatal injuries:

- Falls: ~17% - 77%
- Drowning: ~25% - 50%
- Violence: ~17% - 70%
- Burns: ~ 1% - 98%
- Suicide: 10% - 70%

Treno et al. (1998) argue that such differences might be partly explained by injury cases in ERs being unrepresentative of all injuries in the general population. For example, the less severely injured may not seek professional treatment, may delay seeking treatment or there may be a delay in treatment due to triage policies. Also, the most severely injured are often not interviewed due to an inability to respond or the potential interference of interviews with emergency care needs. More affluent or better insured patients may have other treatment options (e.g., personal physicians) and therefore socio-economically disadvantaged patients, who may have distinct consumption patterns, may be over-represented in ER studies. Nevertheless, though incidence and prevalence estimates may be biased, Treno et al. (1998) found that alcohol–injury associations remained stable even when adjusting for sample selection effects. This suggests that alcohol attributable risks could be obtained from ER studies, but prevalence estimates of injuries and of type of drinkers must be obtained from other sources to obtain a good estimate of the alcohol-related burden of injury in a population.

Certainly, further research into the effects of these potential selection biases is needed. The present chapter, however, takes another perspective. We argue that some of these biases can be easily avoided. For example, the severely injured could be interviewed later. Different degrees of severity of injury could be addressed by combining estimates from different samples (e.g. in emergency departments and level I trauma centers, Li et al., 1997), including fatal injuries from coroner samples. ER studies will always have to be selective as regards the sample selectivity of patients and the non-representativity of ER injuries of all injuries. We argue, however, that the relationship between alcohol consumption and injury is probably best analyzed with ER studies. Though the ER sample's composition of drinkers may not reflect the general population, if the relative risks and odds ratios were obtained separately for different and homogeneous types of drinkers, these risk relationships could then be applied to more representative samples of drinkers (e.g. general population samples) and register-based census data of injuries (e.g., country wide registers of hospitalizations). Alcohol-attributable fractions are derived by combining prevalence estimates with relative risk estimates, but there is no need that the two come from the same data source.

One major factor for the observed differences in findings across ER studies may be that different compositions of types of drinkers (e.g., moderate but regular drinkers or infrequent drinkers with heavy drinking occasions on weekends) are observed across different ER samples. A second major factor in the diversity of study findings is the problem of choosing adequate controls. The present paper pleads for comparative research that addresses these two aspects to obtain more stable estimates for risk-relationships between alcohol use and injury.

Is Diversity of Findings a Threat to Validity?

As the global burden of disease 2000 study (GBD 2000) demonstrated (Rehm et al., 2004), alcohol consumption cannot be considered to have the same effect on injury in all regions of the world (see Table 1). Why should we then assume that different ER studies should show similar effects? The difference in alcohol attributable fractions is related to the prevalence of drinkers in the corresponding region, and to different drinking patterns across regions, i.e. the way alcohol is consumed. It makes a difference on the likelihood of injury whether 14 drinks are consumed regularly per week with 2 drinks a day, or conversely with 7 drinks on each of Friday and Saturday evenings and abstinence during the remainder of the week.

Table 1. Alcohol attributable fractions of injuries, GBD-CRA 2000*

	World			High mortality developing countries AFR-D, AFR-E, AMR-D, EMR-D,SEAR-D		Low mortality developing countries AMR-B, EMR-B, SEAR-B, WPR-B		Developed countries AMR-A, EUR-A, EUR-B, EUR-C, WPR-A	
	Males	Females	Both sexes	Males	Females	Males	Females	Males	Females
Drowning	12	6	10	8	4	10	6	43	25
Falls	9	3	7	5	1	8	3	21	8
Homicide	26	16	24	18	12	28	16	41	32
Self-inflicted injuries	15	5	11	8	2	10	5	27	12
Poisoning	23	9	18	7	3	11	7	43	26
Other intentional injuries	13	7	12	7	3	20	11	32	19
Motor vehicle accidents	25	8	20	19	5	25	8	45	18
Other unintentional injuries	15	5	11	10	4	15	6	32	16

* AFR=Africa, AMR=American, EMR=Eastern Mediterranean, EUR=Europe, SEAR=South East Asia, WPR=Western Pacific

The regional groups are organized as follows: A denotes very low child and low adult mortality, B is low child and low adult mortality, C is low child and high adult mortality, D is high child and high adult mortality, and E is high child and very high adult mortality.

For WHO-regions see e.g. World Health Report (World Health Organization (WHO), 2002)

The main exposure measure in ER studies commonly is acute alcohol intake, e.g. the self-reported amount consumed in the 6 hours before injury, or measures of blood alcohol concentration (BAC). Thus, the patient who consumed 7 drinks before injury can be distinguished from the one who had only 2 drinks. An additional question is however, whether different types of drinkers have the same likelihood of injury if they have consumed the same amount prior to injury. The re-analysis of the Grand Rapids Study (Borkenstein et al., 1964), a case-control study of Michigan drivers (Hurst et al., 1994) showed that risk curves for traffic crashes increased more steeply in less frequent compared with frequent drinkers. Similarly, research using cross-sectional data in the US (Gruenewald and Nephew, 1994; Gruenewald et al., 1996a; Gruenewald et al., 1996b; Treno et al., 1997; Treno and Holder, 1997) showed that the greatest risks for injuries related to drinking and driving were among drinkers consuming large amounts on some occasions, particularly where highest amounts were much larger than usual amounts consumed. This indicates that a) heavy episodic drinking (HED) is more strongly related to injury than volume of drinking and b) that usual drinking patterns may moderate the association between acute intake and injury. There is also some evidence from comparative emergency room studies that different consumption patterns in different ER populations may partly explain differences in risk relationships. For example, the magnitude of alcohol-injury associations has been found to vary across countries (e.g. Cherpitel et al., 1991; Cherpitel et al., 1993b; Cherpitel et al., 1993a) and across regions within countries; e.g. US (Cherpitel, 1997a; Cherpitel, 1997b), Canada (Cherpitel et al., 1999) and Poland (Cherpitel et al., 2005c). It has even been shown that the acculturation into another drinking culture is related to differences in risk relationships (Cherpitel and Borges, 2001). These studies tend to show that differences in risk relations may be related to the different cultural drinking patterns of the so-called "wet" and "dry" regions (Room and Mäkelä, 2000). "Wet" regions are characterized by low abstinence rates, and drinking integrated into everyday life resulting in more frequent drinking with fewer heavy-drinking occasions, while "dry" regions are characterized by high abstinence rates, and a dominant pattern of infrequent but heavy drinking. Higher risk associations are commonly found in dry countries or regions where explosive patterns of heavy drinking occasions are interspersed with less frequent drinking. Recently, meta-analyses of emergency-room studies support such a view, showing that at the aggregate-level a country's predominant drinking pattern affects the association between acute intake and different outcomes such as risk of injury or causal attribution of injury to alcohol (Cherpitel et al., 2003c; Cherpitel et al., 2003a; Cherpitel et al., 2004b; Cherpitel et al., 2005a).

It has been argued that high-volume drinkers are at lower risk of injury than low-volume drinkers at the same levels of acute intake, because of higher tolerance of the former to the effects alcohol (Cherpitel et al., 2003a; Watt et al., 2004). Regular users might have differential risks compared to irregular users with HED, due to the different environment (outside the home, in bars, etc.) in which drinking takes place. Therefore it is necessary to look at usual consumption pattern in addition to acute intake to better understand differences in risk-relationships across ER studies.

Measuring the Interaction Between Usual Consumption Patterns and Acute Intake Prior to Injury

Studies have rarely quantified this moderating effect of usual consumption patterns on acute intake. One reason is that the complexity of the association and the way to model it have not yet been fully understood (Rehm and Gmel, 2000). The association between alcohol consumption and injury requires a three-way interaction model. First, usual consumption patterns have at least two dimensions, namely volume of drinking (e.g., measured average number of drinks a day over the last 12 months) and heavy episodic drinking, also called binge drinking (e.g., measured by drinking 5 or more drinks on an occasion). In a simplified typology, 4 types of drinkers can be distinguished: a) low risk drinkers (low volume, no HED), b) heavy episodic drinkers (low volume, but occasions of HED – stereotypically the weekend drinker with 5+ drinking on weekends and abstinence or low volume during the week), c) high volume drinkers without HED (stereotypically, the daily "meal drinker" with two glasses of wine twice a day), and d) risk accumulators (both high volume and HED). The third dimension to consider in the interaction model is the acute intake of a certain amount of alcohol prior to injury.

To our knowledge, studies in the field have not addressed this three-way interaction, with one exception discussed below. Often the three dimensions (HED, volume, and acute intake) have been addressed in separate models (e.g. Borges et al., 1998b). Closest to what we are suggesting are models in which acute intake and usual drinking patterns were analysed in multiple regression analysis, whereby usual drinking patterns were measured by a combination of frequency of drinking and the maximum amount consumed on one occasion during the last 12 months (e.g. Borges et al., 1998b; Cherpitel and Borges, 2001).

Moreover, most models do not address the interaction between usual consumption patterns and acute intake, but estimate only the main effects. Thus, the effect of acute intake is adjusted for usual drinking patterns, but the differential effects of acute consumption is not estimated for different types of drinkers. Such main effect models can also be quite misleading or even be theoretically impossible. For example, one recent study included volume of drinking (but not HED) and acute intake in a multiple regression model (Watt et al., 2004). After adjusting for a number of socio-demographic variables and injury characteristics, the risk of injury was found to increase for acute intake, but decrease for usual volume of drinking, which was interpreted to mean that regular consumption may have a protective effect on risk of injury due to tolerance. While such an interpretation may be valid, the model is not adequate to support it.

Figure 1 shows what such a main effect model actually estimates. First, main effect models inherently estimate parallel regression lines (in logistic regressions in terms of log odds). Thus, the model "assumes" that the risk of acute intake increases in exactly the same way in all drinker groups (based on usual consumption patterns). Therefore, such models can not estimate differential effects of acute intake in different consumption groups. Second, the models estimate something which is logically implausible, i.e., the risk for abstainers increases with increasing acute intake. Though this may make sense hypothetically if one assumes that the risk of abstainers would increase if they consumed alcohol, and would then be even higher than that for drinkers, it is estimated with empirical data for which such cases (abstainers with acute intake) do not exist. Third, in this model of figure 1 the risk for injury would be lower for alcohol consumers as long as they do not have a risky acute intake, thus low acute intake of alcohol was estimated to be protective compared to abstention. There is little evidence in the literature, however, that acute intake of alcohol, even at moderate doses, is protective for injury.

Figure 1. Main effect model for risk of injury on usual and acute consumption

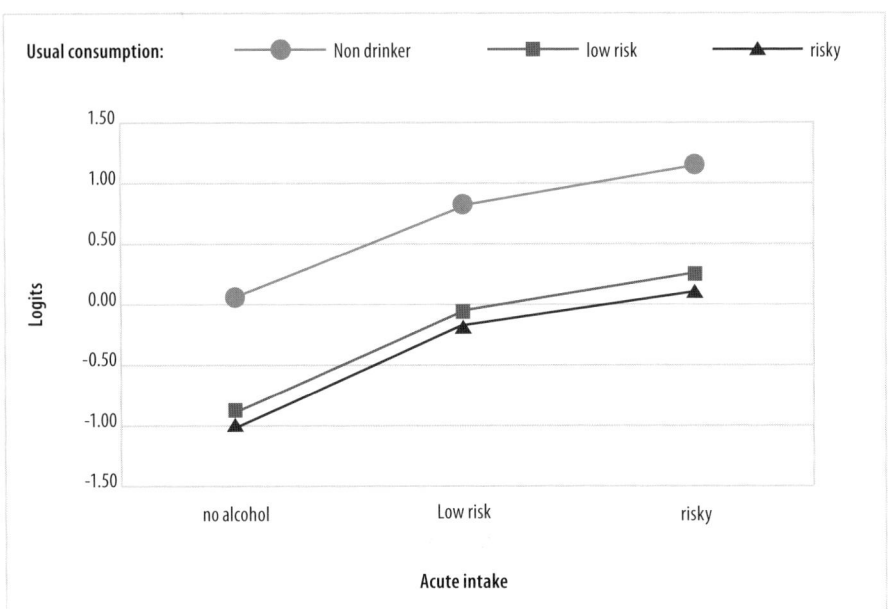

In a recent effort to circumvent these problems of main effect models, groups of drinkers were constructed by crossing volume of drinking with HED and levels of acute intake, leaving out illogical combinations such as abstainers with acute intake (Gmel et al., 2006). This study indicated that a) the risk of injury was higher compared with abstention at any level of acute intake, b) the risk of injury increased in all drinking groups with increasing acute intake, c) the risk of injury among drinkers with no acute intake was similar to that of abstainers, d) for the same amount of acute intake, the risk of injury was lower among high volume drinkers compared with low volume drinkers, and e) the highest risks were found for drinkers with a usual consumption pattern of moderate volume with HED occasions, and who actually had such an occasion prior to injury. Findings from this study suggest that acute intake is more important for injury than usual consumption (people have to drink prior to injury to be at higher risk), but that usual consumption patterns moderate the risk relationship of acute intake with injury.

Conclusion 1

Variability of estimates across different ER studies should not be seen *per se* as a threat to the validity. Simple prevalence estimates of alcohol positive cases are not very informative for cross country comparisons; an Islamic society will have fewer alcohol positive cases than societies where alcohol consumption is integrated into everyday life. Therefore research must move on to estimate risk relationships instead of focussing on pure prevalence studies (percentage of alcohol positive cases). The risk of injury is also related to drinking patterns, not only to the acute intake prior to injury. As these drinking patterns vary across populations, generic risk estimates of acute intake only that do not distinguish between types of drinker must necessarily result in different estimates across studies. A society where frequent moderate intake at home is the norm will probably have a lower risk for alcohol-related injuries than a society where heavy drinking on weekends in bars is the norm. Unfortunately, there are few studies that really differentiate between acute intake and usual patterns, and studies that take into account usual consumption patterns use models that do not empirically test the interaction between usual consumption patterns and acute intake for the risk of injury.

Choice of controls

Still an unsolved issue in ER research is the choice of adequate controls. Common criteria for suitable controls are as follows (Jick et al., 1998; Rothman and Greenland, 1998):

- The time period a subject is eligible to become a control should also be the time period a subject is eligible to become a case.

- A case should be eligible to become a control before the occurrence of the event; a control should be eligible to become a case at the time and location of the occurrence of the event.

- The risk for being involved in an injury should be the same for exposed and unexposed – unless exposure is associated with the outcome.

ER research commonly uses three types of controls, a) injured patients with injuries that are assumed to have theoretically a low alcohol involvement (e.g., animal bites, accidents during recreational activities, occupational accidents, see e.g. Borges et al., 1998b), b) non-injured ER patients with medical emergencies (e.g., digestive bleeding, heart attacks), and c) controls from the general population (e.g., Borges et al., 1998a). In the present discussion we do not include injured patients as controls, because this approach is mainly used to demonstrate the stronger alcohol-relatedness for certain types of injuries (e.g., violent injuries, intentional versus unintentional injuries, etc.) compared with others. The design is thus not suitable to determine absolute risk relationships for each type of injury in general.

Non-injured quasi controls versus general population controls

Often patients attending the ER for other reasons than injuries, the "non-injured", have been used as controls. However, "…it soon become apparent that the non-injured patients and the injured patients alike had higher rates of alcohol use and alcohol-related problems than were found in general population samples" (Cherpitel, 1993). It is therefore assumed that the use of "non-injured" as controls results in a conservative estimate of relative risks.

The studies reviewed by Cherpitel (1993), however, rarely used statistical methods such as logistic regression to adjust for differences in sample composition including age, gender or socio-economic background among injured and non-injured. She came to her conclusion comparing drinking prevalence in ER samples with those in the general population of the county. Certainly a larger-area general population sample might not be the best control group for an emergency room in a special region. For example, in Switzerland, a higher prevalence of positive BACs and BACs of 0.08% and above were found in ERs in the French/Italian-speaking regions compared to the German-speaking regions (Cherpitel, 1993). These regional differences were also found in general population surveys in Switzerland (Gmel and Schmid, 1996). Thus, a general population sample of controls across all regions in Switzerland would result in underestimation of the alcohol-injury association in the German-speaking region and overestimation for French/Italian-speaking ER patients. Cherpitel (1992) compared injured and non-injured patients in an ER sample with a general population sample in the same county who belonged to the same health maintenance organization (HMO). The general population sample had fewer injuries, but the same association with drinking was found for injured compared with non-injured in both samples, and the injured in the general population sample were similar in drinking patterns and drinking problems to injured patients in the ER sample. Additionally, Cherpitel (1995) found a greater proportion of very heavy drinkers in a general population sample of the same HMO compared with a corresponding sample of non-injured cases from an ER and a sample from primary care clinics. Thus, the use of "quasi controls" of non-injured ER patients would not result in a conservative estimate. These findings point to the fact that matching with the general population in the county may not be optimal; at least matching with the catchment area of the ER would be needed, which is commonly the case for non-injured ER patients.

In addition, non-injured patients are usually sampled at the same time as injured cases (see the first criterion, above for choice of controls). Alcohol consumption and related injuries vary over time, with more heavy consumption at night time and on weekends (Gmel et al., 2005). Newer approaches sample controls from the general population of the same catchment area that the ER serves and at time periods comparable to the time of injury occurrence for the ER patient (McLeod et al., 1999). There are problems with the use of general population controls if they are sampled in households, however. Even if the time period is matched with the time of injury of the ER case, a household control being at home is not eligible to be a traffic event case (requirements of exchangeability of cases and controls; see the second criterion under choice of control subjects above). Blomberg and Fell (1979) showed that site-matched controls revealed more conservative risk estimates for pedestrian injuries compared with random-site controls, pointing to the fact that the environment, context and time of injury plays an important role. There may also be healthy-volunteer bias or healthy day bias operating in the general population, that is, household controls more often than cases do not participate in interviews on "bad days" (Maclure and Mittleman, 2000). Such bad days may include, e.g., heavy drinking days. Thus, household controls may not necessary be better controls than non-injured ER patients.

Unfortunately, very few studies have compared different controls (e.g. general population versus non-injured ER patients), and thus it is difficult to decide which controls are more suitable or what the differences in risk estimates are for different control groups. Borges et al. (1998a) in Pachua, Mexico, found no significant differences between non-injured medical ER patients and general population controls of the same region, when analysis was adjusted for socio-demographic characteristics. The strengths of the study were a high completion rate of the general population sample (women 95%, men 80%) and the representativity of the ERs included in the study for Pachua, and thus both the general population survey and the ER sample represented the same region. This finding indicates that, with statistical adjustment of socio-demographic variables that may partly explain differences between cases and controls, non-injured quasi-controls in ERs are not necessarily worse than general population controls.

Conclusion 2

Despite the widespread presumption that non-injured quasi-controls from the same ER are poorer controls than samples from general population surveys, this is far from being established. General population controls also have disadvantages if they do not match on time and location of the catchment area, which might be better controlled in a sample of non-injured patients attending the same ER. More research on which type of controls will lead to more severe biases is needed.

Case-crossover studies: a solution to the control group problem?

To overcome the problems with finding an adequate control group, a relatively new method has entered the field: case-crossover designs. In these designs each case becomes its own control and thus "perfectly" matches characteristics that are time invariant over the study period such as age or gender.

Case-crossover analysis follows the logic of case-control analysis. Simply speaking, in case-crossover analysis cases become the "case intervals" and controls become "control intervals" of the same individual. In the matched-pair interval approach, exposure in the hazard period of the case interval (e.g., 6 hours prior to injury) is matched to exposure in one or several control intervals (e.g., the same six- hour period the previous week). In the usual-frequency approach one estimates the expected exposure time for the hazard interval based on usual frequencies of being exposed (occasions in the past week when people were drinking).

The most fundamental limitation of a case crossover study is inherent in the question it is designed to answer: "Was an event triggered by anything unusual that happened just before?" (Maclure and Mittleman, 2000, p. 193). Case-crossover designs analyze transient effects of a brief exposure on acute outcomes, not effects of chronic exposure. If one also wants to test hypotheses concerning chronic exposure, a traditional case-control study must be conducted simultaneously (Redelmeier and Tibshirani, 1997, and also Maclure and Mittleman, 2000). As acknowledged by the originators of case-crossover studies (Maclure and Mittleman, 2000, p. 200), case-crossover designs may not be optimal if we also wish to look at usual consumption patterns in addition to acute intake. For example, estimates in the matched-pair interval approach discard information on concordant pairs (those drinking in both the hazard and the control interval and those abstinent in both). The former could be very regular drinkers who consume alcohol every day at the same time, biasing case-crossover estimates towards irregular infrequent drinkers, who have been shown to have higher risks for injuries (Hurst et al., 1994; Gmel et al., 2006), and stronger biases in recalling alcohol consumption (Gmel & Daeppen, in press).

Recall bias is particularly relevant for case-crossover studies. It is known that usual quantity-frequency approaches yield lower consumption than diary measures (Feunekes et al., 1999; Gmel and Rehm, 2004), resulting in an underestimation of consumption in control intervals for the usual frequency approach (compared to the diary type of questions for the case interval). Underestimation of alcohol consumption has been found even for short recall periods of, e.g., a week even in retrospective diaries (Ekholm, 2004). This would bias even estimates in the matched-pair approach, because individuals will indicate lower consumption during the six-hour time period the week before the injury event (control interval) compared to the six hours immediately prior to injury (the case interval), due just to errors in recall.

There are also additional problems in using case-crossover designs for the estimation of the alcohol-injury association, such as non-representativeness of control intervals (e.g., the use of control intervals where people are at home for estimating relative risks of traffic injuries; Bateson and Schwartz, 2001; Levy et al., 2001), or spectrum bias (e.g., injuries that require no ER visit in the control intervals might be forgotten in recalling control times compared with more severe injuries that needed an ER visit), and periodicities in exposure (e.g., weekend drinking only), which also leads to particular biases in the case-crossover approach (Vines and Farrington, 2001; Bateson and Schwarz, 2001). Most of these biases favor overestimation of risk relationships.

There are only a few case-crossover studies that allow comparisons of relative risk estimates across different methods or designs (e.g., case-control). These studies, however, tend to show unexplainable high variability (low reliability) in case-crossover estimates, and point to an overestimation of risk-relationships. Borges et al. (2004b), for example, compared matched-pair analysis for three different control times with the usual-frequency approach. Relative risk estimates ranged from 2.71 to 13.52, depending on the particular estimation method. Compared to RR estimates of about 2 for injuries stemming from meta-analytical studies (e.g. Cherpitel et al., 2003a), the latter (13.52), at least, seems to be a strong overestimate.

Vinson et al. (2003a) compared a matched-pair case-crossover design on alcohol use and intentional injury with a case-control design, in which controls came from the same community, with time and day of the interview matched to time and day of the injury for 102 injury cases. Again RR estimates varied between 10 and 34 for different control periods in the case-crossover design (and was 10 in the case-control design). In addition, the authors conducted a control-crossover study (which is the same as case-crossover but with control individuals instead of cases), for which, in the absence of biases, relative risks should be close to 1. Whereas this was nearly found for the control pairs matched with the day before (RR=1.1) a threefold bias (RR=3.2) was obtained for multiple control times using information based on the past 4 weeks. In a second study, Vinson et al. (2003b) analyzed all injuries (n=2517 cases) and found very similar risk estimates for the case-control design (OR=3.1) and the case-crossover design using the day before as the control period (RR = 3.2), with a control-crossover estimate of RR=1.1. These findings point to the possibility that case-crossover designs might yield similar results to case-control designs, if recall errors are minimized (perhaps using a recall of one day only) and sample sizes are sufficiently large.

Conclusion 3

The case-crossover design seems to be highly sensitive to the approach used (matched-pair versus usual-frequency) and the choice of control times in matched- pair analysis. It is likely that case-crossover analysis is prone to recall biases resulting in overestimates for alcohol-injury associations. Case-crossover designs also seem to be sensitive to small sample sizes, particularly when concordant pairs (as in the matched- pair approach) are excluded. It therefore seems prudent to recommend further studies with the case-crossover design that also includes control individuals (e.g. general population controls) to better evaluate the strength and weaknesses of such a design before placing too much confidence in its findings.

Many of the biases, particularly recall bias (Maclure and Mittleman, 2000), may be avoided by incorporating a prospective design, i.e., to sample control periods in time after the event or to use ambi-directional designs (Bateson and Schwartz, 2001; Levy et al., 2001) with control periods before and after the event to control for trends. Lin et al. (2002), in a study on asthma, showed that an ambi-directional design increased the stability of relative risk estimates. To our knowledge there is no prospective case-crossover study on alcohol and injury to date.

Discussion

ER studies cannot be representative of the general population or of all injuries in a population. Patients in an ER at best reflect the population in a small area (e.g., the catchment area of an HMO), and even within this area only a part of the population (e.g. the economically more disadvantaged). Injuries in an ER can not represent all injuries, as minor injuries will not necessarily be professionally treated at all and severe or fatal injuries are commonly excluded in ER studies. In the light of these limits, there are three questions that must be asked.

First, are there alternatives to ER-studies? Certainly, studies in other settings (e.g., autopsy series on fatal injuries) should be coupled with ER studies to yield a more complete picture of the alcohol-injury link. There may also be better designs. Studies where cases and controls are selected directly at the scene (Gmel and Gutjahr, 2001) would avoid some of the control group problems discussed above. Drawbacks to this type of study are the amount of organizational effort (interviewer groups must be quickly at the scene) and costs to implement. Such studies are only practical for certain, prevalent or locally restricted incidents, such as traffic crashes or aggression in bars. The same caveats also apply to prospective studies, where alcohol consumption and injuries are recorded, e.g., by means of diaries (for intimate partner violence see e.g.

Fals-Stewart et al., 2003). We see no alternative to ER studies for studying a wide range of injuries. Unfortunately, due to small sample sizes most studies have been conducted across all causes of injury combined, or have focused on a single type of trauma using only one dimension (either nature of the injury, e.g., fracture, or body part injured, e.g. upper extremities; Felson et al., 1988, Hoidrup et al., 1999; Shapiro et al., 2001; Parkinson et al., 1985; Johnston and McGovern, 2004; Chen et al., 1999; but see Watt et al., 2005). To overcome small sample sizes, collaborative multi-site studies are recommended.

Second, should we be worried about non-representativeness? The answer is, "in principle, yes". Representativeness matters less if we want to show **that** alcohol is associated with injury (which should no longer be a research question); however, we need to know what the exact **magnitude** of the relationship is, e.g., to calculate the alcohol-related burden of disease stemming from injuries. As mentioned above, ER studies cannot be representative of the whole population in a country or society. We will therefore not be able to derive estimates of the percentage of all injuries in a country caused by alcohol consumption from ER studies alone. We argue, however, that the variation in percentages of alcohol-related injuries across ER-studies is partly due to the different composition of samples regarding types of drinkers. Heavy episodic drinkers have higher risks of injuries than moderate drinkers. A culture where drinking takes place more often outside the home and in bars results in more injuries in public places, violent injuries, etc. Thus, the predominant drinking style of the ER region will partly determine how many, and which, injuries are alcohol-related. To obtain alcohol attributable fractions for a larger area than that of the ER (e.g., for a country as a whole) two ingredients are needed: the prevalence of types of drinkers and the relative risks for injuries. Whereas the first can not be obtained from ER studies and needs other approaches such as general population surveys, the second can probably be obtained from ER studies. This, however, would require that ER studies focus more on background variables, such as the location of injury and injury mechanism, and especially on the usual drinking patterns, and not only on the acute intake of alcohol prior to injury. Meta-analytical studies would help to establish associations across sites. Recently, Cherpitel et al. (2003b) showed that estimates across ERs were homogeneous if they were obtained for comparable subgroups of drinkers.

Third, do we know what the best ER study design is? An ER study is essentially a case-control study and thus the principles of case-control studies apply. Assuming that sampling of cases is relatively representative for the ER (e.g., all times of day, all days a week, follow-up of too severely injured) (see Chapter 7 in this volume), the major problem is the best possible choice of controls. It seems to be clear that controls should match the catchment area and the time of injury. Household samples of controls have certain disadvantages (e.g. healthy volunteer bias, being at home and not necessarily sampled at the place of the injury, etc.) as do non-injured medical ER patients (different ages, not sampled at the place of injury, etc). There are not enough ER studies, however, to compare different control groups to see whether statistical adjustment of, e.g., socio-demographic characteristics would remedy some of these problems. Again, meta-analytical studies and comparative studies, as recently carried out in the ERCAAP and WHO multi-site collaborative projects, are needed to quantify eventual biases, and thus to obtain better estimates of risk relationships between alcohol and injury.

References

Bateson TF, Schwartz J (2001). Selection bias and confounding in case-crossover analyses of environmental time-series data. *Epidemiology*, **12**, 654-661.

Blomberg RD, Fell JC (1979). A comparison of alcohol involvement in pedestrians and pedestrian casualties. In *Proceedings of the American Association for Automotive Medicine*. American Association for Automotive Medicine, ed, pp. 1-17. American Association for Automotive Medicine, Louisville, KY.

Borges GLG et al. (1998a). Alcohol consumption in emergency room patients and the general population: a population-based study. *Alcoholism, Clinical and Experimental Research*, **22**, 1986-1991.

Borges GLG, Cherpitel CJ, Mittleman MA (2004a). Risk of injury after alcohol consumption: a case-crossover study in the emergency department. *Social Science and Medicine*, **58**, 1191-200.

Borges GLG et al. (2004b). Episodic alcohol use and risk of nonfatal injury. *American Journal of Epidemiology*, **159**, 565-571.

Borges GLG, Cherpitel CJ, Rosovsky H (1998b). Male drinking and violence-related injury in the emergency room. *Addiction*, **93**, 103-112.

Borkenstein RF et al., eds (1964). T*he Role of the Drinking Driver in Traffic Accidents*. Department of Police Administration, Indiana University, Bloomington, IN.

Chen SC, Lin F Y, Chang KJ (1999). Body region prevalence of injury in alcohol- and non-alcohol-related traffic injuries. *Journal of Trauma*, **47**, 881-884.

Cherpitel CJ (1992). Drinking patterns and problems: a comparison of ER patients in an HMO and in the general population. *Alcoholism, Clinical and Experimental Research*, **16**, 1104-1109.

Cherpitel CJ (1993). Alcohol and injuries: A review of international emergency room studies. *Addiction*, **88**, 923-937.

Cherpitel CJ (1995). Alcohol use among HMO patients in the emergency room, primary care and the general population. *Journal of Studies on Alcohol*, **56**, 272-276.

Cherpitel CJ (1997a) Alcohol and injuries resulting from violence: a comparison of emergency room samples from two regions of the U.S. *Journal of Addictive Diseases*, **16**, 25-40.

Cherpitel CJ (1997b). Alcohol and injury: a comparison of three emergency room samples in two regions. *Journal of Studies on Alcohol*, **58**, 323-331.

Cherpitel CJ et al. (2003a). A cross-national meta-analysis of alcohol and injury: data from the Emergency Room Collaborative Alcohol Analysis Project (ERCAAP). *Addiction*, **98**, 1277-1286.

Cherpitel CJ et al. (2003b). Alcohol-related injury in the ER: a cross-national meta-analysis from the Emergency Room Collaborative Alcohol Analysis Project (ERCAAP). *Journal of Studies on Alcohol*, **64**, 641-649.

Cherpitel CJ, Borges GLG (2001). A comparison of substance use and injury among Mexican American emergency room patients in the United States and Mexicans in Mexico. *Alcoholism, Clinical and Experimental Research*, **25**, 1174-1180.

Cherpitel CJ, Borges GLG, Wilcox HC (2004a). Acute alcohol use and suicidal behavior: a review of the literature. *Alcoholism, Clinical and Experimental Research*, **28**, 18S-28S.

Cherpitel CJ, Flaminio D, Poldrugo F (1993a). Alcohol and casualties in the Emergency room: a US – Italy comparison of weekdays and weekend evenings. *Addiction Research*, **1**, 223-238.

Cherpitel CJ, Giesbrecht N, Macdonald S (1999). Alcohol and injury: a comparison of emergency room populations in two Canadian provinces. *American Journal of Drug and Alcohol Abuse*, **25**, 743-759.

Cherpitel CJ, Pares A, Rodés J (1991). Drinking patterns and problems: a comparison of emergency room populations in the United States and Spain. *Drug and Alcohol Dependence*, **29**, 5-15.

Cherpitel CJ et al. (1993b). Drinking in the injury event: a comparison of emergency room populations in the United States, Mexico, and Spain. *International Journal of the Addictions*, **28**, 931-945.

Cherpitel C J, Ye Y, Bond J (2004b). Alcohol and injury: multi-level analysis from the emergency room collaborative alcohol analysis project (ERCAAP). *Alcohol and Alcoholism*, **39**, 552-558.

Cherpitel CJ, Ye Y, Bond J (2005a). Attributable risk of injury associated with alcohol use: cross-national data from the emergency room collaborative alcohol analysis project. *American Journal of Public Health*, **95**, 266-272.

Cherpitel CJ, al. (2003c). The causal attribution of injury to alcohol consumption: a cross-national meta-analysis from the emergency room collaborative alcohol analysis project. *Alcoholism, Clinical and Experimental Research*, **27**, 1805-1812.

Cherpitel CJ et al. (2005b). Multi-level analysis of alcohol-related injury among emergency department patients: a cross-national study. *Addiction*, **100**, 1840-1850.

Cherpitel CJ et al. (2005c). Risk of injury: a case-crossover analysis of injured emergency service patients in Poland. *Alcoholism, Clinical and Experimental Research*, **29**, 2181-2187.

Ekholm O (2004). Influence of the recall period on self-reported alcohol intake. *European Journal of Clinical Nutrition*, **58**, 60-63.

Fals-Stewart W, Golden J, Schumacher JA (2003). Intimate partner violence and substance use: a longitudinal day-to-day examination. *Addictive Behaviors*, **28**, 1555-1574.

Felson DT et al. (1988). Alcohol consumption and hip fractures: the Framingham Study. *American Journal of Epidemiology*, **128**, 1102-1110.

Feunekes GIJ et al. (1999). Alcohol intake assessment: the sober facts. *American Journal of Epidemiology*, **150**, 105-112.

Gmel G et al. (2006). Alcohol-attributable injuries in admissions to a Swiss emergency room – an analysis of the link between volume of drinking, drinking patterns and pre-attendance drinking. *Alcoholism, Clinical and Experimental Research*, **30**, 501-509.

Gmel G, Daeppen J-B (in press). Recall bias for 7-day recall measurement of alcohol consumption among emergency room patients – implications for case-crossover designs. *Journal of Studies on Alcohol*.

Gmel G, Gutjahr E (2001). Alcohol consumption and social harm: quantitative research methodology. In *Mapping the Social Consequences of Alcohol Consumption*. Klingemann H, Gmel G, eds, pp. 33-52. Kluwer Academic Publishers, Dordrecht.

Gmel G et al. (2005). Drinking patterns and traffic casualties in Switzerland – matching survey data and police records to design preventive action. *Public Health*, **119**, 426-436.

Gmel G, Rehm J (2004). Measuring alcohol consumption. *Contemporary Drug Problems*, **31**, 467-540.

Gmel G, Schmid, H, eds (1996). *Alkoholkonsum in der Schweiz – Ergebnisse der ersten schweizerischen Gesundheitsbefragung*. Verlag Dr Kovac, Hamburg.

Gruenewald PJ et al. (1996a). The geography of availability and driving after drinking. *Addiction*, **91**, 967-983.

Gruenewald PJ, Mitchell PR, Treno AJ (1996b). Drinking and driving: drinking patterns and drinking problems. *Addiction*, **91**, 1637-1649.

Gruenewald PJ, Nephew TM (1994). Drinking in California: theoretical and empirical analyses of alcohol consumption patterns. *Addiction*, **89**, 707-723.

Hingson RW, Howland J (1993). Alcohol and non-traffic unintended injuries. *Addiction*, **88**, 877-883.

Hoidrup S et al. (1999). Alcohol intake, beverage preference, and risk of hip fracture in men and women. Copenhagen Centre for Prospective Population Studies. *American Journal of Epidemiology*, **149**, 993-1001.

Hurst PM, Harte D, Frith WJ (1994). The Grand Rapids dip revisited. *Accident Analysis and Prevention*, **26**, 647-654.

Jick H, García Rodríguez LA, Pérez-Gutthann S (1998). Principles of epidemiological research on adverse and beneficial drug effects. *Lancet*, **352**, 1767-1770.

Johnston JJ, McGovern SJ (2004). Alcohol related falls: an interesting pattern of injuries. *Emergency Medicine Journal*, **21**, 185-188.

Levy D et al. (2001). Referent selection in case-crossover analyses of acute health effects of air pollution. *Epidemiology*, **12**, 186-192.

Li G et al. (1997). Alcohol and injury severity: reappraisal of the continuing controversy. *Journal of Trauma*, **42**, 562-569.

Lin M et al. (2002). The influence of ambient coarse particulate matter on asthma hospitalization in children: case-crossover and time-series analyses. *Environmental Health Perspectives*, **110**, 575-581.

Maclure M, Mittleman MA (2000). Should we use a case-crossover design? *Annual Review of Public Health*, **21**, 193-221.

McLeod R et al. (1999). The relationship between alcohol consumption patterns and injury. *Addiction*, **94**, 1719-1734.

Parkinson D, Stephensen S, Phillips S (1985). Head injuries: a prospective, computerized study. *Canadian Journal of Surgery*, **28**, 79-83.

Redelmeier DA, Tibshirani RJ (1997). Interpretation and bias in case-crossover studies. *Journal of Clinical Epidemiology*, **50**, 1281-1287.

Rehm J, Gmel G (2000). Aggregating dimensions of alcohol consumption to predict medical and social consequences. *Journal of Substance Abuse*, **12**, 155-168.

Rehm J et al. (2003). Alcohol-related morbidity and mortality. *Alcohol Research and Health*, **27**, 39 – 51.

Rehm J et al. (2004). Alcohol use. In *Comparative Quantification of Health Risks. Global and Regional Burden of Disease Attributable to Selected Major Risk Factors* Ezzati M et al., eds, pp. 959-1108. World Health Organization (WHO), Geneva.

Room R, Mäkelä K (2000). Typologies of the cultural position of drinking. *Journal of Studies on Alcohol*, **61**, 475-483.

Rothman KJ, Greenland S, eds (1998). *Modern Epidemiology*. Lippincott-Raven Publishers, Philadelphia, PA.

Shapiro AJ et al. (2001). Facial fractures in a level I trauma centre: the importance of protective devices and alcohol abuse. *Injury*, **32**, 353-635.

Treno AJ, Gruenewald PJ, Johnson FW (1998). Sample selection bias in the emergency room: an examination of the role of alcohol in injury. *Addiction*, **93**, 113-129.

Treno AJ, Gruenewald PJ, Ponicki WR (1997). The contribution of drinking patterns to the relative risk of injury in six communities: a self-report based probability approach. *Journal of Studies on Alcohol*, **58**, 372-381.

Treno AJ, Holder HD (1997). Measurement of alcohol-involved injury in community prevention: the search for a surrogate III. *Alcoholism, Clinical and Experimental Research*, **21**, 1695-1703.

Vines SK, Farrington CP (2001). Within-subject exposure dependency in case-crossover studies. *Statistics in Medicine*, **20**, 3039-3049.

Vinson DC, Borges GLG, Cherpitel CJ (2003a). The risk of intentional injury with acute and chronic alcohol exposures: a case-control and case-crossover study. *Journal of Studies on Alcohol*, **64**, 350-357.

Vinson DC et al. (2003b). A population-based case-crossover and case-control study of alcohol and the risk of injury. *Journal of Studies on Alcohol*, **64**, 358-366.

Watt K et al. (2004). Risk of injury from acute alcohol consumption and the influence of confounders. *Addiction*, **99**, 1262-1273.

Watt K et al. (2005). The relationship between acute alcohol consumption and consequent injury type. *Alcohol and Alcoholism*, **40**, 263-268.

World Health Organization (WHO) (2002). *The World Health Report 2002 – Reducing Risks, Promoting Healthy Life*, Geneva, World Health Organization (WHO).

SECTION III: IDENTIFYING ALCOHOL-RELATED INJURIES IN THE EMERGENCY DEPARTMENT
INTRODUCTON

Tim Stockwell - Centre for Addictions Research of BC | Victoria, BC CANADA

There is increasing international interest in the establishment of data collection systems capable of providing basic epidemiological surveillance of alcohol-related problems. The World Health Organization has recently prepared the second edition of an international alcohol monitoring Guide for member states (WHO, in press) as part of a suite of activities designed to support the development of effective alcohol policies in the wake of a 2005 World Health Assembly resolution (WHO, 2005). In this Guide, and in the two chapters that follow in this section, the importance of monitoring acute alcohol-related harms and related risky patterns of use is stressed, especially given that these both contribute to about half of alcohol's contribution to the global burden of disease (Rehm et al, 2004). Such epidemiological monitoring can support effective policy development variously by raising awareness of alcohol issues in a local or regional context and by contributing to the evaluation of the effectiveness of local policies designed to reduce population rates of alcohol-related harm.

The first chapter by Robin Room notes that alcohol's contribution to injury is normally inferred from a wider international literature and that local data are usually not available. This chapter discusses the potential of using routine recording of alcohol's involvement in emergency department (ED) presentations by encouraging clinicians to apply currently optional elements of the latest International Classification of Diseases (ICD), namely the Y-codes in ICD-10. If compliance issues can be overcome and if maintenance of this compliance is possible, then useful local monitoring of alcohol-related injuries can be achieved.

The second chapter by Stockwell and colleagues discusses a broad array of monitoring options involving the ED, including the use of "surrogate" measures, objective tests of blood alcohol level, the application of etiological fractions, adaptations of routine medical records and the regular sampling of ED attendees for surveys on alcohol and other substance use.

Both chapters stress the need for developing improved methods of monitoring and for the formal evaluation of these so as to better support local, regional and national responses to alcohol-related problems.

References

Rehm J et al. (2004). *Alcohol use*. In: Ezzati M et al., eds. Comparative Quantification of Health Risks. Global and Regional Burden of Disease Attributable to Selected Major Risk Factors. **1**, 959-1108. Geneva: World Health Organization.

World Health Organization (2005). *Public health problems caused by harmful use of alcohol*. Resolution of the General Assembly of WHO, May 25, 2005.
(See: http://www.who.int/nmh/a5818/en/index.html)

World Health Organization (in press). *International guide for monitoring alcohol-related problems, consumption and harm, 2nd Edition*. World Health Organization: Geneva.

CHAPTER 9 :
THE RELATION BETWEEN BLOOD ALCOHOL CONTENT AND CLINICALLY ASSESSED INTOXICATION: LESSONS FROM APPLYING THE ICD-10 Y90 AND Y91 CODES IN THE EMERGENCY ROOM[1]

Robin Room - Turning Point Alcohol and Drug Centre and University of Melbourne | Melbourne, AUSTRALIA

Including Alcohol in Injury Codes in the International Classification of Diseases

A large part of the mortality and disability attributable to alcohol results from injuries; both intentional (suicide and homicide) and unintentional. According to the estimates in WHO's calculation of the Global Burden of Disease in 2000, injuries constitute 46% of the deaths attributable to alcohol and 42% of the Disability-Adjusted Life Years (DALYs) (Rehm et al., 2004; Room et al., 2005). Yet for this very large fraction of the burden attributable to alcohol, the alcohol connection is not usually routinely recorded. Instead, special epidemiological studies, following various designs, have been the primary means by which the picture of alcohol's role in different kinds of casualties has been developed (Aarens et al., 1977). Prominent among these have been studies of series of injury cases coming to Emergency Rooms (ERs), as well as mortality studies.

Although a substantial literature of ER-based studies has accumulated (Cherpitel 1993; Giesbrecht et al. 1989; Romelsjö 1995; Cherpitel, in press), the strongest studies tend to have been in a relatively narrow range of countries, and many of the studies have not included representative samples of patients (Cherpitel & Driggers 2004). For much of the world and much of the time, then, any estimates of alcohol's role in casualties are inferred from studies at another place and time. This way of proceeding may be defensible (though potentially problematic) for estimating alcohol's role in chronic physical conditions such as cirrhosis and esophageal cancer, but it poses obvious problems in estimating alcohol's role in accidents and violence. Not only the pattern of drinking but also the physical and social context affect the relationship with accidents, and for violence there is the additional factor of cultural differences in drunken comportment and alcohol's excuse-value (MacAndrew and Edgerton, 1969). The extent of alcohol's role in a specific kind of casualty thus varies from place to place, according to other factors besides the population's distribution on volume of drinking.

[1] Revised from a paper presented at the International Conference on Alcohol and Injury: New Knowledge from Emergency Room Studies, Berkeley, California, 3-6 October, 2005. This paper draws on the work of the participants in the WHO Collaborative Study on Alcohol and Injuries, and in part on a paper by Cherpitel and colleagues (2005).

For the last 30 years, reviews of the situation have included among their recommendations that an effort be made to include routine coverage of alcohol involvement in reporting systems on casualties and crime (e.g. Aarens et al., 1977:587; Aitken & Zobeck, 1989). In the early 1980s, attention turned to the potential role of changes in the International Classification of Diseases (ICD) in improving the situation. It was noted that, under the ICD rules and codes, there was no way of recording alcohol's involvement as a cause of injury except for alcohol poisoning, and possible solutions were put forward (Room, 1982). At a 1985 conference, it was proposed that "there is perhaps no single step that could be taken with more potential for improving our knowledge of and monitoring of alcohol's role in casualties in every country than to institute a workable way of recording the alcohol dimension in casualties in the 10th revision of the ICD" (Room, 1987). A 1984 note prepared for WHO (Room, 1984) suggested a specific coding scheme "as an extra code in injuries" that would "reflect the degree of intoxication and nature of the evidence", with three levels of positive blood-alcohol (0.1 – 0.9 per mille; 1-1.9; 2 and more – BACs stated in per mille are ten times the mg% measure commonly used in the U.S) and two levels of clinical judgement (highly intoxicated; some intoxication).

The suggestion was further developed in a 1987 proposal for ICD-10 coding from the U.S. Alcohol, Drug and Mental Health Administration (Grant et al., 1987). In this version two codes were proposed, with one recording the blood alcohol content (BAC), and when this was not available, a separate code recording the level of apparent intoxication. For a while, it looked as if instead there would be one code devoted to intoxication by other drugs, and one to alcohol. But eventually, a modification of the proposal by Grant and colleagues was adopted in ICD-10 as the Y90 and Y91 codes, respectively (World Health Organization, 1992). The sub-codes of Y90 are defined by a series of 9 blood alcohol levels: less than 20 mg/100 ml (.02 mg%); 20-39 mg/100 ml; 40-59 mg/100 ml; 60-79 mg/100 ml; 80-99 mg/100 ml; 100-119 mg/100 ml; 120-199 mg/100 ml; 200-239 mg/100 ml; and 240 mg/100 ml or more. There is also a code for the "presence of alcohol in blood, level not specified". Y91, which is intended to be used in the absence of a BAC measure, records an assessment of alcohol involvement determined by level of intoxication, with four levels differentiated: Y91.0 (mild), Y91.1 (moderate), Y91.2 (severe), Y91.3 (very severe), and some criteria for assessment provided (See Table 1). An additional code is used, Y91.9, to indicate alcohol involvement not otherwise specified.

Table 1. Clinical Assessment of Alcohol Intoxication

Clinical signs	Severity / Prominence					
	Very Severe	**Severe**	**Moderate**	**Mild**	**None**	**Not applicable (specify)**
Smell of alcohol on breath						
Conjunctival injection and/or flushed face						
Impairment of speech (e.g. slurring)						
Impairment of motor coordination						
Impairment of attention and/or judgement						
Elated (euphoria) or depressed mood						
Disturbances in behavioural responses						
Disturbances in emotional responses						
Impaired ability to cooperate						
Horizontal gaze nystagmus						

Y91	**Based on the signs above, would you say that the patient is in the state of:** *(Tick appropriate box)*
Y91.3	**Very severe alcohol intoxication** *(Very severe disturbance in functions and responses, very severe difficulty in coordination, or loss of ability to cooperate)*
Y91.2	**Severe alcohol intoxication** *(Severe disturbance in functions and responses, severe difficulty in coordination, or impaired ability to cooperate)*
Y91.1	**Moderate alcohol intoxication** *(Smell of alcohol on breath, moderate behavioural disturbance in functions and responses, or moderate difficulty in coordination)*
Y91.0	**Mild alcohol intoxication** *(Smell of alcohol on breath, slight behavioural disturbance in functions and responses, or slight difficulty in coordination)*

Y91.9	Please specify reason:	*Alcohol involvement, not otherwise specified*	
		Not intoxicated at all	

Do you think that there is any evidence of substance usage other than alcohol?	
	No
	Yes, based on self report
	Yes, based on collateral information
	Yes, based on self report and collateral information
	Not sure

The Status of Y90 and Y91 Codes

While the Tenth Revision of the ICD was "the most radical" revision since 1948 (L'Hours, 1995), the system of international classification of diseases is nevertheless rather conservative, since it aims at wide acceptance by different national recording systems, and since radical changes threaten any ability to track changes over time. Although multiaxial classification systems were proposed during preparation for ICD-10 for the recording of external causes of injuries (L'Hours, 1995), in the end the classification retained the ICD's general "single-variable axis classification". Y90 and Y91 are simply presented as categories which "may be used, if desired, to provide supplementary information concerning causes of morbidity and mortality", with the specification that they are "not to be used for single-condition coding" (WHO, 1992). There is no indication in the ICD-10 presentation of Y90 and Y91 of the relation between them, or of whether one is to be preferred over the other. In the Mental and Behavioural Disorders chapter of ICD-10, where acute intoxication due to alcohol (F10.0) is identified for the first time in the ICD as a specific and separate code, there is a comment that implies that Y91 is viewed as a backup when Y90 cannot be coded: "If desired, the blood alcohol level may be specified by using ICD-10 codes Y90.0-Y90.8. Code Y91.- may be used to specify the clinical severity of intoxication if the blood alcohol level is not available" (WHO, 1992).

The status of Y90 and Y91, then, is as optional extra codes to be used in either morbidity or mortality coding. Implicitly, their placement in the "external causes" chapters orients them to use in injury coding, but the note in F10.0 clearly contemplates that they can be used with any other ICD-10 codes. The same note also implies a preference for Y90 where either could be coded, but this is not specified where the codes themselves are presented.

Previous Studies of the Relation of Clinical Assessment to Blood Alcohol Concentration

The adoption of two parallel codes for recording level of alcohol intake in the ICD system raises the question of the relation between the two codes. What is the level of agreement between clinical judgment and the measured blood alcohol level (BAC), and what factors might affect this relationship? The concordance of these two measures might be expected to vary according to several factors including: 1) the extent of experience of the person making the assessment, 2) the opportunity to observe or test the individual's behavior; 3) the proximity in time of the BAC estimate to clinical assessment; 4) the extent of the subject's physical and/or psychological tolerance of alcohol; 5) cross-cultural variations in the behavior of intoxicated individuals (MacAndrew & Edgerton 1969); and 6) culturally-influenced assumptions about the social location and the recording of intoxication and sobriety on the part of the person making the assessment.

One early study carried out on patients admitted to the hospital with a pre-admission diagnosis of acute alcoholism found 13% to have a zero BAC, with half of these identified as known chronic alcoholics (which presumably accounted for the diagnosis), and another 3% comatose at the time of admission (Jetter, 1938). The odour of alcohol on the breath was the clinical sign reported most frequently (90%), followed by abnormality of gait or inability to walk (75%), abnormality of speech (72%), dilated pupils (71%) and flushed face (61%). A Finnish study in the emergency room, correlating BAC with physician assessment of three levels of intoxication and "no intoxication", found that, among those assessed as not intoxicated, 14% had positive BACs and half of those were above 1.0 per mille, while 12% of those assessed as heavily intoxicated had negative BACs (although the remainder were above 1.0 per mille) (Honkanen, 1977). The physician assessment was found to distinguish better generally between any significant BAC (0.6 per mille or above) and none, than between different levels of positive BAC. A second Finnish ER study, in which assessment was carried out by a separate survey staff, found that, among those assessed as not intoxicated, 11% had a positive BAC, while 90% of those assessed as having some degree of intoxication had BACs of 0.6 per mille and higher. The authors commented that although the degree of intoxication often cannot be estimated clinically, assessment is a valuable and cheap device for studying the strain alcohol places upon emergency services (Honkanen & Ottelin, 1976).

A later U.S. study in the emergency room developed and tested an observational tool designed to assess the presence and level of alcohol intoxication (Teplin & Lutz 1985). This 11-item instrument, called the Alcohol Symptom Checklist, included both physiologic and behavioral manifestations, and was found to be highly correlated (.84) with BAC in the sample in which it was developed, but this line of work subsequently has received relatively little attention.

The limited literature on clinical observation of intoxication in an ER setting and its correlation with BAC thus finds fairly good correlations, but also some striking discrepancies. It should be noted that the judgments in these studies were probably being made after some training and sensitization, and the results might be better than in routine clinical practice. On the other hand, the similar results in the two Finnish studies suggest that a trained lay observer may be able to do about as well as a trained clinician.

Somewhat in contrast to the clinical literature, the mostly US-based literature on estimation of alcohol impairment from behavioural signs by law enforcement officers generally finds a quite high sensitivity and specificity. Considerable effort has been devoted to developing and calibrating "field sobriety tests" to be applied when a driver or a boater is suspected to be above a legal BAC limit (Grossman et al, 1996; Martin, 1998; McKnight et al., 1999), and there is even a study calibrating a field sobriety test battery to measure a BAC level of 0.8 per mille rather than 1.0 per mille (Stuster & Burns, 1998). There are, however, also some more pessimistic reports in this literature (Cole & Nowaczyk, 1994).

It might be noted that there are two substantial differences between the law enforcement and the clinical literature. One is that the aims of the judgment have been somewhat different. In the law enforcement literature, the judgment is aimed directly at estimating the BAC, while in the clinical literature (and also in the case of Y91) the aim is typically a more global judgment of level of intoxication. The other is that in the law enforcement situation, the test battery typically includes such tests as requiring standing on one leg, walking and turning, and clapping hands alternately with the palm and back of the hand while counting. In the context of a busy ER, any such explicit tests of movement and coordination are unlikely to have been applied. Both of these differences might well produce a better concordance between judgment and BAC in the law enforcement situation.

Cross-Cultural Testing of the Relation Between Clinical Judgment and BAC: The WHO Collaborative Study on Alcohol and Injuries [2]

Given the adoption of the Y91 and Y90 codes in the ICD-10, and given the limited knowledge of the empirical relation between such measures, testing the concordance and utility of the codes was among the main aims when the WHO Department of Mental Health and Substance Dependence, in collaboration with the Violence and Injury Department, initiated the WHO Collaborative Study on Alcohol and Injuries in emergency rooms of metropolitan hospitals in 12 countries. In addition to documenting the prevalence and role of alcohol-involvement in injuries, then, the Collaborative Study aimed to explore the concordance of clinical assessment with BAC, based on Y90 and Y91 codes, while testing the feasibility of implementing Y91 coding for assessment and recording of alcohol intoxication in emergency rooms in different societies, and developing and piloting materials to assist ER staff in assessing and coding the degree of alcohol intoxication.

Data from the WHO Collaborative Study on Alcohol and Injuries were collected in 2001-2002 from ERs in Argentina, Brazil, Belarus, Canada, China, Czech Republic, India, Mexico, Mozambique, New Zealand, South Africa, and Sweden. Study methodology across sites was similar to that used in previous ER studies. A probability sample of patients admitted for an injury within six hours of the event at each site was approached as soon as possible with informed consent to participate. Samples were drawn from ER admission forms that reflected consecutive arrival of patients. The total sample across all sites of those 18 years and older was 5243 patients, and represented a 91% completion rate. BAC estimates were obtained using a breath analyzer and patients were given a 25- minute interviewer-administered standard questionnaire. A cohort of interviewers in each site was trained and supervised by study collaborators in their respective locations. A clinical assessment of intoxication made by an ER physician or nurse was also obtained, in most cases prior to the interviewer obtaining the BAC estimate. In instances where BAC was obtained first (10%), the clinician was blind to the BAC estimate. In four sites, Belarus, Brazil, China and India, an additional observational assessment of intoxication was obtained by a second interviewer (Interviewer-Assessment) who had no knowledge of the breath alcohol estimate or of the clinical assessment of intoxication.

Physicians and/or nurses were trained by WHO study staff and site investigators using a module prepared by WHO that included diagnostic criteria for intoxication, behavioral manifestations, description of clinical signs, testing for impairment and other clinical conditions warranting a differential diagnosis. A sufficient number of clinicians were trained at each ER site to assure availability during the periods patients were sampled. Clinicians recorded clinical signs of intoxication and observational assessment codes, based on ICD-10 descriptions of alcohol intoxication at

---

different levels (WHO, 1992; see Table 1). The observation assessment also recorded whether or not the clinician believed a substance other than alcohol might be involved.

Interviewers undertaking the Interviewer Assessment did not receive any special training in assessing for intoxication. The Interviewer Assessment was included to test the ability to evaluate the level of alcohol intoxication without any prior training, using a form specifically designed for this purpose.

In comparing the clinical assessment in Y91 with the BAC levels measured for coding Y90, the following equivalences were assigned:

- No intoxication: ≤ 0.59 per mille

- Mild intoxication: 0.60 – 0.99

- Moderate intoxication :1.00 – 1.99

- Severe/very severe intoxication: 2.00 and more

In general, the study found a moderate degree of concordance between the clinical assessment for Y91 and BAC (Cherpitel et al., 2005). In the cross-site total sample, the Kendall's Tau-B was 0.68, the Pearson's r 0.65, and the weighted Kappa 0.57. At the level of none vs. any intoxication, 1.9% were clinically assessed with some degree of intoxication but with a BAC below 0.60, while 5.7% had a BAC below 0.60 but were assessed with some degree of intoxication. 10.5% were positive both by clinical assessment and by BAC. At higher levels of intoxication, the agreement was less. Only 0.5% were assessed as severe and had a BAC of 2.0 and above, while 1.5% were assessed as severe but had a lower BAC, and 2.0% had a BAC of 2.0 and above but were not assessed as severe. Using a different statistic, the project's Final Report also found moderate agreement between BAC and clinical assessment of intoxication, but that the agreement was considerably less if those with a BAC of 0 were excluded from the calculation (WHO, 2006: section 6.2).

Table 2 shows the concordances between BAC, clinical assessment, and interviewer assessment for each of the sites (Cherpitel et al., 2005). It will be seen that there was considerable variation between sites in the degree of concordance between Y90 and Y91, but the highest concordance for any site was a Tau-B of 0.76. Agreement between the clinical assessment and the interviewer assessment was substantially higher than Y90-Y91 concordance in each of the four sites where this was available. This suggests that the two observational assessments may be closer to measuring the same dimension than either is to measuring BAC. Another interpretation, however, is that BAC

Table 2. Tau-B and SE of Agreement between Clinical Assessment (Y91), BAC (Y90) and Interviewer Assessment (IA) by Country (Source: Cherpitel et al., 2005)

Country	N	Y91 vs. Y90	Interviewer Assessment vs. Y91[a]	Interviewer Assessment vs. Y90[a]
Argentina	424	.65 (.05)		
Belarus	457	.70 (.03)	.81 (.03)	.74 (.03)
Brazil	478	.63 (.06)	.86 (.03)	.60 (.06)
Canada	206	.50 (.31)		
China	453	.70 (.07)	.99 (.003)	.71 (.08)
Czech Republic	455	.35 (.13)		
India	544	.54 (.04)	.71 (.08)[b]	.31 (.11)[b]
Mexico	386	.59 (.06)		
Mozambique	445	.72 (.05)		
New Zealand	119	.68 (.07)		
South Africa	454	.72 (.02)		
Sweden	377	.76 (.05)		

[a] Interviewer assessment was only conducted in Belarus, Brazil, China and India
[b] Only 105 respondents were assessed by interviewer in India study

readings (Y90 codes) may be associated with different clinical signs (Y91) depending on whether BAC is rising or falling, and thereby contribute to the discordance between Y90 and Y91 (along with individual differences in tolerance).

Future Directions for Y90 and Y91

There is an association between clinical judgment or interviewer assessment in terms of the Y91 categories, on the one hand, and BAC on the other. But the association is of varying strength in the different study sites, and is only medium strong at any site. A clinical judgment of intoxication using Y91 is not a satisfactory substitute for Y90, i.e., is not a good way of estimating BAC.

This pushes us back to the question of what we are trying to do with the various codes available in ICD-10.

1. Does it make sense to build in a BAC measure in the ICD system, and if so, is Y90 the right code to use? From the point of view of epidemiology and policy, there are strong reasons for measuring the BAC of injuries in an ER setting. The exemplary case here is the field of drinking driving. The advent of BAC measurement technology eventually transformed our understanding of the relation between drinking and driving casualties, and also transformed the political situation, buttressing the political will to undertake effective countermeasures. Routine BAC measurement of injury cases in ERs, or failing that systematic sampling on a regularly scheduled basis, would be an important contribution to developing knowledge about alcohol's role in different casualties and situations, and to evaluating the success of preventive initiatives. However, it

should be recognized that alcohol is not necessarily involved in causing the injury, even though it is present. By itself, a positive BAC does not say much about alcohol's causal involvement in a particular injury; but routine BAC collection is useful in tracking the rates and proportions of alcohol involvement in casualties in a population. As noted in the following chapter (Chapter 10) in this volume, another issue is that some people will present to the ER several hours after an injury so their BAC may not be a useful indicator of the degree of intoxication at the time of the injury. Where these times are tracked it is possible to exclude such cases and reduce the numbers of false negative BAC readings.

It could be argued that the BAC should be recorded as measured, rather than in summary codes as in Y90. This would allow a maximum of flexibility in analysis, and would improve the accuracy of statistical summaries such as average BAC. But there are also arguments for summary codes. Less exact measures are less likely to become evidence in criminal or other court cases, which mitigates any ethical issues in collecting them and probably means that they are more likely to be collected and recorded. Summary codes also fit more easily in the ICD tradition of categorical measures.

The present specification in Y90 is worth some discussion. It seems appropriate to have a "low BAC" category starting at about 0.2 per mille, conventionally a cut-off for what can reliably be measured as a positive BAC. The next step up could be at any of several levels. A threshold of 0.4 per mille is used in north America for professional drivers, of 0.5 per mille for drinking driving in most European jurisdictions, and 0.8 per mille in the U.S. Beyond the level of 1.0, there are few such guideposts. My suggestion would be to keep down the number of categories, and set the ranges at under 0.2; 0.2 – 0.79; 0.8 – 1.49; 1.5 – 2.49; 2.5 and above.

2. How about developing a clinical instrument to estimate BAC? One option for Y91 would be to specify clearly that it is a fallback from Y90 when that is not available, and to focus on developing an instrument which could be applied in the ER setting with an explicit goal of estimating BAC, at least at a threshold level. This would mean changing the present wording of Y91 towards the kind of criteria used in the U.S. field sobriety tests. Perhaps a standardized test developed for assessing BAC in the context of boating, which used tests applicable in situations with limited mobility (McKnight et al., 1999), might be used or adapted for use in the ER environment. It should be noted that such tests could be applied by a trained admitting or triage nurse or other health worker, rather than a doctor. This would have the advantage of administering the test closer to the time of the injury.

3. The other option for Y91 is to separate it explicitly from Y90 and use it as a measure of ostensive intoxication: how drunk does the person appear and act? There is substantial evidence that drunken comportment varies from one culture to another (MacAndrew and Edgerton, 1969), and that there are variations by culture and also over time within a culture in how much drinking qualifies as "drunk" (Midanik, 1999). The measure might thus be more useful for local tracking and service planning than for international comparisons. Discussion is needed about what the content of such a code might be. One option would be to use it at a threshold level – the equivalent of the "had been drinking" check-box traditionally used on many police reports. Another option would be to focus on physical signs which might require immediate medical attention, or which hold

implications, for instance, for whether the patient should be held overnight or sent home. The present Y91 is worded that way in part – "severe disturbance in functions and responses, severe difficulty in coordination" – although "impaired ability to cooperate" points in another direction.

In considering what to recommend doing with Y91, the existence of F10.0 in the codes should be taken into account, even though it has usually not been considered in an emergency environment. The diagnostic criteria for research for F10.0 are specified in terms of 7 signs of "dysfunctional behaviour", including disinhibition, aggression and impaired judgment, and 7 physiological signs, mostly of lack of coordination but including flushed face and decreased level of consciousness. At least one sign from each list must be present for an F10.0 diagnosis. These lists seem to be based on common clinical wisdom rather than any developmental research and testing.

One way forward might be to use Y91 for signs of impaired coordination and F10.0 for signs of disinhibition, impaired attention and aggressiveness. Whatever is done along this line would have to be compatible with the general rules for F1x.0 codes ("symptoms or signs of intoxication compatible with the known actions of [alcohol] ... and of sufficient severity to produce disturbances in the level of consciousness, cognition, perception, affect, or behaviour that are of clinical importance"; WHO, 1992).

Conclusion

Among its other achievements, the WHO Collaborative Study on Alcohol and Injuries demonstrated that BAC measurement as a routine procedure can be implemented in emergency rooms in a variety of different cultural circumstances. The importance of alcohol's role in injuries argues strongly for building on this base to move towards a situation in which Y90 is measured and coded on a routine basis for injuries in ERs in any society with substantial alcohol use.

With respect to Y91, in my view the WHO Collaborative Study showed that it cannot be used in its present form as a satisfactory substitute for Y90. The alternatives then become to drop Y91 or to reformulate it. One reformulation would be to redesign and reword it so that it is specifically designed to approach a BAC reading as closely as possible, on the model of the standardized field sobriety tests used in north American law enforcement. Another reformulation would be to separate it conceptually from Y90, and use it in one form or another as a clinical judgment of intoxication or of an aspect of intoxication. In deciding on the specific approach to take along the latter path, it would be important to take into account the question of what would be the most useful aspects of drinking to measure and code in the ER situation as information to inform potential immediate clinical responses, brief interventions, and any longer-term clinical follow-up.

References

Aarens M et al. (1977). *Alcohol, Casualties and Crime.* Berkeley CA: Social (now Alcohol) Research Group, Report C-18.

Aitken SS, Zobeck TS (1989). A proposed model establishing an alcohol-related casualty surveillance system. In: Giesbrecht N et al., eds. *Drinking and Casualties: Accidents, Poisonings and Violence in an International Perspective,* pp. 411-424. London & New York: Tavistock/Routledge.

Cherpitel CJ (1993). Alcohol and injuries: A review of international emergency room studies. *Addiction* **88**, 923-937.

Cherpitel CJ (in press). Alcohol and injuries: A review of international emergency room studies since 1995. *Drug and Alcohol Review.*

Cherpitel CJ, Driggers P (2004). *Alcohol and Injuries. A review of international emergency room studies since 1995.* Geneva, Switzerland: World Health Organization.

Cherpitel CJ et al. (2005). Clinical assessment compared to breathalyzer readings in the ER: concordance of ICD-10 Y90 and Y91 codes. *Emergency Medicine Journal* **22**, 689-695.

Cole S, Nowaczyk RH (1994). Field sobriety tests: are they designed for failure? *Perceptual and Motor Skills* **79**, 99-104.

Giesbrecht N et al., eds (1989). *Drinking and Casualties. Accidents, poisonings and violence in an international perspective.* London; New York: Tavistock/Routledge.

Grant B et al. (1987). Proposed coding of alcohol's role in casualties. *Alcohol Health & Research World* **12**, 48-50.

Grossman DC et al. (1996). Validity of police assessment of driver intoxication in motor vehicle crashes leading to hospitalization. *Accident Analysis and Prevention* **28**, 435-442.

Honkanen R (1977). Records based on clinical examination as an indicator of alcohol involvment in injuries at emergency stations. Scandinavian *Journal of Social Medicine* **5**, 91-95.

Honkanen R, Ottelin J (1976). Blood alcohol levels in injury victims at the emergency station of a rural central hospital. *Annales Chirurgiae et Gynaecologiae* **65**, 282-286.

Jetter WW (1938). Studies in alcohol. 1. The diagnosis of acute alcoholic intoxication by a correlation of clinical and chemical findings. *American Journal of the Medical Sciences* **196**, 475-498.

L'Hours ACP (1995). The ICD-10 classification of injuries and external causes. In: *Proceedings of the International Collaborative Effort on Injury Statistics, vol. 1,* pp. 22-1 – 22-16. Hyattsville, MD: Centers for Disease Control and Prevention, National Center for Health Statistics. DHHS Publication No. (PHS) 95-1252.

MacAndrew C, Edgerton RB (1969). *Drunken comportment: A social explanation.* Chicago: Aldine Publishing Co.

Martin CS (1998). Measuring acute alcohol impairment. In: Karch SB, ed. *Drug Abuse Handbook*, pp. 309-326. Boca Raton, FL: CRC Press.

McKnight AJ, Lange JE, McKnight AS. (1999) Development of a standardized boating sobriety test. *Accident Analysis and Prevention* **31**, 147-152.

Midanik LT (1999). Drunkenness, feeling the effects and 5 plus measures. *Addiction* **94**, 887-897.

Rehm J et al. (2004). Alcohol. In: Ezzati M et al., eds. *Comparative Quantification of Health Risks. Global and Regional Burden of Disease Attributable to Selected Major Risk Factors: Volume 1*, pp. 959-1108. Geneva: World Health Organization.

Romelsjö, A. (1995) Alcohol consumption and unintentional injury, suicide, violence, work performance, and inter-generational effects. In: Holder HD, Edwards G, eds. *Alcohol and public policy: evidence and issues*, pp. 114-142. New York, NY: Oxford University Press.

Room R (1982). Notes on alcohol in the Ninth International Classification of Diseases. Berkeley, CA: Alcohol Research Group, Working Paper F85.

Room R (1984). Improving the coding of alcohol and drugs for casualties. Addendum to "Notes on alcohol in the Ninth International Classification of Diseases". Prepared following a WHO Informal Consultation on Proposals for the Classification of Mental Disorders and Psychosocial Factors in ICD-10, April 1984. Berkeley, CA: Alcohol Research Group, Working Paper F85, Addendum.

Room, R. (1987) Models for future work on alcohol's role in casualties. In: Norman Giesbrecht & Honey Fisher, eds. *Alcohol-Related Casualties: Proceedings of an International Symposium Held at the Guild Inn, Toronto, Canada, August 12-16, 1985*, pp. 85-94. Toronto: Addiction Research Foundation.

Room R, Babor T, Rehm J (2005). Alcohol and public health: a review. *Lancet* **365**, 519-530.

Stuster J, Burns M (1998). *Validation of the Standardized Field Sobriety Test Battery at BACS below 0.10 Percent*. Washington, DC: National Highway Traffic Safety Administraion.

Teplin LA, Lutz GW (1985). Measuring alcohol intoxication: the development, reliability and validity of an observational instrument. *Journal of Studies on Alcohol* **46**, 459-466.

World Health Organization. (1992) *International Statistical Classification of Diseases and Related Health Problems* (Tenth revised edn. Vol. 1). Geneva, Switzerland: World Health Organization.

World Health Organization (2006). *WHO Collaborative Study on Alcohol and Injuries: Final Report*. Geneva: WHO.

CHAPTER 10:
SURVEILLANCE AND MONITORING OF ACUTE ALCOHOL-RELATED PROBLEMS IN THE EMERGENCY ROOM

Tim Stockwell, Scott Macdonald, Jodi Sturge - Centre for Addictions Research of BC | Victoria, BC CANADA

Summary

The ability to *monitor* rates of serious alcohol-related harms in the population, both across time and place, is an essential cornerstone of any comprehensive policy to address these harms. Emergency Rooms (ERs) are ideal settings to identify and monitor new emerging trends in risky patterns of alcohol and other substance use that increase the risk of injury, overdose or poisoning and many of the acute harms caused by excessive drinking. The establishment of ongoing population based *surveillance* ER systems to detect such trends can be an invaluable approach for early detection and intervention for a variety of health problems. One challenge to overcome in relation to alcohol-related trauma is the reliable identification of cases which are at least partially caused by alcohol. Five main opportunities for monitoring and surveillance are discussed: (i) surveys of attendees, (ii) objective tests of breath or blood alcohol level, (iii) brief additional questions, flags or codes in routine records, (iv) application of etiologic fractions to diagnostic data, and (v) the development of surrogate measures indicative of high alcohol involvement. While some of these measures have been validated and applied in the evaluation of local harm reduction interventions, they are rarely employed in surveillance and monitoring systems. We argue in this chapter that multi-method approaches can be created in sentinel ER sites as a component of comprehensive monitoring and surveillance systems. Such an approach is perhaps best located within a broad injury and/or poisoning surveillance system.

Introduction

National and international statistics on alcohol-related harms tend to emphasise estimates of total numbers of deaths (e.g. Rehm et al, 2006) or total economic costs (e.g. Collins and Lapsley, 2003) but rarely report trends or variations across place and time. Monitoring such trends can be valuable as a means of guiding the development and evaluation of interventions, whether these are at the national, regional or local level (WHO, in press). While a single estimate of lives lost and economic impacts can raise awareness and build momentum towards new policy initiatives, the monitoring of trends using repeated measures provides a sharper focus on whether prevention and treatment policies are being well directed and effective. When such monitoring achieves a continuous coverage or includes very frequent assessments, this is often termed "surveillance" (Hirshon, 2000). Continuous monitoring of alcohol and other substance use in the ER has the potential to identify new and emerging patterns of risk for serious injury, overdose and poisoning events in a timely way that may reduce or prevent future occurrences.

There is growing international interest in the broadening of national alcohol monitoring beyond general population surveys to "harder" measures of estimated rates of alcohol-caused mortality and morbidity (WHO, 2000 and in press; Chikritzhs et al, 2003). General population surveys tend to be periodic (e.g. every 3 to 10 years), suffer from substantial underreporting of drinking (Stockwell et al, 2005), have variable (mostly falling) rates of compliance and rely on self-attributions of causation for alcohol-related problems (e.g. Adlaf et al, 2005). By contrast, harm indicators derived from coroner and health statistics offer comprehensive coverage of rates of serious harms over both time and place. The major challenge with these sources is the identification of cases that are specifically alcohol caused or those with a high probability of being so. This challenge is especially apparent in relation to presentations to ERs: while these are more numerous than either deaths or hospital episodes, information specifically linking ER presentations with previous alcohol consumption is mostly absent. This chapter will discuss the potential for different types of data collection in the ER to contribute to comprehensive national alcohol and other drug monitoring in a significant way.

Monitoring and surveillance systems, whether at the national, regional or local level, have tended to focus more on illicit drug use and related harms than on alcohol. Well developed drug surveillance systems exist in several countries such as the Illicit Drug Reporting System (IDRS) in Australia (Topp et al, 2004), the Drug Abuse Warning Network (DAWN) in the US (Kraman, 2004), the South African Community Epidemiological Network on Drug Use (SACENDU) (Parry et al, 1997), the Canadian Community Epidemiological Network on Drug Use (CCENDU) (e.g. Buxton et al, 2005) and the US Arrestee Drug Abuse Monitoring (ADAM) (Ashcroft et al, 2003). Each of these systems has only a limited focus on alcohol despite global evidence of substantially greater harm associated with alcohol misuse (Rehm et al, 2004). One exception is the Australian National Alcohol Indicators Project (Chikritzhs et al, 2003) which has reported on trends in risky patterns of alcohol use and per capita consumption. This reports serious alcohol-related harms across all Australian jurisdictions as well as within major subpopulations defined by age and ethnic background (see www.ndri.curtin.edu.au). This monitoring system has been explicitly developed on the basis of the *International Guide to Monitoring Alcohol Consumption and Related Harms* (WHO, 2000 and in press) and has been applied in the evaluation of major policy initiatives such as alcohol tax changes (Stockwell et al, 2001; Chikritzhs et al, 2005).

As is apparent from many of the contributions to this volume, there has been an upsurge of data collection initiatives in the ER which now span a few dozen countries and mostly use comparable protocols. This has facilitated cross-cultural analyses of the extent to which alcohol contributes causally to various kinds of injury presentations (e.g. Macdonald et al, 2005) and also provides possible directions for monitoring and surveillance (Young et al, 2004).

Advantages of the ER for surveillance and monitoring

The great potential of the ER for alcohol surveillance is indicated by the fact that a substantial proportion of alcohol-related morbidity and mortality is often related to the short term effects of heavy episodic alcohol use. It has been estimated that 68% of cases of in-patient hospital care in Australia caused by alcohol misuse were from the acute effects of alcohol (Chikritzhs et al, 2003). In addition, over two-thirds of these episodes were from acute causes, mostly some form of injury (Chikritzhs et al, 2003). Another advantage of an ER setting for monitoring purposes is the high volume of people presenting with conditions related to their alcohol and other substance use. In Canada it was reported that as many as 13% of Canadians visited an ER in 2003, most of whom were not admitted and not therefore otherwise entered into the hospital record system (Carriere, 2004).

One significant advantage of ER data is that people often seek medical attention for acute injuries and illnesses provided they can physically reach treatment sites. For example several studies have found that many people that present violence related injuries to the ER do not report these incidents to the police (Brinkman et al, 2000). This is not to say that there are no biases in these data sets (see below), however, in the main, the existing biases limit the direct comparability *between sites* rather than within *one site* over time.

Challenges of the ER for surveillance and monitoring

There is a challenge to identify reliable subsets of presentations with a high probability of alcohol involvement in order to form the basis for an indicator of alcohol-related harm. These ER presentations are highly variable among hospitals and reflect the unique geographic composition of the catchment area, density, hours of opening, availability of transportation and waiting times for treatment – as well as changes in these in one location over time. ER staff are generally too busy to do more than respond to the immediate presenting problem and will often be selective of whom they will ask about alcohol use (Brinkman et al, 2000). An additional challenge is that even standard medical diagnostic information is often not readily accessible in an electronic format from ER departments. This at least has been the experience of the first author in both the Australian and Canadian context. While it is possible to hire interviewers to collect alcohol and other drug use information by self-report and breathalyser from persons presenting, this is usually too costly a procedure for routine surveillance and monitoring but may be possible with recurrent sampling.

Available Strategies for ER Monitoring

Special surveys

Following the lead of Cherpitel and colleagues work over two decades examining alcohol-related injuries in the ER and reviewed extensively in this volume, over 30 countries have developed and implemented consistent protocols to assess alcohol and other substance use through interviews focusing on the six-hour period leading up to an injury or acute illness event. These protocols typically serve an analytic purpose rather than that of monitoring. They typically involve approaching a representative sample of ER patients round-the-clock for several weeks. Interviews last between 15 and 30 minutes and include a request to take a breathalyser sample. Response rates have typically been in the region of between 70 and 80% (Cherpitel, 1993).

While alcohol-related presentations to the ER have been used as one indicator in the evaluation of local community intervention in the US (e.g. Puttnam et al, 1991; Treno and Holder, 1997) and in Australia (e.g. Burns et al, 1995) these have not involved the use of such interviews. In one case the prohibitive cost of doing so was a major reason for the creation of an alternative surrogate measure (Treno and Holder, 1997).

The DAWN surveillance tool, used in US ERs, introduced some limited coverage of alcohol-related presentations in 2003: visits involving the misuse of alcohol for minors are recorded and also for adults when this is judged to occur in conjunction with another drug (Ball et al., 2005). DAWN involves the rating of case notes of ER attendees and has coverage rates ranging from 45% to 77% (SAMHSA, 2006). Clearly this procedure is dependent on the clinical staff detecting the presence of alcohol and reporting this in the case notes.

With obvious cost considerations for ongoing monitoring, options for ER surveillance relying on special surveys might be affordable in some jurisdictions provided careful time sampling is applied. Key ER departments in a particular jurisdiction would need to be selected as "sentinel sites", chosen on the basis of serving a significant catchment area in a city or region of interest. Establishing consistent data collection protocols would enable tracking variations in levels and types of alcohol-related harm over time. Time sampling (e.g. see Room et al, 1987) would need to allow for seasonal variation in drinking patterns and levels of related harms and focus on high risk times (e.g. between 10 p.m. and 6 a.m. on weekends).

Another approach to collecting data from ER patients is through self-administered question-naires. This approach allows for the collection of information from a large number of people at much lower costs than interviews and avoids the possibility of interview biases that could skew the results. This approach is more appropriate in jurisdictions with a high literacy rate, such as developed countries, because those who cannot read would be excluded from the study, intro-ducing possible biases in the data.

Objective tests for alcohol use

The issue of objective tests in the ER is beset with ethical and technical difficulties, whether it be a routine sample of urine, blood, breath or saliva. In relation to assessing presence of alcohol, the evidence, to date, is that self-reported consumption is a more sensitive test. This is because by the time an individual is interviewed in the ER and they submit to a test, blood alcohol levels may have ready fallen to zero – as well alcohol consumption may have occurred since the injury or illness event (McLeod et al, 2000). There is no doubt, however, that high levels of compliance with a simple breath test can be achieved, being a more cost-effective objective method. A clear advantage for objective tests is that despite the lack of sensitivity in many cases, a relatively standard procedure can be applied for taking the measurement and hence utilising the results to develop indicators over time for a particular area or sentinel site. Restricting data used in these indicators to those tests conducted within a relatively short time of the injury event would increase the sensitivity of the measure. Some ER departments already implement routine urine, blood or breath alcohol assessments and it is possible that such clinical assessments are easier to institutionalize in acute medical care settings than are interview protocols (e.g. Treno and Holder, 1997). There are also clini-cal reasons for assessing blood-alcohol levels in the management of many acute conditions which may also help to cement routine alcohol screening in an ER department. Anecdotally, some policy makers and health-care workers have objected to the application of breathalyser testing in health care settings because of the association of this procedure with police and law enforcement. There is a concern that the implied lack of trust in this procedure may interfere with the doctor-patient or nurse-patient relationship. One possible way of diminishing the conflict here is to use measurements, the accuracy and specificity of which fall below legal standards; for example, saliva sticks that change color (Guang-chou et al, 1992) or handheld breathalyzers (which look like a torch/ flashlight and sometimes used by traffic police for screening using random breath testing) that measure alcohol in the air around the subject. For basic epidemiological monitoring and surveillance complete accuracy is not a great concern, although inaccurate measurement is of no use to the legal system.

A predominant research issue with ER studies is that informed consent is typically required for ethical reasons from patients to be interviewed or to obtain breath samples for research purposes. Consent can introduce a bias in that those who refuse to participate are more likely to have consumed alcohol than those who consent. This bias can be especially strong for events where criminal liability may be involved (i.e. drinking and driving) but also can occur due to social desirability effects. One method of avoiding non-response bias is to take samples without consent to use for research purposes. This situation can occur when blood samples are required for medical purposes, often in cases of severe trauma where the patient is unconscious. In some jurisdictions, procedures can be implemented where the results of blood tests can be used for epidemiological research studies, while preserving the identity of the subjects. The National Trauma Registry in Canada is an example of this type of injury surveillance system (Public Health Agency of Canada, 2005). The trauma registry includes 40 participating hospitals and utilizes external causes of injury (E codes), blood alcohol levels (BAL) from designated hospitals and the cause of death extracted from coroners records.

Alcohol flags, use of Y-codes and other adaptations of medical record forms

A variety of simple devices have been tried both in healthcare and police settings with the objective of making it easy for emergency personnel to indicate if alcohol was a factor in a presenting problem. Examples include a simple box to be ticked, a couple of questions to be administered by a triage nurse or the use of the under-utilised "Y-codes" in ICD-10. The latter has been discussed in some detail in the preceding chapter in this volume. In each case one major problem is that an individual has to make a difficult judgment in a busy stressful environment regarding whether alcohol is likely to have played a contributory role in a presenting complaint. This judgment has many subjective elements regarding interpretation of verbal reports and clinical signs (Brinkman et al, 2000).

Brinkman et al (2000) cite examples where the use of such flags and special codes often decline over time leading to a particular ambiguity that the absence of a tick can variously mean a failure to consider the question, not being certain of the answer or just the absence of alcohol as a risk factor. Indeed, with one or two exceptions, the experience of a WHO exercise to increase compliance with Y-codes confirmed that rates of their use vary considerably between sites and overtime, thus compromising the interpretation of data as indicators (see Chapter 9 in this volume). Y-codes require ongoing training of staff and are plagued by compliance issues and inconsistent interpretation of the codes. We recommend that consideration be given to these devices only in situations where the completion of the flag, box or code is compulsory so that missing data problems are reduced.

The application of etiological fractions to ER attendance data

In the absence of reliable local estimates of alcohol's contribution to injury, well developed and widely used methods now exist for estimating the proportions of cases presenting to hospital departments with particular ICD-9 or ICD-10 diagnoses in which alcohol was a causal factor, otherwise known as "etiologic fractions" (see WHO, 2000 and in press). Estimates of these are derived from meta-analyses of national and/or international studies linking levels of alcohol consumption to different injury and illness outcomes and take account of different rates of hazardous alcohol use in different age and gender groups in a particular population (e.g. English et al, 1995; Rehm et al, 2004). In theory, it is possible to apply these fractions to ER attendance data. In practice, it is often not the case that the necessary diagnostic information is both assessed routinely in the ER and, where it is, that it is then also entered into an electronic database. This, however, is changing as more information is recorded electronically in different healthcare environments.

International ER studies are increasingly providing a rigorous basis for the calculation of relative risks and alcohol etiological fractions (e.g. Cherpitel et al, 2005a; McLeod et al, 2000). Some studies have simply estimated relative risks for all injury events rather than breaking them down into smaller diagnostic categories – such as those found in the International Classification of Diseases (ICD). In practice, this amounts to the weighting of each injury presentation to an ER according to age and gender using meta-analyses such as Cherpitel et al, (2005b) to estimate the probability of alcohol consumption in each case. It cannot be assumed that relative risks and etiologic fractions are universally transferable across all cultures and settings (WHO, 2000) and hence the meta analyses used for these estimates should, as far as possible, be based on the most recent studies of broadly comparable populations. Again, local data (when available) provides the most reliable estimates of relevant etiologic fractions, conventionally taken to be the proportion of cases with injuries presenting with a BAC above 0.08 mg/100ml (e.g. see English et al, 1995).

Other surrogate measures of alcohol-related injury

A recent analysis of the international ERCAAP dataset found that 75% of young, single males presenting an injury between 12 midnight and 4am had recently consumed some alcohol (Young et al, 2004). Variations were also analysed and confirmed for other sociodemographic subgroups of ER attendance at different times of day and week. The authors recommended that these data could be used to derive surrogate measures for local rates of alcohol-related injuries. The WHO International Guide on Monitoring Alcohol Consumption and Related Harm (WHO, 2000) recommends this as a surrogate indicator of alcohol-related harm. It should be noted that variations in trading hours will impact upon the application of this surrogate measure in one local area and may also require variations in the specific time period employed.

A similar 'surrogate' measure involving weekend injury presentations to ERs was successfully used to evaluate the US Community Trials Project (Holder and Treno, 1997). These authors noted the high prevalence of alcohol-related injuries in the US and evaluated alternative means of monitoring levels of these, both for epidemiological and evaluation purposes. They identified characteristics of certain hospital discharge cases indicative of alcohol-involvement and used these to develop a highly cost-effective monitoring tool method compared with the use of regular population surveys. They identified advantages such as the availability of data for many years baseline prior to an intervention being implemented, good statistical power when interventions cover a large population, very low costs and good accessibility of the data for researchers. A weighting method was developed and applied to reflect the probability of any individual injury presentation being alcohol-related based on estimates derived from a sizeable hospital database that included BAC measures on admission. Weighted hospital discharge data were then used as part of the evaluation of the Community Trials Project and, along with other measures, demonstrated a significant impact on overall injury rates (Treno & Holder, 1997). The final model demonstrated a significant increase in the probability of injuries being alcohol-related variously for males, patients age 21 to 34, patients presenting on Fridays and Saturdays, those presenting between midnight and 4am, and also those presenting with head injuries. Disadvantages of this method include a significant delay before the key data usually become available and that, for smaller geographic areas and short time periods, numbers of discharged patients with injuries may not be large enough for analysis. Furthermore, the highest probabilities of alcohol-involvement available from the model were relatively modest, being only in the region of 40% (e.g. for a young male, admitted on a Saturday with a head injury).

Young et al (2004), however, extended the Treno and Holder (1997) approach by applying it to the "walking wounded", i.e., ER attendees who are not admitted. They provided analyses of different subsets of attendees who varied in the extent to which their injuries were alcohol-related as evidenced by self-report or BAC, using the international ERCAAP dataset. A trade-off was identified between having a high-volume of cases but with a lower proportion being alcohol-related (e.g. 46% of all persons attending between the hours of 10 pm and 6 am) versus a relatively small volume of more narrowly defined cases containing a much higher proportion with prior alcohol consumption (e.g. 75% of all young single males presenting between midnight and 4 am). The authors recommend developing surrogate measures that optimise the specificity of alcohol involvement with a sufficient volume of data to generate requisite statistical power for any required analyses of these surrogate measures of local rates of alcohol-related harm (Young et al, 2004).

Conclusions and Recommendations

The above five general approaches to monitoring acute alcohol-related harm in the ER are not mutually exclusive and an optimal strategy might involve encouraging elements of each to be applied in a consistent manner – depending on the availability of resources and access to requisite data. The single most specific, cost-effective and objective indicator would be the number of injuries involving young single males presenting between midnight and 4 am on Friday and Saturday nights (Young et al, 2004). This could also be presented as a proportion of all presentations to that ER on Fridays and Saturdays to control for seasonal variations in the local population. Such an indicator could be used to contribute to the evaluation of a local alcohol harm reduction initiative. Total number (or proportion) of positive BAC cases would also be a specific, but less cost-effective, indicator given the expense of conducting breath tests, calibrating the breathalysers and compiling the data. Applying etiologic fraction weights to all admissions depending on age, sex and presenting problem would also be a useful, reasonably cost-effective though less specific approach. Its value would be increased greatly if local data collection was used to estimate alcohol etiologic fractions for injuries presenting to a particular ER (e.g. Treno and Holder, 1997). In the latter case, it would be ideal to base etiologic fraction estimates on BAC levels obtained no more than two hours after the injury event and also after determining no alcohol consumption had occurred since the injury event. As discussed by Room (Chapter 9 in this volume), further work is needed to simplify the use of the new Y-codes in ICD-10 where they are related to categories of blood alcohol level and/or ratings of degree of intoxication so as to increase the likelihood that they are used reliably.

As one component of a national surveillance and monitoring system, it would be feasible to establish sentinel sites in major cities where the above indicators of acute alcohol-related harm could be collected routinely. Their added value to more traditional alcohol harm indicators based on alcohol-related mortality and morbidity data lies in the much higher frequency of ER presentations than of either hospital admissions or deaths. When combined with sample interviews at high risk times (i.e. late weekend nights), it would also be possible to gather information about the combined use of alcohol with other psychoactive substances, both licit and illicit (Sturge et al, 2006). Combined use of alcohol with other substances such as opioids or other central nervous system depressants is known to present special risks for drug overdose or injury indicating potential value in monitoring trends in patterns of combined substance used in a comprehensive alcohol and other drug surveillance system (Sturge et al, 2006).

References

Adlaf EM, Begin P, Sawka E, eds (2005). *Canadian Addiction Survey (CAS): A national survey of Canadians' use of alcohol and other drugs: Prevalence of use and related harms: Detailed Report.* Ottawa: Canadian Centre on Substance Abuse.

Ashcroft J, Daniels DJ, Hart SV (2003). 2000 *Arrestee Drug Abuse Monitoring: Annual Report.* Washington, DC: U.S. Department of Justice, Office of Justice Programs, National Institute of Justice.

Ball JK, Ducharme L, Green J (2005). The DAWN Report – New DAWN: Why It Cannot Be Compared with Old DAWN.
(Available at: http://dawninfo.samhsa.gov/files/DAWN_TDR_new_old.pdf)

Brinkman S et al. (2000). "An indicator approach to the measurement of alcohol-related violence." In: Williams P, ed. *Alcohol young people and Violence, Australian.* Australian Institute of Criminology, Research and Public Policy, Canberra.

Burns L et al. (1995). Policing pubs: what happens to crime? *Drug and Alcohol Review* (14):369-376.

Buxton J (2005). Vancouver Drug Epidemiology, Vancouver Site Report for the Canadian Community Epidemiology Network on Drug Use.
(Available at: http://www.vancouver.ca/fourpillars/pdf/report_vancouver_2005.pdf)

Carriere G. Use of Hospital Emergency Rooms. *Health Reports* 16, 1 (October 2004), Statistics Canada, catalogue no. 82-003.

Cherpitel CJ (1993). Alcohol and injuries: A review of international emergency room studies. *Addiction* **88**: 923-937.

Cherpitel CJ, Ye Y, Bond J (2005a). Attributable risk of injury associated with alcohol use: a cross-national meta-analysis from the Emergency Room Collaborative Alcohol Analysis Project. Am. J. *Public Health* **95**:266-272.

Cherpitel CJ et al. (2005b). Multi-level analysis of alcohol-related injury among emergency room patients: A cross-national study. *Addiction* **100**, 1840-1850.

Chikritzhs T et al. (2003). *Australian Alcohol Indicators, 1990-2001: Pattern of alcohol use and related harms for Australian states and territories.* National Drug Research Institute, Curtin University of Technology, Perth, West Australia. ISBN 1 74067 300 X.

Chikritzhs T, Stockwell T, Pascal R (2005). "The impact of the Northern Territory's Living With Alcohol Program, 1992-2002: revisiting the evaluation." *Addiction.* **100**, 1625-1636.

Collins D, Lapsley H (2002). *Counting the cost: estimates of the social costs of drug abuse in Australia in 1998/9.* Canberra: Commonwealth Department of Human Services and Health, Monograph Series Number 49.

Cummings GE et al. Health Promotion and Disease Prevention in the Emergency Department: A feasibility study. Can J Emerg Med 2006;8:100-5.

English D et al. (1995). *The Quantification of Drug Caused Mortality in Australia 1992*, Commonwealth Department of Human Services and Health, Canberra.

Guang-chou T, Bhushan K, Isreal Y (1992). Characteristics of a New Urine, Serum, and Saliva Alcohol Reagent Strip. *Alcoholism: Clinical and Experimental Research*, **16**, 222.

Hirshon JM. The rationale for developing public health surveillance systems based on emergency department data. *Acad Emerg Med* 2000;7:1428-32.

Kraman P (2004). *Drug Abuse in America: Rural Meth*. Lexington: The Council of State Governments.

McLeod R et al. (2000). *The influence of alcohol and drug use, setting and activity on the risk of injury – a case-control study*. National Drug Research Institute, Curtin University of Technology, Perth, Western Australia.

Parry CDH et al. (1997). The South African Community Epidemiology Network on Drug Use (SACENDU): description, findings (1997-99) and policy implications. *Addiction* **97**:969-976.

Public Health Agency of Canada. *Inventory of Injury Data Sources and Surveillance Activities*. Centre for Surveillance Coordination. 2005 (ISBN: H121-3/2005E-PDF 0-662-40052-6)

Putnam S, Rockett I, Campbell M (1993). Methodological issues in community-based alcohol-related injury prevention projects: attribution of program effects, in *Experiences with Community Action Projects: Research in the Prevention of Alcohol and other Drug Problems*. Greenfield TK, Zimmerman R, eds. US Department of Health and Human Services: Rockville, Maryland.

Rehm J et al. (2004). *Alcohol Use*. In: Ezzati M et al., eds. Comparative Quantification of Health Risks. Global and Regional Burden of Disease Attributable to Selected Major Risk Factors. 1: 959-1108. Geneva: World Health Organization.

Rehm J et al. in collaboration with Adlaf E, Recel M, Single E (2006). *The Costs of Substance Abuse in Canada 2002*. Ottawa, ON: Canadian Centre on Substance Abuse.

Room R (1987). Models for future work on alcohol's role in casualties. In: Norman Giesbrecht, Honey Fisher, eds. *Alcohol-Related Casualties: Proceedings of an International Symposium held at the Guild Inn, Toronto, Canada, August 12-16, 1985*. Toronto: Addiction Research Foundation, pp 85-94.

Substance Abuse and Mental Health Services Administration, Office of Applied Studies. Drug Abuse Warning Network, 2004: *National Estimates of Drug-Related Emergency Department Visits*. DAWN Series D-28, DHHS Publication No. (SMA) 06-4143, Rockville, MD, 2006.

Stockwell TR et al. (2001). "The public health and safety benefits of the Northern Territory's Living With Alcohol program." *Drug and Alcohol Review*, 20, (2), pp. 167-180.

Stockwell T, Sturge J, Macdonald S (2005). *Patterns of risky alcohol consumption in British Columbia: Analysis of the 2004 Canadian Addiction Survey.* Statistical Bulletin number 1: Centre for Addictions Research of BC, University of Victoria, British Columbia.

Sturge J, Stockwell T, Macdonald S (2006). A comprehensive alcohol and other drug epidemiological monitoring system for Canada: A pilot project in British Columbia and Ontario, Canada. Paper presented at the 32nd annual symposium of the Kettil Bruun Society for Social and Epidemiological Research on Alcohol, Maastricht, the Netherlands, May 28-June 2, 2006.

Topp L et al. (2004). Adapting the Illicit Drug Reporting System (IDRS) to examine the feasibility of monitoring trends in the markets for 'party drugs'. *Drug and Alcohol Dependence* **73**(2):189-197.

Treno AJ, Holder HD (1997). Measurement of alcohol-involved injury in community prevention; the search for a surrogate. *Alcoholism: clinical and experimental research* **21**(9), 1695-1703.

World Health Organization (2000). *International guide for monitoring alcohol-related problems, consumption and harm*, World Health Organization: Geneva.

World Health Organization (in press). *International guide for monitoring alcohol-related problems, consumption and harm, 2nd Edition.* World Health Organization: Geneva.

Young D et al. (2004). Emergency room injury presentations as an indicator of alcohol-related problems in the community: A multilevel analysis of an international study. *Journal of Studies son Alcohol* **65**(5), 605-612.

SECTION IV: SCREENING AND BRIEF INTERVENTION IN THE EMERGENCY DEPARTMENTS AND TRAUMA CENTERS
INTRODUCTION

Daniel W. Hungerford - Centers for Disease Control and Prevention | Atlanta, GA US

The seven papers in this section present in a limited space a wide range of experience with alcohol screening and brief intervention (SBI) in acute care medical settings – emergency departments (EDs) and trauma centers (TCs). Although the papers taken collectively can hardly be considered comprehensive, they are a reasonable introduction to SBI in acute care clinical settings and will help readers motivated to learn more delve into the history of work in these settings and its current status.

The first paper, by D'Onofrio and Degutis, reviews current efficacy studies in EDs and TCs, the limitations of current research, and the feasibility of SBI in real-world clinical settings. It also considers the challenges of translating research results into routine practices outside the research setting. The next five papers – Bernstein and Bernstein, Broderick, Woolard et al., Touquet and Brown, and Gentilello et al. – discuss the practical details of implementation. In the process, they describe what kind of help should be offered to acute care patients with alcohol problems, how to implement SBI in real-world clinical settings, and how the operational characteristics of their programs developed over time. The papers by the Bernsteins and Broderick describe the difficulties of addressing the entire severity spectrum of alcohol problems as opposed to focusing exclusively on addicted patients, the ones with the most severe problems. Broderick notes the difficulties in communicating to ED staff that SBI was designed to help primarily patients who have alcohol-related problems but are not addicted. Woolard et al. describes two unique projects – not truly SBI – that focus on impaired drivers treated in the ED. One increases ED-based reporting of impaired drivers to the Department of Motor Vehicles, which has the power to revoke driving licenses. The other is a court-mandated program that provides group counseling for adolescent drivers convicted of driving under the influence of alcohol and observation sessions in the ED and TC during weekend evenings. The third project involves an intervention method that could increase the efficiency of SBI and acceptance by ED staff. It is currently evaluating the impact on future injury of brief interventions provided over the telephone after patients with alcohol problems are discharged from the ED. Touquet and Brown address the question of who should provide SBI. They describe a protocol that efficiently integrates physicians and nurses into the SBI endeavor but does not overwhelm them with a major new initiative. Gentilello and colleagues describe how TCs are different from EDs, and how those differences influence the way SBI is implemented, efficacy, and the feasibility of dissemination. They contend that current efforts in TCs deserve high priority support.

In the last short chapter, Hungerford considers the potential impact SBI could have if it were broadly implemented in EDs and TCs on alcohol problems in society as a whole.

CHAPTER 11 :
EVIDENCE-BASED EMERGENCY DEPARTMENT SCREENING AND BRIEF INTERVENTION FOR ALCOHOL PROBLEMS

Gail D'Onofrio, Linda C. Degutis - Yale University | New Haven, CT US

Introduction

While screening and brief intervention for unhealthy alcohol use has been shown to be effective in primary care settings (Whitlock et at 2004; U.S. Preventative Services Taskforce, 2004), the evidence to date in emergency department (ED) settings is limited. This chapter will review what may be described as the first intervention in the ED, and the randomized controlled trials in the ED setting published to date. In addition, one inpatient study that randomized trauma center patients initially seen in the ED will be discussed.

Chafetz and colleagues (1962) published what might be considered the first report of a brief intervention for alcohol problems in the ED. Their intervention consisted of a referral to a specialized alcoholism treatment clinic as well as information on strategies for assisting with social issues, such as food and housing. They demonstrated the effectiveness of their intervention in motivating alcohol-dependent patients to engage in treatment. These investigators reported that 65% of patients with alcohol dependence who received the intervention and a direct referral from the ED to an alcoholism treatment clinic kept their initial appointment at the clinic, compared with 5.4% of the control group who received only a referral.

Randomized Controlled Studies in the ED Performed in the U. S.

Although a number of studies have subsequently evaluated the effectiveness of brief intervention in the ED, they are difficult to compare for a number of reasons. These studies have used limited populations (e.g., only young adults or injured patients), involved interventions of varying lengths, and had methodological limitations (D'Onofrio and Degutis 2002). To date, only five studies in the U.S. have analyzed the effectiveness of brief interventions in ED patients using a control group and randomly assigning participants to an intervention.

Alcohol interventions in adolescent populations

Monti and colleagues (1999) compared the effectiveness of "standard care" with that of a brief motivational interview (MI) in reducing alcohol-related consequences and alcohol use among ED patients ages 18 and 19. Standard care was described as "consistent with general practice for treating alcohol-involved teens in an urgent care setting" and included a handout on avoiding drinking and driving, as well as a list of local treatment agencies. The 94 participants were recruited for the study because they had positive blood alcohol concentrations (BACs) or an alcohol-related injury (i.e., reported drinking prior to the injury that required treatment). Follow-up assessments were conducted by phone after 3 months and through face-to-face interviews after 6 months.

Both groups of patients decreased their alcohol consumption, but patients who participated in the MI showed greater improvements in the following outcomes, all of which were statistically significant, compared to those in the control group which received standard care: 1) lower incidence of drinking and driving (62% vs. 85%); 2) fewer citations for moving violations during the follow-up period (3% vs. 23%); 3) fewer alcohol-related injuries during the follow-up period (21% vs. 50%); and 4) fewer alcohol-related social and legal problems in the 6 months following treatment, such as problems with dates, friends, parents, school, or the police (mean.89 vs. 1.44).

The generalizability of these results to routine ED practice may be limited because the population was restricted to injured adolescents, all interventions were performed by trained social workers hired for the project, and the refusal rate was relatively high. Monti's results are similar to those from BI studies in primary care settings – alcohol consumption was reduced in both groups, and negative consequences were reduced only in the treatment group.

Spirito and colleagues (2004) studied adolescents ages 13 to 17 who were treated in an ED for an alcohol-related event. The adolescents were eligible to participate in the study if they had evidence of alcohol in their blood, breath, or saliva (N = 142), or if they reported drinking alcohol in the 6 hours before the injury that required treatment in the ED (N = 10). The participants underwent a battery of assessments that took an average of 45 minutes to complete. They reported their drinking behavior over the past 12 months and completed the Adolescent Drinking Questionnaire (which assesses behavior over the past 3 months), the Young Adult Drinking and Driving Questionnaire, and the Adolescent Injury Checklist. Furthermore, at baseline the investigators administered the Adolescent Drinking Inventory (ADI) to identify adolescents with alcohol problems that warranted a treatment referral, and for use in the personal feedback component of the intervention condition. The ADI is a 24-item measure of severity of alcohol involvement, with a score of > 15 indicating that referral for alcohol problems is needed. Participants were then randomly assigned to receive standard care or a motivational interview.

Researchers interviewed the adolescents by phone after 3 months and interviewed them in person after 6 and 12 months. The investigators found that adolescents in both groups drank less alcohol during the 12-month follow-up period. However, adolescents in the MI group with a baseline ADI score indicating problematic alcohol use improved significantly in two outcomes, average number of drinking days per month (frequency) and frequency of high-volume drinking (binging). Based on these findings, the investigators recommended that adolescents who are treated in the ED for an alcohol-related injury should be screened for pre-existing alcohol problems and should receive a brief intervention if the screen is positive.

Maio and colleagues (2005) tested a laptop computer-based interactive program addressing alcohol misuse compared with standard care in adolescents aged 14-18 with minor injuries. Telephone follow-up of both groups occurred at 3 and 12 months post ED visit. The interactive computer program used the setting of a house party where alcohol was available in order to guide the patient in making decisions not to use alcohol, and to make other choices that decreased the risk of negative consequences. The characters in the role play provide feedback and information to the participant. Alcohol misuse was measured by binge drinking, defined as the number of times in the previous 3 months they drank 5 or more drinks on 1 occasion at baseline and follow-up, and the Amidx, a validated 10-item measure quantifying negative consequences of drinking alcohol, with values ranging from 0-60; higher scores mean more misuse. A total of 655 patients completed the program with only a 15% refusal rate. Few agreed that they drank 6 hours prior to the injury, 2.5%. Forty-three percent admitted to drinking, 26% to binge drinking and 22% to riding in the past three months with a driver who had been drinking. There was no significant difference between groups for either outcome measure. A major limitation is that the entire group had low Amidx scores at baseline, indicating a low rate of alcohol misuse.

Adult injured hazardous and harmful drinkers

Longabaugh and colleagues (2001) evaluated the effects of a brief motivational intervention in injured drinkers age 18 or older who visited the ED of an urban teaching hospital. Patients were eligible for the study if they screened positive for hazardous or harmful drinking (i.e., had breath alcohol concentrations greater than 0.003 g/dL, reported having ingested alcohol in the 6 hours before the injury, or scored positive on the AUDIT screening test. Participants were randomly assigned to one of three groups: 1) standard care (SC) (N = 188), 2) brief intervention (BI) consisting of a 40- to 60-minute session provided by non-ED staff, i.e., a social worker or graduate student (N = 182), 3) BI with a booster (BIB) that entailed a scheduled return visit 7 to 10 days after the initial BI (N = 169).

At 1 year after the intervention, participants in all three groups reported having reduced their days of heavy drinking, although the differences between the groups were not significant, similar to the findings of the study by Monti, et al (1999). Moreover, the BIB group reported significantly fewer alcohol-related negative consequences (e.g., hangovers and lost work time) and alcohol-related injuries than did the SC group. However, the average number of injuries the participants had sustained in the year preceding the ED visit was low (i.e., an average of 1.6 injuries), as was the incidence of new injuries. Therefore, it is difficult to demonstrate significant changes, which limits the interpretation of the findings. Nevertheless, the investigators concluded that a booster added to a brief intervention with injured ED patients who engage in hazardous and harmful drinking may be helpful in reducing negative consequences and alcohol-related injuries.

D'Onofrio and colleagues (2005a) studied brief interventions for harmful and hazardous drinkers in the ED provided by emergency practitioners. Participants were identified by screening all patients and enrolling those who reported drinking in excess of the National Institute of Alcohol Abuse and Alcoholism (NIAAA) guidelines for low risk drinking. Potentially dependent drinkers were excluded. At 12 month follow-up, with 93% retention, both the control (brief advice) and the intervention groups had significantly reduced their drinking; however, there was no significant difference between the groups.

Hospitalized trauma patients

Two studies have explored interventions in hospitalized trauma patients. Gentilello and colleagues (1999) included hospitalized trauma patients who were found to have an alcohol use disorder (AUD) based on screening and/or testing. They did not include patients who were discharged after treatment in the ED. The participating patients represented the full spectrum of unhealthy drinking. Patients were randomly assigned to either the control group or the intervention group, which received a single 30-minute motivational interview conducted by a doctoral-level psychologist in the inpatient setting. Follow-up with the participants was conducted after 6 and 12 months. Although only 54 % of participants were available for follow-up at 12 months, the investigators found that the intervention group decreased their weekly alcohol consumption by significantly more (21.8 drinks) than the control group (6.7 drinks). The decrease was greatest in patients with mild to moderate AUD problems at the beginning of the study. The beneficial effects of the intervention appeared to be persistent. At 3-year follow-up period, the investigators found a 47% reduction in injuries requiring an ED visit or readmission to the trauma service in the intervention group; however, this was not statistically significant. Because of the low follow-up rate, it is not possible to generalize the findings of this study, as one cannot determine whether the patients who were not followed also decreased their alcohol consumption.

One difference between this and the other studies that may have contributed to the positive findings was the control group, which received minimal screening and assessments as would patients who receive the prevailing current standard of ED care. Therefore, the control protocol was less likely to have acted as an intervention and thereby minimize the differences in results for the experimental and control groups. Also, in contrast to the other studies, this investigation included patients who had sustained injuries significant enough to warrant admission to the hospital, which in itself may lead to a so-called teachable moment and may contribute to the patients' motivation to change their drinking behavior. Limitations of the study included a relatively high refusal rate (34% of eligible patients did not participate) and a relatively low follow-up rate of 54%. It is important to note that these results may not be generalizable to the ED setting, as hospitalized trauma patients differ in many ways from patients who are treated and released from the ED. In addition, the practice setting of an in-patient care unit differs considerably from that of an ED.

Sommers et al. (2006) studied patients who were admitted to the hospital after an alcohol-related motor vehicle crash. Patients (n=187) were randomized into two types of brief intervention, brief advice or brief counseling. At 12 months, 100 were followed up. Both groups significantly reduced their drinking and had decreases in traffic citations, however there was no difference by condition.

Studies of Brief Intervention Outside the U.S.

To date there are a few randomized controlled studies completed outside of the United States. Crawford and colleagues (2004) from London, England screened patients for misuse of alcohol as defined by the Paddington Alcohol Test (see Touquet and Brown, Chapter 12 in this volume). Men who drank more than 8 units of alcohol in any one session at least once a week and women who drank more than 6, in addition to any patient who believed their ED visit was related to alcohol, were judged to be misusing alcohol and eligible for inclusion in a single blinded pragmatic randomized controlled trial. A total of 599 patients were assigned to either a control group, receiving an informational leaflet, or an intervention group that received the leaflet plus an appointment with an alcohol health worker. Interestingly, at 6 months there was a significant decrease in alcohol consumption in the intervention group to a mean of 59.7 units of alcohol consumption per week compared to 83.1 in the control group (p=0.02); however this was not related to whether or not they attended an appointment with the alcohol health worker. At 12 months there was no significant difference in alcohol consumption between the two groups. In addition, those patients who were assigned to the intervention and referral group had 0.5 fewer visits to the ED over the following 12 months (1.2 vs. 1.7, p=0.046). It is unclear exactly why the referral group had a decrease in alcohol consumption even if they did not attend the brief intervention session, but it may be hypothesized that this group received more explicit messaging during the ED visit than those that received just a leaflet. This has implications that even minimal messages may be of value.

Tait and colleagues (2005) from Australia published a randomized controlled trial that studied the results of an ED-based intervention for adolescents aged 12-19 who presented to an ED with a history of alcohol and/or other drug use. Of 127 adolescents who were recruited, 60 were randomized to the intervention and 67 received usual care. At 12 months 87(69%) were re-interviewed. Significantly more of the intervention group than the usual care group (12 vs. 4) had attended a treatment agency. In addition, the intervention group showed a greater reduction in the proportion of alcohol and other drug ED presentations over time.

Rodriguez-Martos Dauer and colleagues (2005) from Spain compared the effectiveness of a brief motivational intervention with a minimal intervention for reducing alcohol consumption in adult, alcohol-positive traffic casualties presenting to an emergency room. After 1 year 67% of patients had reduced their consumption, and the percentage of heavy drinkers had dropped by 47%. Binge drinking also dropped significantly, however there were no differences between conditions.

Finally, Neumann and colleagues from Germany (2006) conducted a randomized controlled trial of sub-critically injured patients in an ED who screened positive for an alcohol use disorder on the AUDIT using a laptop computer. The intervention group received a computer generated customized printout based on the patient's own alcohol use pattern, level of motivation and personal factors which was provided in the form of feedback and advice. At 6 months the intervention group had a 35.7% decrease in alcohol intake as compared to a 20.5% decrease in controls (p<.006). At 12 months there was still a significant decrease in the intervention vs. the control group, 22.8% vs. 10.9% (p<.023).

Limitations of Published Studies

Several methodological issues may have influenced the results and limited the generalizability of the studies described. First, all had high refusal rates – that is, as many as 47% of patients in the studies refused to participate. Refusal rates of this magnitude can introduce significant bias (e.g., only patients who have less severe problems or are motivated to change their drinking behavior may agree to participate). Refusal rates in studies involving adolescents may be particularly high because parents may need to give consent for their children to participate, and adolescents may not want their parents (or others) to know about their alcohol consumption for fear of getting in trouble due to underage drinking.

Second, the standard care conditions in the studies did not truly represent the standard of care. Other studies have demonstrated that emergency practitioners rarely screen their patients for alcohol problems or provide any intervention (D'Onofrio et al, 2005b). However, in several studies the patients in the SC groups received at least brief advice and a pamphlet – *minimal interventions that go beyond the standard of care commonly seen in ED settings.*

Third, the screening and assessment of the participants may have acted as an intervention, as indicated by the fact that participants in all study groups decreased their alcohol use. Assessment questionnaires that take 30-45 minutes to complete may have an impact similar to that of brief interventions of similar duration. Furthermore, questionnaires such as the Adolescent Injury Checklist, Adolescent Drinking Inventory, and the Drinking Inventory of Consequences emphasize the consequences of alcohol misuse and, in themselves, may provide feedback that motivates people to think about their behavior; such feedback is one of the key components of the intervention being tested.

Fourth, to detect statistically significant differences between control and intervention groups, adequate numbers of patients must participate and be randomly assigned to the different conditions. The studies discussed did not report power calculation to determine if their sample size was sufficient to detect differences. This makes interpretation of the studies' findings difficult, particularly because none of the studies found a significant difference between the intervention and control groups with respect to alcohol consumption. Without power calculations, it is difficult to determine whether this lack of difference is genuine, or merely a result of inadequate sample sizes. Although power calculations are recommended and useful, they may be challenging because some of the negative consequences of alcohol consumption may be low at baseline, e.g., drinking and driving or the number of alcohol-related injuries.

Internal and External Validity

The fidelity of the interventions is an important concern. One might assume that interventions delivered by the same practitioner to all patients would be consistent in content and form, which was likely to have been the case in the study by Gentilello, et al (1999). Other interventions might follow a specific framework, but reports of study findings generally do not describe how adherence to the intervention protocol was monitored. This poses problems for the interpretation of both internal and external validity, as well as the acceptance of the interventions in the practice setting, especially since a range of intervention strategies was used.

Selection bias is one of the potential threats to internal validity in these studies. This can be minimized in several ways. While few studies can allow enrollment 24 hours per day, 7 days per week, it is possible to schedule enrollment times so that there is maximum opportunity to screen and enroll a heterogeneous group of patients. In addition, researchers can track patients who are not screened or enrolled in order to ensure that they are similar demographically to enrolled patients. Limiting screening to a pre-selected group of patients based on diagnosis or demographics limits the generalizability of results.

It is important that patients who did not complete a particular study protocol be included in the analysis, using an intent-to-treat model. Post-hoc elimination of patients threatens validity, as does modification of study procedures and enrollment processes during the course of the study. Changing assessment instruments or interventions contributes to bias, and should be avoided. These types of issues can be minimized by piloting procedures prior to beginning enrollment in the study. Loss to follow-up can also be reduced through the use of surrogate contact information for patients.

High refusal rates can affect generalizability of the study. Self-selection bias, which is possible whenever the group of patients being studied has any form of control over whether to participate, creates problems for interpretation of results. Patients' decisions to participate may be correlated with traits that affect the study, making the sample non-representative. For example, people with strong opinions or substantial knowledge may be more willing to spend time answering a survey than those who do not.

Without large multi-site trials, external validity is threatened, as there is no opportunity to test an intervention across various settings. Since it is not clear that the emergency department-based studies used the same experimental procedures, the question of external validity is raised.

Applicability of Study Findings in Routine ED Practice

All studies have suggested that providing some form of brief intervention to selected ED patients may decrease their alcohol consumption and alcohol-related negative consequences. The specific message that should be delivered to patients, however, is not so clear, because the standard care groups – that received some brief advice, information or assessment containing motivational statements – also experienced positive outcomes. Also, because these studies used research staff for the interventions (i.e., social workers, graduate students, or doctoral-level psychologists), not typically available in most EDs, it is unclear how the findings can be translated into the real-world ED setting. Other research on the feasibility of screening and brief interventions in the ED setting can shed additional light on this question.

Degutis (1998) demonstrated that screening adult ED patients with tools such as quantity/frequency questions and the four-item CAGE questionnaire was feasible in a real-world ED setting. Screenings were performed by ED staff and 2,349 patients were screened. An additional aspect of the screening was the use of saliva testing for estimating blood alcohol concentration (BAC). In this study, the refusal rate was less than 2%. Of the patients who were screened, 6% had BACs of > 20 mg/Dl. Similarly, Hungerford and colleagues (2003), using a study population of young adults ages 18 to 39, reported that screening and intervention could be integrated into the ED setting. In this study, research staff screened a convenience sample of ED patients who were waiting for treatment. The investigators found that 87% of the young adult drinkers consented to screening. Of these, 43% screened positive for alcohol problems on the AUDIT, and of those with positive screens, 94% received counseling. The high prevalence of alcohol problems and the broad acceptance of screening and brief intervention in this sample indicated that even though the study used research staff not present in the real-world ED setting, screening is feasible in the ED, and this setting is a promising venue for screening and brief intervention. (See Table 1 for screening instruments which are commonly used to identify those with alcohol use disorders; see also Chapter 12.4 in this volume).

Table 1. Commonly Used Screening Instruments to Identify Alcohol Use Disorders

CAGE: A positive response to one or two items is considered a positive screening result.

Have you ever felt you ought to **c**ut down on your drinking?

Have people **a**nnoyed you by criticizing your drinking?

Have you ever felt bad or **g**uilty about your drinking?

Have you ever had a drink first thing in the morning to steady your nerves or to get rid of a hangover? (**E**yeopener)?

Developed to identify alcohol dependence on a lifetime basis, but may be asked for the last 12 months to identify current dependence.
Ewing JA (1984). Detecting alcoholism: The CAGE questionnaire. *JAMA* **252**, 1905–1907.

Rapid Alcohol Problems Screen (RAPS4): A positive response to any item is considered a positive screening result.

During the last year have you had a feeling of guilt or remorse after drinking? (**R**emorse)

During the last year has a friend or family member ever told you about things you said or did while you were drinking that you could not remember? (**A**mnesia/blackout)

During the last year have you failed to do what was normally expected of you because of drinking? (**P**erform)

Do you sometimes take a drink in the morning when you first get up? (**S**tarter/eyeopener)

Developed to detect current alcohol use disorder in the emergency department.
Cherpitel CJ (1995). Screening for alcohol problems in the emergency room: a rapid alcohol problems screen. *Drug Alcohol Depend*, **40**, 133-137.
Cherpitel CJ (2000). A brief screening instrument for problem drinking in the emergency room: the RAPS4. *J. Stud. Alcohol*, **61**, 447-449.

The Alcohol Use Disorders Identification Test (AUDIT): A weighted score of 8 or more for men up to age 60 or 4 or more for women, adolescents, and men over the age of 60 is considered a positive screening result.

How often do you have a drink containing alcohol?

How many drinks containing alcohol do you have on a typical day when you are drinking?

How often do you have six or more drinks on one occasion?

How often during the last year have you found that you were not able to stop drinking once you had started?

How often during the last year have you failed to do what was normally expected of you because of drinking?

How often during the last year have you needed a first drink in the morning to get yourself going after a heavy drinking session?

How often during the last year have you had a feeling of guilt or remorse after drinking?

How often during the last year have you been unable to remember what happened the night before because of your drinking?

Have you or someone else been injured because of your drinking?

Has a relative, friend, doctor, or other health care worker been concerned about your drinking or suggested you cut down?

Developed to identify current problem drinking in primary care settings.

Scoring

Question 1: Never (0), Monthly or less (1), 2-4 times a month (2), 2-3 times a week (3), 4 or more times a week (4).

Question 2: 1 or 2 (0), 3 or 4 (1), 5 or 6 (2), 7 to 9 (3), 10 or more (4)

Questions 3 – 8: Never (0), less than monthly (1), monthly (2), weekly (3), daily or almost daily (4)

Questions 9 &10: 2 points for yes, but not in the last year; 4 points for yes, during the last year.

Saunders JB et al. (1993). Development of the Alcohol Use Disorders Identification Test (AUDIT): WHO Collaborative Project on Early Detection of Persons with Harmful Alcohol Consumption–II. *Addiction* **88**, 791–804.

Babor TF et al. (2001). AUDIT: The Alcohol Use Disorders Identification Test: Guidelines for Use in Primary Health Care. Geneva, Switzerland: World Health Organization (Available at: http://www.who.int/substance_abuse/publications/alcohol/en/index.html. Accessed June 27, 2005).

TWEAK: A weighted score of 2 or more indicates an at-risk drinker.

How many drinks can you hold? (**T**olerance)

Have close friends or relatives **w**orried or complained about your drinking in the past year?

Do you sometimes take a drink in the morning when you first get up? (**E**yeopener)

Has a friend or family member ever told you about things you said or did while you were drinking that you could not remember? (**A**mnesia)

Do you sometimes feel the need to **c**ut down on your drinking?

Developed to identify at-risk drinking among pre-natal patients.

Scoring:

Tolerance: scores 2 points if the patient can hold more than 5 drinks without falling asleep or passing out.

Worry: scores 2 points

All other questions: score 1 point each.

Russell M et al. (1994). Screening for Pregnancy Risk-Drinking. *Alcohol Clin Exp Res* **18**, 1156-1161.

A survey of emergency practitioners (O'Rourke et al. 2006) found that these clinicians considered performing a brief intervention for harmful and hazardous drinkers feasible and acceptable in their everyday practice. Other investigators demonstrated that emergency medicine residents who received training in screening and brief intervention in a skills-based workshop increased their knowledge and practice of these procedures (D'Onofrio et al. 2002). Fifty-eight percent of medical records of patients treated by trained residents contained evidence of screening and intervention, compared with 17% of records of patients treated by a control group of similar residents who did not receive training.

Translation into Practice

Moving from an experimental approach to implementation in a real-world ED setting creates many challenges. Challenges in implementation result not only from translating experimental methods into practice, but also from existing difficulties in implementing a new practice in any setting. Additional practice barriers related to implementation of interventions for alcohol use exist and include biases, discomfort with performing the intervention, insufficient knowledge about the intervention, inadequate reimbursement, lack of incentive, concern about patient reaction, an unsupportive environment and time pressures (Degutis, 1998; Miller, et al, 2006; Hoyt 2005).

There are several considerations that need to be addressed in moving brief interventions from the experimental setting to practice in the emergency department. One of the primary considerations is who has the responsibility for performing the intervention. This has implications for practitioner education, as well as practitioner performance and quality improvement. As other studies have demonstrated, physicians and nurses have received little basic education in the area of intervention for alcohol problems, and have very little education with respect to brief interventions or motivational interviewing. Because of this, the barrier of lack of education among practitioners in intervention techniques must be overcome, or the responsibility for the intervention needs to be assigned to practitioners who have education and experience in delivering the BI. Some practitioners have suggested that social workers perform the interventions, but in a busy ED, it is unlikely that there will be sufficient social work staff to perform the number of interventions that are required. Other EDs have addressed this through the implementation of programs such as Project ASSERT (Bernstein, et al, 1997; see also Chapter 12 in this volume), which integrates ancillary personnel into the ED setting in order to provide interventions for alcohol and other drug problems.

Summary and Recommendations

Studies of brief interventions for alcohol use in the emergency department setting have used various methods of delivering interventions. These have contributed to both the measured effectiveness of the interventions and the generalizability of outcomes, as well as the translation of research findings to practice. Methodological challenges including subject selection, fidelity of interventions, and statistical power limit the generalizability of results, however. This chapter has discussed the issues related to the methods used to deliver brief interventions in ED studies, the effect on the interpretation of results, and implications for translation of the research findings into practice. In addition, it is clear that further studies of brief interventions in the ED setting will require careful consideration of the limitations identified in previous work, so as to truly test the effectiveness of the components of the intervention.

Finally, if ED physician and nursing staff are to provide these interventions, they need to have education, as well as resources on which to rely, for referral of patients when necessary. There are numerous strategies available for providing education, including in-person as well as computer-based approaches. Standardized curricula are available and readily accessible, but have not yet been translated into multiple languages. The provision of education to practitioners, and implementation of interventions in the clinical setting hold promise for decreasing alcohol-related problems in emergency department patients.

References

Bernstein E, Bernstein J, Levenson S. Project ASSERT: an ED-based intervention to increase access to primary care, preventive services, and the substance abuse treatment system. (1997) *Annals of Emergency Medicine* **30**:181-9.

Chafetz ME et al. (1962) Establishing treatment relations with alcoholics. *Journal of Nervous and Mental Disorders* **134**, 395-409

Crawford MJ et al. (2004). Screening and referral for brief intervention of alcohol-misusing patients in an emergency department: a pragmatic randomised controlled trial. [see comment]. *Lancet* **364**, 1334-9.

Dauer et al. (2006). Brief intervention in alcohol-positive traffic casualties: is it worth the effort? *Alcohol & Alcoholism* **41**, 76-83.

Degutis LC (1998). Screening for alcohol problems in emergency department patients with minor injury: results and recommendations for practice and policy. *Contemporary Drug Problems* **25**, 463-476.

D'Onofrio G, Degutis LC (2002). Preventive care in the emergency department: screening and brief intervention for alcohol problems in the emergency department: a systematic review. *Academic Emergency Medicine* **9**, 627-38.

D'Onofrio G et al. (2002). Improving emergency medicine residents' approach to patients with alcohol problems: a controlled educational trial. *Annals of Emergency Medicine.* **40**,50-62.

D'Onofrio G et al. (2005a). Emergency Practitioner Performed Brief Intervention for Harmful and Hazardous Drinkers in the Emergency Department. *Alcoholism: Clinical and Experimental Research* **29**(5) (Supplement): S37: 179A. (abstract)

D'Onofrio G et al. (2005b). Development and implementation of an emergency practitioner-performed brief intervention for hazardous and harmful drinkers in the emergency department. *Academic Emergency Medicine* **12**, 249-56.

Gentilello LM et al. (2005). Alcohol interventions for trauma patients treated in emergency departments and hospitals: a cost benefit analysis. *Annals of Surgery* **241**, 541-50.

Hoyt DB. Are we the problem? Overcoming obstacles to implementing intervention programs. *The Journal of Trauma, Injury, Infection and Critical Care.* **59**, S135-S136.

Longabaugh R et al. (2001). Evaluating the effects of a brief motivational intervention for injured drinkers in the emergency department. *Journal of Studies on Alcohol* **62**, 806-16.

Maio RF et al. (2005). A randomized controlled trial of an emergency department-based interactive computer program to prevent alcohol misuse among injured adolescents. [see comment]. *Annals of Emergency Medicine* **45**, 420-9.

Miller WR et al. (2006). Addressing substance abuse in health care settings. *Alcoholism: Clinical and Experimental Research* **30**, 292-302.

Monti PM et al. (1999). Brief intervention for harm reduction with alcohol-positive older adolescents in a hospital emergency department. *Journal of Consulting and Clinical Psychology* **67**, 989-994.

Neumann T et al. (2006). The effect of computerized tailored brief advice on at-risk drinking in subcritically injured trauma patients. *Journal of Trauma Injury, Infection, and Critical Care* **61**, 805-814.

Nordqvist C et al. (2005). Can screening and simple written advice reduce excessive alcohol consumption among emergency care patients? *Alcohol & Alcoholism* **40**, 401-8.

O'Rourke M et al. (2006). Alcohol-related problems: emergency physicians' current practice and attitudes. *Journal of Emergency Medicine*. **302**, 63-8.

Saitz R et al. (2006). Challenges applying alcohol brief intervention in diverse practice settings: populations, outcomes, and costs. *Alcoholism: Clinical & Experimental Research* **30**, 332-8.

Sommers MS et al. (2006). Effectiveness of brief interventions after alcohol-related vehicular injury: a randomized controlled trial. *Journal of Trauma Injury, Infection and Critical Care* **61**, 523-533.

Spirito A et al. (2004). A randomized clinical trial of a brief motivational intervention for alcohol-positive adolescents treated in an emergency department. *Journal of Pediatrics* **145**, 396–402.

Tait RJ et al. (2005). Emergency department-based intervention with adolescent substance users: 12-month outcomes. *Drug and Alcohol Dependence* **79**, 359-363.

U.S. Preventive Services Task Force (2004). Screening and Behavioral Counseling Interventions in Primary Care To Reduce Alcohol Misuse: Recommendation Statement. *Ann Intern Med*. **140**, 554-556.

Whitlock EP et al. (2004). Behavioral Counseling Interventions in Primary Care To Reduce Risky/Harmful Alcohol Use by Adults: A Summary of the Evidence for the U.S. Preventive Services Task Force. *Ann Intern Med*. **140**, 557-568

CHAPTER 12 :
IMPLEMENTING BRIEF INTERVENTIONS:
A SERIES OF FIVE PAPERS

12.1 – Evolution of an Emergency Department-Based Collaborative Intervention for Excessive and Dependent Drinking:
from one Institution to Nationwide Dissemination, 1991-2006

Edward Bernstein | Boston University School of Medicine
Judith A. Bernstein | Boston University School of Public Health

Mixed Blessings: Opportunity and Challenge in the Nation's Emergency Departments

ED patients are 1.5–3.0 times more likely to report excessive drinking or alcohol-related consequences than primary care patients (Cherpitel, 1999). More than 10% met formal criteria for alcohol dependence (Lowenstein, 1998), and 25% screened positive on the Alcohol Use Disorders Identification Test (AUDIT) in a study that used a probability sample (Cherpitel, 1995). In a five year follow-up study, alcohol-intoxicated ED patients had twice the mortality rate as a non-intoxicated comparison group (Davidson, 1997). In another study, providers identified less than one third of the 31% who were dependent drinkers as having a current problem, and less than one-fourth of those identified received treatment referrals, despite the location of an assessment and placement facility for specialized treatment adjacent to the ED (Bernstein et al., 1996).

Emergency physicians have ample opportunity to encounter ED patients at risk for injury and other alcohol-related consequences. However, both preventive and chronic care are typically considered outside the mission of emergency medicine despite the fact that alcohol dependence or excessive drinking is at the root of many presenting problems. Injured patients are often stitched up and discharged without addressing either excessive or dependent alcohol use, and the excessive drinker who lacks the stigmata of 'alcohol on breath' is very likely to leave the ED undetected.

In this chapter we share the evolution of a strategy to incorporate public health principles into ED services to improve quality of care for patients with high risk and/or dependent drinking.

Project ASSERT: From Interesting Idea to Demonstration Project

Project ASSERT was established to improve **A**lcohol, **S**ubstance Abuse **S**ervices, **E**ducation and **R**eferral to **T**reatment. Since 1994 it has served more than 50 000 patients at the Boston Medical Center ED, where Health Promotion Advocates (HPAs) screen for substance abuse and offer brief intervention and access to primary care, preventive services, and substance abuse treatment. Project ASSERT was derived from evidence supporting the role of community health workers as case finders, culture brokers, educators and access facilitators in underserved areas

(Swider 2002, Brownstein et al., 2006), and motivational interviewers in a strategy for behavior change (Miller & Rollnick, 1991). Fifty years ago a landmark study at Massachusetts General Hospital provided inspiration for change. In a controlled trial, Dr Chafetz enrolled 200 middle-aged, homeless, dependent drinkers to test a non-confrontational brief intervention delivered by trained residents and social workers. As a result, 42% of the intervention group compared to 1% of the controls kept five alcohol treatment appointments (Chafetz, 1962).

If the intervention worked so well with alcoholics from Boston's notorious Scollay Square, why not give it a try in a comparable ED? Project ASSERT was established in 1994 at Boston City Hospital with a demonstration grant from the US Center for Substance Abuse Treatment (CSAT) (Bernstein et al., 1997).

Phase One: the CSAT grant years, 1994-1997 – 25, 541 ED patients served

Working with residents and social workers, Dr Chafetz improved the care of patients presenting with alcohol- and drug-related illnesses and injuries. Why, then, did Project ASSERT employ HPAs (Health Promotion Advocates) – community outreach workers hired to do "in reach" in the ED under the direction of clinical staff? The HPA role was established in recognition of time constraints, overcrowding, and resource limitations in the modern ED; the need for a stable core of dedicated, experienced personnel with substance abuse treatment and community resource contacts; and protected time to educate and motivate patients to make healthier choices.

– Case finding strategy

HPAs performed universal screening, brief intervention and referral to treatment (SBIRT) at the patient bedside 16 hours daily and interviewed patients about access to primary care, preventive clinical screening, seat belt use, smoking, substance abuse, and experience of violence and depression. Alcohol and drug screening questions were embedded in the screening interview and included 1) last-year illicit drug use, 2) consumption of alcohol in the last 24 hours and admission of a drinking problem, 3) episodes of binge drinking, and 4) report of alcohol- or drug-associated injury within the last year. This health promotion approach, education and referral to medical services, was readily accepted by patients who otherwise might have felt stigmatized by an exclusive emphasis on substance use.

– ED-based brief intervention strategy

When patients screened positive for excessive and dependent drinking or illicit drugs, the HPAs utilized a motivational interviewing strategy, *Brief Negotiation Interview* or BNI developed with Dr Stephen Rollnick. The BNI is a behavior change strategy to establish trust, reduce resistance, and promote choice, and contains the following elements: 1)asking patient permission to discuss alcohol or drug use, 2) exploring pros and cons of substance use, 3) promoting reflection about discrepancies between current life circumstances and future goals, 4) assessing readiness to change, 5) identifying patient strengths and prior successes, 6) providing a menu of options, and 7) negotiating a specific action plan.

Among 7,118 patients screened by Project ASSERT during 12 months in 1995–1996, we found 41% positive for excessive or dependent drinking or drug use (31% female, 61% Black, 11% Hispanic), 61% without a regular doctor, 80% smokers, 8% reporting alcohol- or drug-related injury, and 24% mildly/moderately depressed. Among enrollees, 10% accepted referral to detox, 41% to outpatient, acupuncture, or central intake for placement into specialty treatment, 34% to Alcoholics Anonymous (AA) or Narcotics Anonymous (NA), and 47% to primary care. At 3-month follow-up, half had kept referral appointments, and more than half reported reduced drinking, drug use and related consequences (Bernstein et al., 1997).

When Boston City Hospital and Boston University Medical Center Hospital merged in 1996, the future of ASSERT was uncertain. With supporting data, petitions, and testimony from ED nurses, physicians, and patients, Project ASSERT won the Mayor of Boston's Customer Service Award. The Boston City Council passed a resolution recommending funding, and shortly thereafter, Project ASSERT became a line item in the ED budget.

Phase Two: the evolution of the Project ASSERT model from 1997 to the present

The hospital change from city to private funding, and the ASSERT funding change from a grant to institutional, created new challenges. Central Intake, which provided outpatient addiction services, was relocated in a hard-to-find spot several blocks from the ED, with reduced staff, hours, and accessibility. These changes forced Project ASSERT to implement an aggressive, time-consuming referral process requiring hourly calls to a long list of facilities to locate beds and negotiate on behalf of patients with private or managed care insurance for prior approval. In addition, HPAs had to track down medical staff to complete medical and psychiatric clearance examinations when required. ED physicians documented Project ASSERT consultations in the medical record and the facility billed for these services. When beds were unavailable, HPAs provided low-impact case management until placement in treatment was found.

From 1999-2005, Project ASSERT provided services to 27,101 patients: 32% female; 46% Black; 16% Hispanic; 27% without primary care, 28% unable to afford medications, 61% smoked, 24% always used a seat belt, 31% with an alcohol- or drug-related injury, and 20% mildly or moderately depressed. Among the 15,786 who drank in excess of National Institute of Alcohol Abuse and Alcoholism guidelines (NIAAA, 2005) or used drugs within 30 days, 44% were placed in detox, 9% in outpatient programs, and 42% referred to AA or NA. An average of 3 out of 4 patients requesting inpatient detox programs were placed daily. The patients for whom no bed was available were encouraged to return until a bed opened up. Moreover, 11, 315 patients who screened negative for substance abuse were referred for primary care, and 42% received an array of other mental health and preventive referrals.

Compared to Phase One, more referrals in Phase Two were made directly to detox and primary care and fewer to central intake and outpatient addiction programs. Overall, Project ASSERT allocated greater resources for crisis care for dependent drinking than for excessive drinking. The majority of Phase Two patients were referred to Project ASSERT directly by ED staff. There was simply only limited time for screening room-to-room for excessive drinking.

– HPAs are necessary but not sufficient.

The Project ASSERT model requires active participation of clinical staff on a number of levels. ED providers have gotten better at detecting and referring patients to Project ASSERT when the visit is obviously alcohol-related. However, unless providers utilize validated screening questions instead of relying on 'smell of alcohol on breath' or profiling obvious alcohol-related visits, they will continue to miss hidden dependent drinking, and they will only rarely detect excessive drinking. Recognizing that our original intent to broaden the base of treatment for alcohol problems by providing real-time universal screening and intervention across the spectrum of excessive to dependent drinking had not yet been realized, we increased our efforts to enlist ED providers in alcohol screening and motivational intervention.

– The 14-site ED Study: A vehicle for developing a truly collaborative SBIRT model

As a result of participation in National Alcohol Screening Day, the National Institute of Health's National Institute of Alcohol Abuse and Alcoholism (NIAAA) awarded a series of grants* to develop a standardized SBIRT curriculum for ED providers (including physicians, registered nurses, nurse practitioners, emergency medical technicians, and physician assistants), and to conduct a controlled trial of SBIRT with ED patients. The curriculum consisted of 1) a brief slide show describing evidence for the efficacy of SBIRT, 2) videos in which ED providers demonstrate intervention skills with simulated patients, 3) scripted scenarios for practicing skills, 4) pocket-sized plastic cards with NIAAA screening guidelines, a graphic display of typical drinks, and the intervention algorithm, and 5) a web site designed for independent learning (www.ed.bmc.org/sbirt). Prior to training, and at 3 and 12 months post-training, a survey instrument developed by D'Onofrio and colleagues. (D'Onofrio et al. 2001) was administered to 402 ED clinicians (21% nurses, 60% physicians, 7 % nurse practitioners or physician assistants, and 12% social workers), with 74% reporting <10 hours prior professional alcohol-related education ever. At the 3-month follow-up interview, scores significantly improved over baseline for self-reported confidence in ability and responsibility to intervene and actual utilization of SBIRT skills. However, scores decreased at the 12 month interview. Practitioners appear to need a more supportive infrastructure to maintain SBIRT-related skills and attitudes. Significant barriers were time limitations, lack of referral resources and reimbursement. We hypothesize that resources such as computerized screening, addition of ancillary support personnel like the HPAs, and a booster training session might increase SBIRT utilization. A core group of SBIRT champions at each site were critical to successful SBIRT implementation.

– Going Live with ED-based SBIRT

A pre- and post-training comparison tested whether the trained providers across the 14 sites could utilize SBIRT skills to effect drinking behavior change among their patients (Academic ED SBIRT Research Collaborative, 2007). From April–August 2004, about 26% of the 7,746 ED

---

* The Academic Emergency Medicine Alcohol SBIRT Research Collaborative: Boston University, Brown University, Charles R. Drew University, Denver Health Medical Center, Emory University, Howard University, Tufts University, University of California, San Diego, University of Medicine and Dentistry of New Jersey, University of Michigan, University of New Mexico, University of Southern California, University of Virginia, and Yale University.

patients screened were drinking above the NIAAA guidelines. Among screen-positive patients, 1104 enrolled – 566 were recruited during the six-week control period prior to training, and 538 enrolled during the intervention period immediately following training. Both control and intervention groups received a brochure listing local treatment resources. A trained clinical provider delivered the brief ED-based BNI to the intervention group. At the 3-month follow-up interview, 63% of patients (n = 699) called in to the Interactive Voice Recognition system at the University of Connecticut. The dependent drinkers in the intervention and control groups did not differ significantly on outcome variables at follow up. However, among excessive drinkers, the intervention group reported an average of 3.25 fewer drinks per week than controls and 0.72 fewer drinks per heavy drinking occasion, and compared to 18% of the control group, 28% of the intervention group was no longer drinking in excess of NIAAA guidelines.

Phase Three: a paradigm shift is needed

In the 1960s there was a paradigm shift in primary care as physicians began to accept responsibility to help patients adopt a healthy life style, i.e., changes in diet, smoking and exercise to reduce cholesterol levels and cardiovascular risk. It is now standard of care for physicians to address these issues. Emergency Medicine is in the early stages of a similar shift regarding unhealthy drinking and drug use. The American College of Emergency Physicians (ACEP) and the Emergency Nurses Association (ENA) have strong policy statements supporting SBIRT. Recently the American College of Surgeons made it an essential requirement that level I trauma centers screen and have the capability to provide an intervention for patients identified as problem drinkers (Committee on Trauma 2006). There are now a number of tool kits, web sites and strategies to assist ED providers to acquire SBIRT skills (www.ed.bmc.org/sbirt)

Documented success with our peer educator protocol (Bernstein et al., 2005) has sparked two large-scale efforts to disseminate the Project ASSERT collaborative model. In 2005, New York City's Department of Health and Mental Hygiene provided funding to train clinical staff at five EDs, support infrastructure development, and hire HPAs. The Massachusetts Department of Public Health has recently funded a five year project in collaboration with Massachusetts ACEP and ENA to implement Project ASSERT at six EDs throughout the state. These two initiatives include systems changes to increase collaboration between emergency care providers, behavioral health and substance abuse specialists, and primary care practitioners, and to build long term sustainability.

From more than 12 years of Project ASSERT practice, and training more than a thousand ED providers, we have learned that a collaborative model works best when HPAs are fully integrated into the ED staff and consistently available at critical hours when patients with problems are likely to present. We encourage ED clinicians, when time is available, to address unhealthy alcohol and drug use, but changes in clinical practice cannot be sustained without 1) institutional support, 2) changes in reimbursement structures to allow the ED to bill for addiction services or increased intensity of the visit, 3) parity with medical services supporting physical health for reimbursement for mental health and substance abuse services, and 4) access to quality mental health and substance abuse treatment.

References

Academic ED SBIRT Research Collaborative (2007). An evidence-based alcohol screening, brief intervention and referral to treatment (SBIRT) curriculum for emergency department (ED) providers improves skills and utilization. *Subst Abuse* **28**(4), 79-92.

Bernstein E et al. (1996). Emergency Department detection and referral rates for patients with drinking problems. *Subst Abuse* **17**, 69-76.

Bernstein E, Bernstein J, Levenson S (1997). Project ASSERT: An ED-based intervention to increase access to primary care, preventive services and the substance abuse treatment system. *Ann Emerg Med* **30**, 181-189.

Bernstein J, et al. (2005). Brief motivational intervention at a clinic reduces cocaine and heroin use. *Drug Alc Depend* **77**, 49-59

Chafetz ME, et al. (1962). Establishing treatment relations with alcoholics. *J Nerv Ment Dis,***134**, 395-409.

Cherpitel CJ. (1999). Drinking patterns and problems: a comparison of primary care with the emergency room. *Subst Abuse* **20**, 85-95.

Cherpitel CJ. (1995). Screening for alcohol problems in the emergency department. *Ann Emerg Med* **26**, 158-166

Committee on Trauma, American College of Surgeons. (2006). *Resources for Optimal Care of the Injured Patient.* Chicago, Illinois: American College of Surgeons.

Davidson P, et al. (1997). Intoxicated ED patients: A five year follow-up of morbidity and mortality. *Ann Emerg Med* **30**, 593-597.

D'Onofrio G, et al. (2002). Improving emergency medicine residents' approach to patients with alcohol problems: A controlled educational trial. *Ann Emerg Med* **40**, 50-62.

Lowenstein SR, et al. (1998). Behavioral risk factors in Emergency Department Patients: A multi-site survey. *Acad Emerg Med* **5**, 781-787.

Miller WR, Rollnick S. (1990). *Motivational Interviewing: Preparing People to Change Addictive Behaviors.* New York: Guilford.

National Institute for Alcohol Abuse and Alcoholism (2005) Helping patients who drink too much: A clinician's guide. http://pubs.niaaa.nih.gov/publications/Practitioner/CliniciansGuide2005/clinicians_guide.htm, accessed 6 June 2006)

Swider SM. Outcome effectiveness of community health workers: An integrative literature review. (2002) *Public Health Nursing* **19**(1), 11-20.

12.2 – Implementing Brief Alcohol Intervention in the Emergency Department

Kerry B. Broderick - University of Colorado | Denver Colorado, CO US

Background

Denver Health Medical Center [DHMC] is a 275 bed hospital, including a Level I trauma center. It is a major academic hospital in Denver, Colorado with 60 residents in its emergency medicine residency program, and sees approximately 57,000 emergency patients per year. Approximately 20% of visits to the emergency department (ED) involve patients who have alcohol problems as a major or contributing illness. The DHMC ED provides medical control for the Denver Comprehensive Addiction Rehabilitation and Evaluation Services program, a city wide acute detoxification facility that has 22 female and 66 male beds, and also provides education for health workers on issues related to substance use. DHMC nurses, physicians and ancillary staff see the effects of alcohol continually in the ED, but traditionally were given no formal training in screening for substance abuse and alcohol problems or performing brief interventions.

I began as an emergency nurse in Chicago, IL where many patients had alcohol-related problems. However, these problems were essentially ignored. The medical staff sutured lacerations, splinted fractures, and hydrated kidneys, but paid little attention to their origin (alcohol dependence and hazardous drinking) while treating the symptoms. While an attending physician at the Erie County Medical Center (ECMC) in Buffalo, New York, I developed discharge instructions for patients with alcohol and substance problems. Despite the fact that ECMC had a very active alcohol service with alcohol counselors placed in the ED almost 24 hours a day, there were no discharge instructions for these patients. This intrigued me, and I wondered what policies other facilities had, or more importantly, did not have. So, in 1998 I completed a research project that surveyed the routine practices of emergency physicians (EPs) regarding the use of substance abuse discharge sheets. Two of the top reasons EP's did not discuss substance abuse with their patients were lack of time and relevant education. (Broderick, 2000)

Phase 1: Planning for Implementation

– Planting the seed

In 2000, the Robert Wood Johnson Foundation's Demand Treatment program awarded the city of Denver, Colorado $ 60 000 for an alcohol intervention and services program. Program partners included the Mayor's Office on Drug Control and Denver Health Medical Center. The program initiated a campaign to educate public and medical personnel regarding the deleterious effects of alcohol and to heighten awareness regarding alcohol problems. In October 2002, the Demand Treatment grant financed a trip to DHMC by Drs. Edward and Judith Bernstein to lead a grand rounds forum on ED-based screening, brief intervention, and referral to treatment (SBIRT) (see Chapter 12 in this volume). Invitations targeted staff from social work, emergency nursing, and emergency psychiatry in addition to emergency physicians. This presentation and forum planted the seed that more training was necessary. It was at this stage of the grant that I became involved. In spring of 2003, the Demand Treatment program partners decided to use a portion of grant funds toward educating health care workers at DHMC about SBIRT.

– Gaining Momentum

Although capable of delivering an educational workshop on SBIRT, concern that my co-workers might view this as a personal crusade and not assign proper value to it led me to contract with an out-of-agency educator with expertise in providing education about alcohol interventions to health care practitioners. Together we developed a curriculum specifically for emergency health care providers.

– Budget

With the funds allocated through the grant, we developed a budget for SBIRT training in various departments of the hospital, using the hourly wages of emergency medicine residents, attending physicians, nurses, social workers and psychiatric workers who would be attending the workshop. This included funds for the educators' time to develop the curriculum, three 3-hour workshops, and 8-hours for the educator to work directly with the staff to perform SBIRT with ED patients. We felt that hands-on teaching with actual patients would help increase staff utilization of SBIRT methods.

– Developing a training module

We developed a 3-hour training module patterned after the Bernstein/ D'Onofrio model, (D'Onofrio, 2002). It consisted of a pre-test and a one-hour lecture on alcohol and SBIRT. The lecture included a video tape that modeled the brief intervention (BI). We then split the class into smaller workshops that used patient scenarios from the Bernstein Module. Each workshop included a facilitator already trained in BI methods. Training ended with a wrap-up session.

– Buy in

Bringing to the table the budget and training module, I met with ED leaders – physicians, nursing, social work and psychiatry – to gather support for the effort to educate appropriate personnel about SBIRT. To me, asking for significant support from these leaders meant asking for manda-tory training, which would give heightened validity to the training among the staff. The project agreed to train all staff who attended and reimburse each hospital department for the time its staff spent for training, using the hourly wages of attending staff to prorate the available budget among participating departments. Because the project's budget could not reimburse the total cost of staff time, each department had to allocate some of its own financial resources for the training, which helped them to become, literally, invested in the process.

Phase 2: Training

To accomplish the mandatory training, we scheduled three separate trainings over a period of six weeks to accommodate staff vacation and scheduling conflicts. Training times were also staggered to accommodate shift workers, with one training session held during the routine emergency medicine residency conference time which is a 3-hour educational time slot. The initial budget for the SBIRT training, which was completed in September 2003, covered training for 60 nurses, 48 physicians, 6 social workers, and 15 psychiatry staff; actually undergoing training, however, were 30 nurses, 33 physicians, 27 social workers, and 38 psychiatry staff.

Rather than focusing primarily on dependent drinkers, training stressed that brief interventions could have a greater impact on at-risk drinkers – those patients who were just beginning to interface with the health care system because of substance use problems. Most of these patients do not need a referral to specialized treatment, but should receive a brief intervention in the ED. As SBIRT training and implementation progressed, a concern with the limited referral resources continually surfaced among medical students and health care providers (HCP). We continued to stress that the majority of patients are impacted by the very brief ED-based intervention and do not need specialized treatment; they just need their HCP to have a brief conversation with them.

The workshop educator spent eight hours in the ED assisting staff with providing SBIRT. This proved challenging as ED staff was resistant to providing SBIRT, especially while being observed. The educator was not adequately assertive in being able to get the ED staff into patients' rooms to implement SBIRT. In retrospect, it is clear that more clinical time should have been devoted to this component of the training. This was a very important learning event for us.

Phase 3: Implementation

– Sustainability

In an effort to encourage health care providers to continue providing SBIRT, we added two new items to our paper ED-patient medical record. The first item was a check box to note that an alcohol intervention had been completed, and the second was a check box to note that a referral had been made. These were included directly under the review of systems portion of the medical record that health care providers are responsible for completing. The utilization of these check boxes varies and constant reminders are necessary to get the health care providers to complete them. An electronic medical record may prove useful as these check boxes could then be made mandatory fields.

– Continuing the Momentum

In October of 2003, we applied for and were subsequently granted funding by the National Institutes of Health, National Institute on Alcohol Abuse and Alcoholism for research on screening and brief intervention at Denver Health Medical Center, one of 14 such academic emergency department grants nationwide. This grant came at a crucial time and helped sustain the momentum from the original training at DHMC. The new study was a clinical trial, and it attracted 42 health care providers to enroll for the required training and helped to cement understanding and use of SBIRT in the ED. Eighty patients were enrolled in the trial – 40 in the control group and 40 in the BI group. The study was completed in 2006.

Each year the emergency medicine residents receive a Grand Round's lecture on alcohol and SBIRT, and each month the medical and physician assistant students who rotate through our department receive the same presentation. Additionally, throughout the year, residents and students in the ED experience mini presentations on SBIRT while on patient rounds. As in most EDs, the culture in our ED and paramedic division was one in which the word 'drunk' was frequently used. Gathering support from some of our faculty, we encouraged use of 'intoxicated patient' and 'intoxication'. The rationale was that 'drunk' is a derogatory term and does not really convey the patient's medical condition. Many of our patients are intoxicated on a variety of substances and have numerous medical co-morbidities. We hoped this change in language would parallel a change in how these patients, as well as alcohol dependence and hazardous drinking, were viewed. This change has brought increased awareness among the ED staff and paramedics of the importance of alcohol-related problems and respect for patients with these problems. Although alcohol dependence and alcohol abuse are listed as diagnostic categories in the American Psychiatric Association's Diagnostic and Statistical Manual of Mental Disorders, patients who drink excessively but do not meet criteria for either diagnosis are more prevalent in EDs and should be identified and receive an intervention. Even though a brief intervention is generally not adequate for patients with alcohol dependence, it is adequate for the patient who is not dependent but drinking excessively. In educating HCPs, I use the term alcohol problems to include both types of patient – those who are dependent and those who are not but who are drinking excessively. The diagnostic label assigned to a patient by medical personnel is important because it influences whether and how they act when they encounter the condition.

– National Alcohol Screening Day

Each year Denver Health Medical Center participates in National Alcohol Screening Day® (NASD) (http://www.mentalhealthscreening.org/events/nasd/), an annual event that provides information about alcohol and health as well as free, anonymous screening for alcohol-use disorders. Event sites are located in community, college, primary health care, military and employment settings. The program is designed to provide outreach, screening and education about alcohol's effects on health for the general public. DHMC provides 24-hour screening and intervention in the ED, and screening during business hours in the adult walk-in and adult medicine clinics on the Denver Health Hospital campus. In 2006, 36 staff volunteered for NASD and screened 480 patients. This annual program helps to heighten awareness with patients and staff of the importance of identifying and helping patients with drinking problems.

Phase 4: Future Plans

In 2007 Colorado was awarded federal funds for SBIRT, through the state governor's office as a means to involve policy makers in implementing change at both the judicial and insurance levels.

This program will include DHMC emergency department and obstetric and gynecology staff and patients. The women, infant and children (WIC) clinics, as well as youth, through DHMC's school-based clinics will also be involved, and some rural sites will be used for comparisons.

We have developed a new educational web-based module, which includes the recently released video, "The Emergency Practitioner and the Unhealthy Drinker: Motivating Patients for Change" from the National Highway Traffic Safety Administration. This video is available through the American College of Emergency Physicians (www.acep.org). Staff will complete modular questions on the web, which facilitates offering the program to outside sites without a large outlay of time. This will allow us to track those who complete the module and make sure they pass the modular test, and will be followed-up with a short session to go over questions and model behavior.

In order for SBIRT to become truly sustainable, staff and the institution must be held accountable to provide it. I firmly believe that SBIRT will not become routine in medical settings until health insurers – both private and public – reimburse for this important clinical preventive service.

References

Broderick K (2000). Alcohol/Substance Abuse regarding Screening and Discharge Instruction Utilization in Emergency Physician Practices. *Annals of Emergency Medicine*. **36**(4), S71.

D'Onofrio G et al. (2002). Improving emergency medicine residents' approach to patients with alcohol problems: a controlled educational trial. *Annals of Emergency Medicine*. **40**(1), 50-62.

12.3 – Changes in Clinical Practice Regarding Alcohol and Motor Vehicle Crashes

Robert H. Woolard - Texas Tech University | El Paso, Texas

Michael J. Mello, Janette Baird, Ted Nirenberg - Rhode Island Hospital | Providence, RI USA

Introduction

In Rhode Island, the practice of emergency medicine (EM) is changing with regard to addressing alcohol in the emergency department (ED). Many ED staff have been trained in motivational intervention (MI) in programs offered on several occasions during 13 years of research and regular post-graduate seminars. Staff is encouraged to counsel patients during their treatment. However, only some physicians and nurses offer brief alcohol counseling to their patients and most do not routinely do so. Competing clinical demands prevent most ED staff from the routine practice of MI. In contrast, screening and referral has increased. Several initiatives have led to increased screening for alcohol use problems among ED patients. While research projects and alcohol education projects have been operational, the ED staff has shown both a willingness and ability to engage in brief screening and referral for treatment for ED patients with alcohol use problems. Many ED staff are routinely identifying patients' alcohol use through a templated record system that prompts a single question about alcohol use. Through one initiative patients with alcohol problems, particularly impaired drivers, are reported to a medical advisory board by completion of a brief one page form which leads to a referral for counseling. Some ED staff have taken on an injury prevention role by advising youths referred to observe trauma in the ED by the courts.

Through our research programs, which began in 1996 at the Rhode Island (RI) Hospital ED, many ED staff has become aware of the success of brief counseling. Currently, three programs influence and change our practice of emergency medicine in RI: Screening for Our Safety (SOS), Reducing Youthful Dangerous Driving (RYDD); and Decreasing Injuries with Alcohol (DIAL). Two of these programs (SOS and RYDD) have introduced interventions for alcohol and injury into our routine emergency practice. All three programs focus on motor vehicle crash (MVC) injuries and one, particularly, on young drivers.

Background

Injury is one of the major health consequences of alcohol consumption and frequently requires treatment in an ED. The Emergency Room Collaborative Alcohol Analysis Project (ERCAAP), a meta-analysis of 15 studies involving 7 countries, found that heavy drinking (above NIAAA guidelines) is highly predictive of an alcohol-related injury (Cherpitel et al, 2003). At RI Hospital, 21% of injured ED patients were found to have a blood alcohol concentration (BAC) of .10g/dl or greater on arrival at the ED (Becker et al, 1995). In those presenting to another university ED, the prevalence of alcohol abuse and alcohol dependence among MVC patients was 23%, with almost half having negative BACs (Maio et al, 1997). Alcohol-related problems cost the U.S. billions of dollars each year, including the cost of lost earnings and the cost of medical and alcohol treatment (D'Onofrio et al, 1998).

For decades, a strong connection between alcohol use and MVCs has been documented. In 2002 in the US, 258 000 injuries resulted from MVCs where alcohol was involved (US Department of Transportation, http://www-fars/nhtsa.dot.gov/). The Fatality Analysis Reporting System (FARS) has allowed the National Highway Traffic Safety Administration (NHTSA) to conclude that in 2005, there were 16 885 deaths in alcohol-related car crashes (US Department of Transportation, http://www-nrd.nhtsa.doc.gov). In Rhode Island, half of all fatal MVCs are alcohol-related (ranking near the highest state in the nation). In almost half (49%), the blood alcohol level was .08 g/dl or greater. Among adolescents involvement in fatal MVCs has increased significantly, with a 12% increase from 1992 to 2002. In the US, despite the 21-year-old minimum drinking age laws in most states, 24% of 15-20 year old drivers who died in a MVC were intoxicated (US Department of Transportation, http://www-fars/nhtsa.dot.gov/).

Confronting alcohol-related injuries daily, ED staff is often discouraged by the lack of effective programs to prevent future alcohol-related injuries. A MVC patient's ED visit presents the potential for a "teachable moment," that could be a life-saving opportunity not just for that patient, but also for other drivers and passengers. Using the principles of motivational interviewing, an injury can become a motivator, and create a "window of opportunity" to reduce drinking after an ED visit (Longabaugh et al, 1995) and after a MVC (Sommers et al, 2000; Mello et al., 2005). In the ED, MI can be adapted to explore the pros and cons of alcohol use and driving, to motivate the patient to change behavior, and to help the patient to reduce the risk of future alcohol-related injuries and negative consequences as demonstrated in the Rhode Island Early Intervention Study (Longabaugh et al, 2001).

Screening for Our Safety (SOS)

After a MVC, impaired drivers who are taken to the ED frequently escape prosecution (Biffl et al, 2004), and are less likely than those not taken to the hospital to receive court-mandated treatment for their alcohol problems. NHTSA, in a 2002 report, concluded that implementing ED protocols for screening and intervention would increase the likelihood of these patients receiving treatment for alcohol problems (US Department of Transportation, 2002). Changes in Rhode Island, due to SOS, have begun to close loopholes and bring more patients needing treatment into counseling.

Rhode Island, along with several other states, allows physicians to report specific medical problems that would impair driving skills to a group of physicians on the Medical Advisory Board (MAB) of the Department of Motor Vehicles (DMV), circumventing involvement with the judicial or law enforcement sectors. This alerts the MAB-DMV to the fact the individual has a condition that may make it unsafe to drive. In the past, seizures and syncopal events have led to reporting by physicians of their patients to the MAB-DMV. Currently, similar reporting of patients with alcohol problems results in a medical evaluation with possible restriction or revocation of driving privileges until the alcohol problem is corrected by completion of an alcohol treatment program. RI physicians are more comfortable with this type of reporting for alcohol-impaired drivers than reporting to law enforcement (Mello et al, 2003). The Screening for Our Safety project disseminated and promoted this mechanism for emergency physicians to report problem drinkers.

SOS created an online educational program for emergency medicine physicians concerning alcohol injury and the need to screen ED patients for alcohol use problems, and presented potential interventions for those who screen positive. More than half of all RI emergency medicine physicians participated in the educational intervention. After the SOS program, changes in physicians' practice were evident. Before SOS, 47% stated they discussed with their patients the role that alcohol may have played in their injury, while afterwards, 58% reported these discussions (Mello, et al, 2003). Before SOS, only 65% of participants knew that RI physicians could report alcohol-impaired drivers to the DMV; after SOS 100% knew this and 91% reported that they planned to increase their reporting of impaired drivers to the DMV (Mello et al, 2006). In the three months prior to the SOS, 17 reports of impaired driving were sent to the DMV, while in the three months following SOS, 42 reports were sent (Mello et al, 2006). Referrals were collaborated by an ED chart review from two hospitals (including the state trauma center). Review found that in the month before SOS 14% of positive screens were referred for intervention, but in the month after SOS 32% were referred (Mello et al, 2006).

The SOS educational curriculum was accessed by the emergency medicine physicians statewide. With SOS, we found that participating physicians screen and refer patients more often after training. Many emergency physicians are now making this a routine part of their practice. SOS also provides a mechanism for organizing physicians statewide to further address impaired driving, with 37% of the physicians willing to give personal contact information to advocate for improvement of this issue (Mello et al, 2006). In sum the SOS program has successfully educated more than half of the ED physicians in the state and engaged them in changing their practice.

The Reducing Youthful Dangerous Driving (RYDD) Program

In Rhode Island, some ED staff members are involved as preceptors in Reducing Youthful Dangerous Driving, a program of our Department of Emergency Medicine Injury Prevention Center to reduce impaired driving and other risky behaviors. RYDD offers counseling to 16–20 year olds referred from Rhode Island courts after their conviction for driving under the influence of alcohol (DUI) or other high-risk driving offenses. These high risk driving behaviors result in motor vehicle violations and are predictors of future MVC's. RYDD participants spend time observing in the RIH trauma center and ED, where they are exposed to the devastating negative consequences of high-risk driving behaviors. During these observational sessions the RYDD participants are accompanied by medical staff who explains the medical scenarios occurring in real time. Observing others suffering negative consequences allows RYDD participants to more fully understand the impact of risk behaviors and especially consequent alcohol-related trauma. When not directly involved in RYDD, ED staff is aware of the many youth observers who benefit from several hours of counseling and ED observation through the RYDD program.

The RYDD Program is divided into six sessions, four of which are interactive group MI sessions, and two are observation sessions in the ED/trauma center. In the group sessions, the RYDD participants discuss their actions leading up to court; the negative consequences including fines, license suspension and community service; their expectations of their trauma center experience including MVCs and injuries, their thoughts about safety; and the pros and cons of the high-risk behaviors that brought them to the court. Participants also explore their motivation for and against high-risk driving. All participants are encouraged to create a change plan addressing their high-risk driving behaviors.

An emergency physician prepares the group for their ED/trauma center visit by presenting slides of actual car crashes in RI. The RYDD participants observe for two four-hour sessions in the ED/trauma center on Friday or Saturday nights. The ED medical and nursing staff engages these participants during their ED/trauma center observation sessions, while they are exposed to a diversity of trauma including serious injuries related to MVCs, alcohol use and high-risk behaviors.

Efforts are underway to promote RYDD throughout the state courts. Exposure to RYDD has changed the view of many practitioners, motivating them to screen and refer patients to counseling, especially those involved in MVC with DUI. Data from this program show reduced high-risk behaviors, such as drinking and driving, and feedback from the participants suggests that both the personal qualities of the ED staff and the ED observational experience are important to motivate change. In RI, through programs like RYDD, ED staff knows first hand the positive outcomes that can be attained with effective counseling.

Decreasing Injuries with Alcohol (DIAL)

For more than ten years in three EDs affiliated with Brown Medical School, counseling concerning injury and alcohol has been offered to ED patients by research staff under research protocols. In the Rhode Island Early Intervention Study (REIS), patients with histories of hazardous drinking responded to brief motivational intervention (BMI) with reduced injuries at one-year follow-up. (Longabaugh et al, 2001). REIS patients with MVC injury were found to have the greatest positive treatment effects (Mello et al, 2005). While the ED visit is proximate to the injury, it is not necessarily the only opportunity to address alcohol's negative consequences. Counseling soon after the ED visit may be more practical. The ED is a stressful setting where the patient may be fatigued or under the influence of alcohol, and such distracting factors could limit the effectiveness of the intervention. During a follow-up intervention within a few days of the ED visit, the patient may be sober, more motivated and alert. Since the patient may still have emotional and physical reactions to the injury, the "teachable moment" should persist. In a more comfortable setting, the patient may be more open to counseling. Referral to follow-up counseling from the ED has been shown to be effective (Crawford et al, 2004).

In DIAL, we evaluated the effect of telephone counseling on changing behavior and reducing future injury. An intervention, consisting of two follow-up phone calls after a primary care physician visit, has been found to result in significant reductions in alcohol use and binge drinking (Fleming et al, 2002). This effect occurred within six months and was maintained over 48-months with reductions in both days of hospitalization and ED visits. Telephone and computer follow-up have also been found to result in improved outcomes for preventive care, diabetes, and arthritis, as well as ED visits (Balas et al, 1997). Telephone interventions and counseling can increase adherence to medication and treatment protocols and diagnostic procedures (Duan et al, 2000; Westfall and Narducci, 1997; Maisiak et al, 1996; Miller et al, 1997). Telephone interviews may also provide more accurate responses to sensitive questions than in-person interviews, and telephone surveys have yielded significantly higher reported rates of alcohol-related harm compared to in-person surveys (Midanik, et al, 2001).

In our DIAL study, we have found that patients can be contacted by telephone within days of ED discharge, and are open to discussions about their alcohol use and behavioral change plans, suggesting that telephone intervention following an ED visit is practical and can be accomplished effectively. Dial was a randomized clinical trial, funded by the U.S. Centers for Disease Control and Prevention, to determine the outcome of a telephone intervention to reduce alcohol use and high-risk behaviors following an ED visit. ED staff in RI is most enthusiastic about referring patients to MI counseling by telephone. ED clinical care is not delayed nor is discharge hindered when MI is given as follow-up care. Our experience is that in overcrowded EDs, the overburdened ED staff is more willing to identify and refer patients than to take on the additional task of delivering a MI in the ED. Telephone MI is an attractive intervention option that could have wide spread appeal because it requires less effort from the ED staff, is easy to implement, can be utilized in all types of EDs, allows access to counselors fluent in other languages, and may cost less to provide.

Conclusion

Our ED staff is now identifying patients who use alcohol and seeking programs such as DIAL, to which they can refer patients with alcohol problems. Other than under research protocols, few programs are available which will continue the counseling care needed by these patients with alcohol problems. Some of our ED staff provide brief advice and refer their patients to further treatment when time is available, many staff have become preceptors in counseling programs like RYDD, and more staff screen and refer impaired MVC drivers with DUI to the Medical Advisory Board at the DMV. As emergency departments continue to develop clinical options, screening and intervention will become even better integrated into routine emergency clinical care. Use of simple standard screening questions as a follow-up to an alcohol use question must be added to the ED staff routine. More services to provide needed intervention must be developed and funded to encourage staff to refer. In Rhode Island, our focus on MVC patients and impaired drivers has led to changes in our routine practice of emergency medicine, which promise to reduce injury.

References

Balas EA et al. (1997). Electronic communication with patients: evaluation of distance medicine technology. *JAMA* **278**(2), 152-9.

Becker B et al. (1995). Alcohol Use Among Subcritically Injured Emergency Department Patients. *Academic Emergency Medicine* **2**(9), 784-790.

Biffl WL et al. (2004). Legal prosecution of alcohol-impaired drivers admitted to a level I trauma center in Rhode Island. *J Trauma* **56**(1), 24-29.

Cherpitel CJ et al. (2003). Alcohol-Related Injury in the ER: a Cross-National Meta-Analysis from the Emergency Room Collaborative Alcohol Analysis Project (ERCAAP). *Journal of Studies on Alcohol* **64**, 641-649.

Crawford MJ et al. (2004). Screening and Referral for Brief Intervention of Alcohol-Misusing Patients in an Emergency Department: A Pragmatic Randomized Controlled Trial. *Lancet* **364**, 1334-1339.

D'Onofrio G et al. (1998). Patients with Alcohol Problems in the Emergency Department, Part I: Improving Detection. *Acad Emerg Med*, 1200-1209.

Duan N et al. (2000). Maintaining mammography adherence through telephone counseling in a church-based trial. *Am J.Public Health* **90**(9), 1468-71

Fleming MF et al. (2002). Brief physician advice for problem drinkers: long-term efficacy and benefit-cost analysis. *Alcohol Clin. Exp. Res* **26**(1), 36-43.

Longabaugh R et al., (1995). Injury as a motivator to reduce drinking. *Acad Emerg Med* **2**, 817-825.

Longabaugh R et al. (2001). Evaluating the effects of a brief motivational intervention for injured drinkers in the emergency department. *J. Stud. Alcohol* **62**(6), 806-16.

Maio R et al. (1997). Alcohol Abuse/Dependence in Motor Vehicle Crash Victims Presenting to the Emergency Department. *Acad Emerg Med.* **4**, 256.

Maisiak R et al. (1996). Health outcomes of two telephone interventions for patients with rheumatoid arthritis or osteoarthritis. *Arthritis & Rheum* **39**(8), 1391-99.

Mello MJ et al. (2003). Physicians' attitudes regarding reporting alcohol-impaired drivers. *Subst Abuse* **24**(4), 233-242.

Mello MJ et al. (2006). Alcohol screening in the emergency department: Screening for our Safety (SOS). American Automobile Association Foundation for Traffic Safety, Washington DC, December, 1-54.

Mello MJ et al. (2005). Emergency department brief motivational interventions for alcohol with motor vehicle crash patients. *Ann Emerg Med* **45**(6), 620-625.

Midanik LT et al. (2001). Reports of alcohol-related harm: Telephone versus face-to-face interviews. *J. Stud. Alcohol* **62**(1), 74-78.

Miller SM et al. (1997). Enhancing adherence following abnormal pap smears among low-income minority women: A preventive telephone counseling strategy. *J. Natl. Cancer Inst* **89**(10), 703-08.

Sommers MS et al. (2000). Attribution of injury to alcohol involvement in young adults seriously injured in alcohol-related motor vehicle crashes. *Am J Crit Care* **9**(1), 28-35.

US Department of Transportation. Fatality Analysis Reporting System (FARS) National Center for Statistics and Analysis.(http://www-fars.nhtsa.dot.gov/)

US Department of Transportation (2002). National Technical Information: Identification and Referral of Impaired Drivers Through Emergency Department Protocols. DOT HS 809 412, National Highway Traffic Safety Administration (NHTSA), Springfield, VA.

Westfall GR, Narducci WA (1997). A community-pharmacy-based callback program for antibiotic therapy. *J. Amer. Pharm Assoc.* **37**(3), 330-34.

12.4 – Pragmatic Implementation of Brief Interventions: An Alcohol Nurse Specialist for Every Acute Hospital

Robin Touquet, Adrian Brown - St. Mary's Hospital | London, UK

Background

Patient alcohol misuse is a well-recognised problem for all in the caring professions (Royal College of Physicians of London, 2001), but especially so for emergency departments (EDs) – which never close. They are places of safety for those that have made themselves vulnerable. The initial effects of alcohol are to depress the cerebral inhibitory centres, increasing risk-taking behaviour that can all too easily precipitate attendance at the ED. The top ten alcohol-related presentations at EDs have been identified as fall, collapse, head injury, assault, accident, unwell, non-specific gastro-intestinal, psychiatric, cardiac, and repeat attendance (Smith et al, 1996; Patton et al, 2004; Fig. 1). Repeated attendance by intermittent binge drinkers (hazardous drinkers) and dependent drinkers – often those at the extreme of homelessness – can generate negative attitudes from hospital staff.

Alcohol misuse is among the commonest causes of premature illness and death in the developed world (WHO, 2002). All health care interactions present opportunities for patients to develop insight into the consequences of their drinking. (Rollnick et al, 2005). No problem is more acute or unpredictable – often frightening – than one that necessitates a visit to the ED. The cause is often unpleasant, the patient wishing to avoid repetition; hence the opportunity for a teachable moment (Williams et al, 2005). ED staff must develop insight and understanding of this opportunity. Staff must have confidence that their response will be time well spent rather than increasing workload and stress within the peripatetic cauldron of the ED. Hence staff should be educated about the potential benefits of brief intervention (Huntley et al, 2004; Touquet & Brown, 2006; Touquet & Paton, 2006).

Since 1994 when the first alcohol nurse specialist was appointed to undertake alcohol health work in the ED at St Mary's Hospital in London (Smith et al, 1996; Wright et al 1998), the St Mary's system of work has evolved (Patton et al, 2004), and consists of three components. The first is selective screening of patients by ED medical staff using the Paddington Alcohol Test (PAT, Fig. 1) and providing screen-positive patients with brief advice. The second involves referral to an alcohol nurse specialist who carries out substance misuse assessment and a brief intervention. And, the final component involves feedback about the case to the referring member of staff.

Figure 1

PADDINGTON ALCOHOL TEST
(**PAT** Feb. 2006)

PATIENT IDENTIFICATION
NAME
D.O.B.

Consider PAT for ALL of the <u>TOP 10</u> reasons for attendance.

1. FALL (inc. trip)	**2. COLLAPSE** (inc. fits)	**3. HEAD INJURY**	**4. ASSAULT**
5. ACCIDENT	**6. UNWELL**	**7. NON-SPECIFIC GASTRO-INTESTINAL**	**8. CARDIAC** (inc. chest pain)
9. PSYCHIATRIC (inc. Deliberate self harm & overdose, please specify)	**10. REPEAT ATTENDER**	**Other** (specify):	

Proceed only after dealing with patient's 'agenda,' i.e. patient's reason for attendance.

1 Do you drink alcohol?	**YES** (continue)	**NO** (end)

2 What is the most you will drink in any one day?	(units) =	☐

If necessary, please use the following guide to estimate total daily units.

[A table of various types of alcohol is inserted here to aid practitioners in calculating amounts]

Advise patient "The medically recommended daily limits of alcohol are 4 units per day for a man or 3 units per day for a woman".

3   How often do you drink more than twice the recommended amount?

☐ Everyday	Dependant Drinker	(PAT + ve?)
☐ _____ times per week	Hazardous Drinker	(PAT + ve?)
☐ Never or less than weekly	GO TO QUESTION 4	

4 Do you feel your attendance here is related to alcohol?	YES (PAT + ve) NO

If **Pat + ve give feedback** e.g. "We advise you that this drinking is harming your health".
If drinking daily, but not excessively, advise about drink free days.

5 We would like to offer you advice about your alcohol consumption; would you be willing to see our alcohol nurse specialist?	YES (PAT + ve) NO

If "YES" to #5 give AHW appointment card and make appointment in diary (@ 10am)
Please note if patient admitted to ward
....

REFERRER'S SIGNATURE:
DATE:

Selective Focused Screening with Brief Advice

The PAT is a brief, four-question clinical tool used by doctors and nurses to identify patients who need an intervention for alcohol problems. (Smith et al, 1996). Its brevity enables the PAT to be a pragmatic tool, rather than one designed for research use. Nonetheless, it is applied in a structured way intended to be acceptable to staff and patients – being concise and to the point, whilst non-judgemental (Patton et al, 2004). The PAT is applied at the end of a consultation, for the top ten alcohol-associated presentations (see Figure 1), often collapse or falls (Huntley et al, 2001). It is therefore embedded in the clinical process to help patients develop insight, confidence and indeed gratitude. This gratitude is gained by first managing patients' presenting problems, which are likely to be their priority.

When introducing the PAT, the patient is told it is routine practice. If the patient says he or she does not drink alcohol, the PAT ends at question 1, unless the practitioner suspects otherwise. For the patient who does drink alcohol, the fourth question begins the process of brief advice by asking, "Do you feel your attendance here is related to alcohol?"

PAT-positive patients are given feedback about the potential and actual harmful effects of their drinking to help generate a desire for change (Patton et al, 2003). Lastly, they are asked, "Would you like to see our alcohol nurse specialist?" Since May 2005, about one third have accepted an appointment, and 60% of those were seen by the alcohol nurse specialist. This very brief process is made easy for staff by the PAT's inclusion in all case-notes, with instructions printed on the back. The appointment book for the alcohol nurse specialist is readily available, and he maintains a high profile within the department from 9 AM to 5 PM. The availability of an alcohol nurse specialist relieves staff stress, because now it is possible to provide services for previously neglected problems. This also enhances its sustainability. (see Costs, Benefits, Sustainability below)

Brief Intervention

The brief intervention is a single, 30–45 minute session conducted by the alcohol nurse specialist. It combines an initial assessment (Table 1) with feedback and counselling within a motivational interviewing framework (Rollnick et al, 1992). The session can be an introduction to specialised treatment services. However, unlike dependent drinkers, hazardous drinkers generally do not require further treatment options, or they may decline such a referral. It has been suggested that people already contemplating change respond to brief advice regarding alcohol-related problems (Leontieva et al, 2005). Furthermore, such advice helps a significant proportion of patients move from the pre-contemplation stage, where they do not admit they have a problem with alcohol, to the contemplation stage, where they recognise they have a problem (Prochaska et al, 1992).

Table 1: The Major Components of the brief intervention

Take an Alcohol History

Focus on current consumption - amount and frequency of drinking

Explore drinking career. Note incidence of withdrawal symptoms, previous treatment, and periods of abstinence. Emphasize the latter as the ability to change.

Explore Alcohol-Related Consequences

Current health and psychiatric status, referring to reason for admission.

Current social issues – close family background, relationships & children, housing arrangements, employment, legal issues.

Significant history – although the intervention focuses on the current situation and the patient's priorities.

Provide Feedback about the effects of drinking

Provide health education, e.g.,

- describe units of alcohol and recommended healthy limits
- comment on the patient's consumption
- ask if patients associate their alcohol consumption with the consequences they have experienced

Make recommendations about action planning

- what step the patient might take next
- how to cut down or detox safely
- treatment options.

Offer written information about

- abstinence or cutting down drinking
- how to contact local services which offer appropriate help
- individual responsibilities and choices (Miller, 1983)

Alcohol Health Work

Management of potential alcohol misuse is a two-part process: brief advice as part of screening followed, where appropriate and where accepted by the patient, with brief intervention, which includes assessment and planning. However, the work of the alcohol nurse specialist at St Mary's is much broader than just providing brief interventions. It involves changing attitudes, culture and practices in the whole ED and the wider hospital environment, tasks we refer to as 'alcohol health work'. The alcohol nurse specialist has a supportive role for all doctors and nurses, providing specialized knowledge and practical experience gained from working with alcohol patients through various stages of treatment, detoxification, and rehabilitation. Some understanding of the treatment options and outcomes can help ED staff to improve their interactions with patients, to address what they can manage themselves, and to identify which issues to refer to the alcohol nurse specialist. Within the St Mary's model, alcohol health work is described in Table 2.

Table 2: Alcohol Health Work at St Mary's – mnemonic: **'BEAFS'.**

Brief intervention	includes liaison with local treatment services
Education of ED staff	about screening, withdrawal symptoms, detoxification regimes, physical effects of excessive drinking, and brief advice staff can provide
Audit of the process	improving practice by tracking PATs and referrals to alcohol nurse specialist and highlighting missed cases to ED staff
Feedback to ED staff	to enhance staff compliance by reporting the rate at which patients act on referrals
Screening improved	with associated benefits: for the patient of reduction in drinking; for staff of quicker resolution of emergency department presentations, and reduced re-attendance

Cost, Benefits and Sustainability

In his seminal paper on obstacles to screening and interventions for alcohol problems in trauma centers, Gentilello (2005) has eloquently summarized the current situation of providing brief intervention for alcohol misusing patients presenting to trauma centers in the USA. In the UK, an increasing number of EDs and hospitals are establishing alcohol liaison links or employing alcohol nurse specialists to carry out this work. There are financial and resource implications but the benefits for hospital and patient outweigh the costs.

Gentilello states that to be successful it is necessary to integrate alcohol specialists into hospital practice rather than solely at specialist alcohol clinics (Gentilello, 2005). This is exactly the position in the UK (Royal College of Physicians, 2001). Every hospital needs an 'alcohol czar' – a senior clinician to provide advice and leadership to ensure that advantage is taken of the teachable moment provided by an ED visit. Every acute hospital with an ED should provide alcohol health work full time, ideally by an alcohol nurse specialist. The resulting benefits – including stress reduction for staff dealing with 'difficult' patients – are then embedded in the service for sustainability (Patton et al, 2003; Williams et al, 2005; Touquet & Brown, 2006).

The largest pragmatic randomised controlled trial ever reported showed that for every two referrals for alcohol health work by ED staff, there is one less re-attendance during the following twelve months. Furthermore, there was a mean reduction in consumption among those who were drinking above the UK recommended limits (Crawford et al, 2004). Cost savings are substantiated both from this randomised controlled trial (Barrett et al, 2006) and the UK Alcohol Treatment Trial (UKATT, 2005) conducted in primary care practices. The latter demonstrated that providing brief interventions for alcohol misuse gave cost savings of five times the expenditure on health, social and criminal justice services, echoing Gentilello's work (Gentilello et al, 2005). Evidence now shows us how we can respond to reduce this misuse (Touquet & Brown, 2006; Touquet & Paton, 2006).

Benefits are often deferred, rather than perceived at the point of delivery – a doctor advising a patient that alcohol misuse is harming their health will rarely see the positive effects of change by that patient. Therefore, education of all staff is necessary (Huntley et al., 2004) as well as repeated audit (Huntley et al., 2001).

Costs are directly financial (the cost of employing an alcohol nurse specialist) and secondary, i.e. use of time and resources. Benefits fall into two categories: the patient makes connections between drinking and health problems, reduces their alcohol consumption and receives health gains; the hospital sees reduction in attendances and develops a staff-base confident in dealing with such problems. The posts of alcohol nurse specialist and alcohol czar can be seen as financially sustainable because the financial and time costs are more than recouped in savings from reduced admissions due to accidents and health problems caused by drinking.

'Drunks' are not popular in EDs. However, if a high profile service for these patients is seen to be effective in reducing re-attendances in addition to reducing injuries and chronic health problems secondary to alcohol misuse, it will generate positive attitudes among staff. Once they see that something so simple can help improve patients' health with such a minimal amount of time spent, such services become sustainable.

References

Barrett B et al.(2006). Cost-effectiveness of screening and referral to an AHW in alcohol misusing patients attending an AED. A decision-making approach. *Drug Alcohol Depend*. **81**, 47–54.

Crawford MJ et al.(2004) .Screening and referral for brief intervention of alcohol misusing patients in an emergency department: a pragmatic randomised controlled trial. *Lancet* **364**, 1334–9.

Gentilello LM et al. (2005). Alcohol interventions for trauma patients treated in Eds and Hospitals. *Ann Surg*. **241**, 541-50.

Gentilello LM (2005) Confronting the obstacles to screening and Interventions for alcohol problems in trauma centers. *The Journal of Trauma, Injury, Infection and Critical Care*; Supplement. Session 4 **59**(3), S137-143.

Huntley JS, Patton R, Touquet R (2004). Attitudes towards alcohol of emergency department doctors trained in the detection of alcohol misuse. *Ann R Coll Surg Engl* **86**, 329–333.

Huntley JS et al. (2001). Improving detection of alcohol misuse in patients presenting to an accident and emergency department. *Emerg Med J* **18**, 99–104.

Leontieva L et al. (2005). Readiness to change problematic drinking assessed in the emergency department as a predictor of change. *Journal of Critical Care* **20**, 251– 256.

Patton R, Crawford M, Touquet R (2003). The effect of health consequences feedback on patient's acceptance of advice about alcohol consumption. *Emerg Med J* **20**, 451–2.

Patton R et al. (2004). The Paddington Alcohol Test: a short report. *Alcohol Alcohol* **39**, 266–8.

Prochaska JO, DiClemente CC, Norcross JC (1992). In search of how people change. Applications to addictive behaviors. *Am Psychol* **47**(9), 1102- 14.

Rollnick S et al. (2005) Consultations about changing behaviour. *BMJ* **331**, 961–3.

Rollnick S, Heather N, Bell A (1992). Negotiating behavior change in medical settings: the development of brief motivational interviewing. *J Ment Health* **1**:25–37.

Royal College of Physicians (2001). Alcohol – *Can the NHS Afford It?*. London: Royal College of Physicians.

Smith SGT et al. (1996). Detection of alcohol misusing patients in accident and emergency departments: the Paddington alcohol test (PAT). *J Accid Emerg* **13**, 308-312.

Touquet R, Brown A (2006). Alcohol misuse: Positive response. Alcohol Health Work for every acute hospital saves money and reduces repeat attendances. *Emerg Med Australas* **18**, 103-107.

Touquet R, Paton A (2006). Tackling alcohol misuse at the front line (editorial). *British Medical Journal* **333**, 510-11

UKATT Research Team. (2005) Cost effectiveness of treatment for alcohol problems: findings of the randomized UK alcohol treatment trial. *BMJ* **331**, 544–8.

World Health Organisation (2002). The world health report 2002 – reducing risks, promoting healthy life. WHO: Geneva.

Williams S et al. (2005). The half-life of the 'teachable moment' for alcohol misusing patients in the emergency department. *Drug Alcohol Depend* **77**, 205–8.

Wright S et al. (1998). Intervention by an alcohol health worker in an accident and emergency department. *Alcohol Alcohol* **33**, 651-6.

12.5 – Alcohol Interventions in Trauma Centers and Emergency Departments: Same Place, Different Services

Larry M. Gentilello - Southwestern Medical School | Dallas, TX US

Carol R. Schermer - Loyola University Medical Center | Chicago, IL US

Daniel W. Hungerford - Centers for Disease Control and Prevention | Atlanta, GA US

Introduction

Many elements of screening and brief intervention (SBI) for alcohol problems are implemented in a similar way across different medical settings. However, to understand how to implement SBI in a trauma center (TC), it is necessary to understand how trauma centers operate, and in particular, how they differ from emergency departments (EDs). The first section of this paper will provide this background. The second section will describe how the differences influence the ways in which SBI is implemented in TCs. The final sections of the paper will review the effectiveness of SBI in helping TC patients with their alcohol problems and the reasons it is critically important to implement SBI programs in trauma centers now.

How is a trauma center different from an emergency department?

In 2004 over 41 million patients in the U.S. required ED care for treatment of an injury. Almost 90% were treated in the ED and discharged without being admitted to the hospital (McCaig and Nawar, 2006). However, patients who had more severe injuries, perhaps a million or two, went to an emergency department situated in one of the more than 1,300 hospitals (MacKenzie et al, 2003) that are designated as trauma centers. Trauma centers are hospitals that have been verified by a state, local authority, or by the American College of Surgeons as having the personnel, equipment, and other resources needed for optimal care of the injured patient. Trauma centers are ranked Level I to Level IV based on the specialty services they can provide, how rapidly they can mobilize resources, what types of research they perform, and the number of trauma patients they treat annually. In general a Level I trauma center has surgeons available immediately and resources to provide emergency general surgical and subspecialty care, including thoracic, orthopedic and neurosurgical care.

Patients who sustain minor injuries usually travel to an ED under their own power, by car, or using other simple means of transportation. However, when a major incident occurs such as a serious car crash, shooting or stabbing, fall from a great height, or other type of serious event, pre-hospital emergency medical technicians or paramedics are summoned to the scene to evaluate the patient. If the patient is suspected to have sustained a life-threatening injury, the medics activate the trauma system by communicating with the closest appropriately designated trauma center. The patient is then transferred to that trauma center, even if several EDs are passed along the way.

At the trauma center the patient is met by members of the trauma service or team. The trauma service is run by general surgeons who have additional training in trauma and critical care. An attending trauma surgeon must be present in the hospital 24 hours per day, and must be present to receive the patient prior to arrival of the ambulance or aeromedical helicopter. The trauma service is in charge of the initial evaluation, resuscitation, and management of the patient. After admission to the hospital ward or to the intensive care unit, the trauma service functions as the patient's primary physician. A trauma service not only provides the immediate surgical and medical care, but also organizes multidisciplinary care that is provided by other hospital services – orthopedic, neurosurgical, and services such as physical therapy, social work, and rehabilitation.

In other words, in contrast to the brief contact context of the ED, the trauma service cares for patients from the time they enter the emergency department, through the intensive care unit onto the regular hospital floor, in clinic settings after hospital discharge and if needed, after rehabilitation. This process of facilitating and providing complete care to the injured patient allows the trauma staff to establish a therapeutic relationship with the patients they serve. The average duration of inpatient hospitalization for admitted trauma patients is 5.6 days (Peitzman et al, 1999).

This process is distinctly different from what happens in the ED, where emergency physicians typically evaluate and treat patients with low-acuity injuries, who do not require hospitalization, and who can be managed in the ED and discharged. The patients they care for spend at most a few hours under their care. For ED patients with more complicated injuries, the emergency physician may ask for a consult from the trauma service, after which the patient's care may be taken over by the trauma service or remain with the emergency physician. Generally, injured patients make up about a third of the total ED patient load; the remainder is made up of patients with medical problems such as chest pain, abdominal pain, or infections. In contrast, on a trauma service, all patients are injured.

Another difference is the prevalence of patients who have alcohol problems, which have been shown to be a leading risk factor for injuries and their recurrence. Roughly one quarter of injured patients treated in an emergency department are found to have an alcohol problem (Gentilello et al, 2005). In contrast, an analysis of data from six regional trauma centers involving 4,063 patients demonstrated that over 40% of admitted patients were under the influence of alcohol, and 85-90% of these patients met criteria for high-risk drinking or alcohol dependence (Soderstrom et al, 1992; Rivara et al, 1993; Dischinger et al 2001; Reed et al, 2005). After trauma center discharge the unexpected death rate for patients with a positive blood alcohol test is almost 200% greater than for patients who were not drinking at the time of injury, with 37%-77% of these deaths occurring as a result of repeat injuries (Dischinger et al 2001; Sims et al 1989).

How do these differences influence implementation of SBI in TCs?

Although the overall number of patients seen in TCs is smaller than those seen in primary care practices or EDs, TCs are an ideal place to aggressively pursue implementation of SBI. SBI has already been shown to be effective, feasible, and cost-effective in TCs (Schermer, 2005), and trauma center interventions have been associated with a decreased risk of subsequent drinking driving arrests (Schermer, 2006). Moreover, the prevalence of alcohol use disorders in TC populations is extraordinarily high, and trauma surgeons overwhelmingly recognize the importance of alcohol problems as a risk factor for injury, so they strongly support implementing SBI services in trauma centers (Schermer et al, 2003; Gentilello et al, 2005).

Trauma centers, therefore, appear to be optimal clinical settings in which to implement SBI programs. Trauma patients have a longer duration of hospitalization and a provider who they see throughout their hospitalization, and who organizes and provides their care. They have time to detoxify from any acute effects of alcohol or drug use, and they have repeated exposures to the same physician over the course of hospitalization. In addition, by the nature of their injuries that resulted in admission to the hospital, they usually have more severe injuries, and perhaps have experienced a bigger negative impact on their lives that can be attributed to their alcohol use.

Another advantage of the length of hospital stay for trauma patients is that not as much manpower is needed to perform the brief alcohol interventions. A feasibility study of SBI performed at three busy trauma centers showed that only a half-time employee was necessary to perform interventions at even the busiest centers. The provision of SBI in the ED would require someone to perform SBI 24 hours per day. In contrast, the in-patient setting of a trauma center allows interventions to be performed on a more scheduled basis, at an optimal time during the day. Most patients hospitalized for injury are also seen by a person from care management or social work to facilitate their discharge, rehabilitation and social needs. These visits with the social worker are ideal for additional discussions of alcohol consumption, related problems, and how behavior change plays a factor in subsequent health risks.

Although typical obstacles to disseminating SBI are present in TCs and among trauma staff, they appear to be less problematic than in other medical settings because of the cohesive nature of the profession. For example, designated TCs have unique organizational characteristics that provide advantages that favor successful implementation of SBI. They must undergo a rigorous site visit to demonstrate that they meet certain criteria with respect to specialty availability, equipment, facilities, and range of services. Most states have adopted the criteria developed by the American College of Surgeons Committee on Trauma, as outlined in the monograph *Resources for Optimal Care of the Injured Patient* (Committee on Trauma, 2006). This book provides a detailed list of the essential elements required to meet trauma center criteria. As of 2002 there were 214 Level I trauma centers, and 406 Level II trauma centers in the United States. Almost no other type of medical service is franchised by a State regulatory authority and held accountable for performance in such a rigorous manner.

The needs of a typical large city are usually met by only one or two Level I trauma Centers. In suburban areas, Level II trauma centers serve a similar purpose. The requirement to adhere to the American College of Surgeons trauma center criteria enables advances in medical care, including newly developed evidence-based alcohol screening and intervention protocols, to be incorporated into care in a systematic way, whereas incorporation of new medical advances into other specialties within a given geographic region usually requires changing the practice patterns of hundreds of physicians. SBI need only be adopted by a few trauma centers in order to benefit the majority of seriously injured patients in a given region, since most are likely to be cared for within the trauma system.

Conclusion

For the past twenty years considerable research funding and clinical effort has focused on integrating SBI into the broader health care system, with relatively little focus on trauma care. While not misplaced, this effort should now be increasingly focused on the opportunity provided by the recently adopted requirement for alcohol screening and interventions in Level I trauma centers. Investigations are needed to compare the effectiveness of different psychotherapeutic training modalities including combinations of treatment manuals, workshops, and methods of ongoing coaching and supervision.

Instead of dividing efforts to change medical care practices in multiple clinical arenas, implementing new practices in a single clinical setting is the most successful strategy for creating broad support for adoption of SBI. To date, organizational behavior research on program change, theories of organizational behavior, and theoretical and empirical work on diffusion of innovation support using this strategy. The quickest route to increasing the use of SBI across medical settings is by successfully implementing it in trauma centers. Stakeholders should work toward this goal by influencing perceptions, attitudes, and decisions at multiple levels – trauma staff, trauma center, and professional society levels. By providing the trauma community with technical information, data characterizing implementation problems in other settings, and evaluation materials about outcomes, we can ensure the success of SBI in trauma centers nationwide.

References

Committee on Trauma. Resources for Optimal Care of the Injured Patient. (2006) Chicago, IL: American College of Surgeons.

Dischinger PC et al. (2001). A longitudinal study of former trauma center patients: the association between toxicology status and subsequent injury mortality. *J Trauma* **51**,877–886.

Gentilello LM, Donato A, Nolan S (2005). Effect of the uniform accident and sickness policy provision law on alcohol screening and intervention in trauma centers. *J Trauma*. **59**, 624–631.

Gentilello LM et al. (2005). Alcohol interventions for trauma patients treated in emergency departments and hospitals: a cost benefit analysis. *Ann Surg* **241**, 541–550.

MacKenzie EJ et al. (2003). National inventory of hospital trauma centers. *JAMA*. **289**(12),1515–1522

McCaig LF, Nawar EN. (2006). National Hospital Ambulatory Medical Care Survey: 2004 emergency department summary. (2006) *Advanced data from vital and health statistics*; no 372. Hyattsville, MD: National Center for Health Statistics.

Peitzman AB et al. (1999). Trauma center maturation: quantification of process and outcome. *Ann Surg*. **230**, 87–94.

Reed DN Jr et al. (2005). Use of a single question to screen trauma patients for alcohol dependence. *J Trauma* **59**, 619–623.

Schermer CR et al. National survey of trauma surgeons' use of alcohol screening and intervention. *J Trauma*. 2003;55:849–856.

Schermer CR (2005). Feasibility of alcohol screening and brief intervention. *J Trauma*. **59**, S119–S124.

Schermer CR et al. (2006). Trauma center brief interventions for alcohol disorders decrease subsequent driving under the influence arrests. *J Trauma*. **60**, 29–34.

Sims DW et al. (1989). Urban trauma: a chronic recurrent disease. *J Trauma*. **29**, 940–946.

Soderstrom CA et al. (1992). Psychoactive substance dependence among trauma center patients. *JAMA* **267**, 2756-2759.

CHAPTER 13 :
POTENTIAL IMPACT OF SCREENING AND BRIEF INTERVENTION PROGRAMS IN EMERGENCY CARE SETTINGS

Daniel W. Hungerford - Centers for Disease Control and Prevention | Atlanta, GA

In 2000, the World Health organization reported that out of 26 risk factors evaluated, alcohol problems were the fifth most harmful, accounting for 4% of the global burden of disease (Alcohol & Public Policy Group, 2003). It is difficult to overstate the manifold ways in which human lives are affected. The human cost can be understood as more than increased individual vulnerability to injury, disease, disability and death. Alcohol problems burden the economy through losses to productivity and increased use of the medical, legal and penal systems and tear the social fabric and destroy families through neglect, violence, and loss of livelihood.

In the U.S. in 2004, there were over 110 million visits to emergency departments (EDs); one out of every ten ambulatory medical care visits was to an ED. About 13% of those visits resulted in a hospital admission. More than 41 million visits were for injuries (McCaig et al, 2006), some of which were severe enough to require care by surgeons in the trauma service – perhaps as many as two million. The proportion of patients in acute care settings who have alcohol-related problems is much higher than in the general population. Even though no reliable national estimates exist, studies in single institutions indicate the prevalence can be as high as 25% in EDs and 50% in trauma centers (TCs) (Cherpitel and Clark, 1995; Jurkovich et al, 1993; Mackersie et al, 1995). Therefore, emergency departments (EDs) and trauma centers (TCs) are important venues in which this problem can be addressed. They already act as the canary for alcohol problems in society's coal mine.

Some patients go to emergency and outpatient departments for their care more frequently than average – those under age 45, males, African Americans, Medicaid beneficiaries, and the uninsured (McCaig et al, 2006). These segments of the population often do not receive regular care in other medical settings. Therefore, if their alcohol problems are not addressed in the TC or ED, they may not be addressed at all.

As in most medical settings, the prevailing practice in EDs and TCs is either to ignore alcohol problems or to refer addicted individuals to specialty treatment. However, the latter tactic is too narrow. Although treatment works (O'Brien and McLellan, 1996), only about 15% percent of individuals in the U.S. who need treatment get it (Substance Abuse and Mental Health Services Administration, 2006). The remainder fail to get treatment for a variety of reasons: they don't believe they need it, they aren't ready, they can't afford it, they can't get access, or available treatment isn't appropriate.

There is another major reason that specialty treatment is not the major solution to society's alcohol problems. In the last 40 years, epidemiologic studies have shown that most alcohol-related harm cannot be attributed to addicted individuals but to drinkers who are not addicted (Hungerford, 2005). For every alcohol-addicted individual in the U.S., there are more than six individuals who are not addicted but are drinking excessively (Grant et al, 2004; Dawson et al 2004). Historically this realization and the search for more effective and efficient treatment methods motivated research that led to clinical-preventive-service strategies: public-health-style screening and brief counseling (Hungerford, 2005). In the U.S., these strategies are often lumped under the labels *screening and brief intervention (SBI) or screening, brief intervention, and referral to treatment (SBIRT)*. Public health screening involves using a uniform method with a predefined population to identify individuals at elevated risk for alcohol problems (including excessive drinking), a method unlike prevailing case-finding methods that assess individuals who medical staff suspect might be addicted. SBI programs typically provide brief interventions for individuals who are at elevated risk of harm because of their drinking patterns, but are unlikely to be addicted. SBIRT programs typically provide the same service, but devote added attention to making sure that individuals with more severe problems and addicted individuals get additional assessment and appropriate definitive treatment.

Like traditional specialty treatment, SBI and SBIRT strategies focus on individuals; like prevention programs, these strategies address alcohol problems early, before individuals become addicted. Unlike either treatment or prevention strategies, these strategies are opportunistic; that is, individuals are engaged when they present for medical or other services in clinical settings that are not traditionally associated with alcohol treatment or prevention efforts. Hence, with their more inclusive screening method and their broader definition of the problem – elevated risk rather than addiction, they help high-risk individuals not typically covered by prevailing methods and have the potential for a larger societal impact.

Although important SBI research questions remain – how to improve its efficacy, efficiency, and reliability in various clinical settings and with specific demographic groups (Saitz et al, 2006), controlled research in both EDs and TCs has demonstrated that screening and brief intervention programs can be effective and confer a favorable cost-benefit ratio (Gentilello et al, 1999; Longabaugh et al, 2001; Spirito et al, 2004; Crawford et al, 2004; Zatzick et al, 2004; Bernstein et al, 2005; Gentilello et al 2005; Neuman et al, 2006; Schermer et al, 2006). Acute care settings provide a readymade, enriched sample of patients experiencing the whole range of society's alcohol-related problems, both patients with obvious severe problems and patients with less severe problems, i.e., the ones who are not currently being identified. As epidemiologists know, screening is less expensive in a population with a high prevalence of the condition of interest. Decision makers should understand that they cannot expect to address alcohol problems adequately without addressing them in EDs and TCs. Furthermore, they cannot expect the professional denizens of the acute care world to address this widespread problem by themselves on top of their already-difficult task of treating urgent medical problems and injuries. Acute care settings perform the role of the canary in the coal mine for society; they signal the existence of alcohol problems and provide an opportunity to address them. However, to address alcohol problems properly, EDs and TCs do not need the owners of the mine – society's decision makers – to add another task to their to-do list. Instead, they need decision makers to understand that acute care clinical settings are productive locations in which to address society's alcohol problems. When decision makers understand this, they will create the political will to provide the necessary financial and human resources.

References

Alcohol & Public Policy Group. (2003) Alcohol: No Ordinary Commodity. A summary of the book. *Addiction* **98**, 1343–1350.

Bernstein J et al. (2005). Brief motivational intervention at a clinic visit reduces cocaine and heroin use. *Drug and Alcohol Dependence* **77**, 49–59.

Cherpitel CJ, Clark WB (1995). Ethnic differences in performance of screening instruments for identifying harmful drinking and alcohol dependence in the emergency room. *Alcoholism: Clinical and Experimental Research* **19**, 628–634.

Crawford MJ et al. (2004). Screening and referral for brief intervention of alcohol-misusing patients in an emergency department: a pragmatic randomised controlled trial. *Lancet* **364**(9442), 1334–1339.

Dawson DA et al. (2004). Toward the Attainment of Low-Risk Drinking Goals: A 10-Year Progress Report. *Alcohol: Clinical and Experimental Research* **28**, 1371-1378.

D'Onofrio G, Becker B, Woolard RH (2006). The impact of alcohol, tobacco, and other drug use and abuse in the emergency department. *Emergency Medicine Clinics of North America* **24**, 925-967.

Gentilello LM et al. (1999). Alcohol interventions in a trauma center as a means of reducing the risk of injury recurrence. **Annals of Surgery** *230*, 473-480.

Gentilello LM et al. (2005). Alcohol interventions for trauma patients treated in emergency departments and hospitals: a cost benefit analysis. *Annals of Surgery* **241**, 541–550.

Grant BF et al. (2004). The 12-month prevalence and trends in DSM-IV alcohol abuse and dependence: United States, 1991–1992 and 2001–2002. *Drug and Alcohol Dependence* **74**, 223–234.

Hungerford DW (2005). Interventions in trauma centers for substance use disorders: new insights on an old malady. *Journal of Trauma* **59**, S10–S17.

Jurkovich GJ et al. (1993). The effect of acute alcohol intoxication and chronic alcohol abuse on outcome from trauma. *Journal of the American Medical Association* **270**, 51-56.

Longabaugh R et al. (2001). Evaluating the Effects of a Brief Motivational Intervention for Injured Drinkers in the Emergency Department. *Journal of Studies on Alcohol* **62**, 806–816.

Mackersie RC et al. (1995). High-risk behavior and the public burden for funding the costs of acute injury. *Archives of Surgery* **130**, 844–849

McCaig LF, Nawar EN (2006). National Hospital Ambulatory Medical Care Survey: 2004 emergency department summary. *Advance Data from Vital and Health Statistics*, no. 372. National Center for Health Statistics, Hyattsville, MD.

Mello MJ et al. (2005). Emergency department brief motivational interventions for alcohol with motor vehicle crash patients. *Annals of Emergency Medicine* **45**, 620–625.

Neumann T et al. (2006). The effect of computerized tailored brief advice on at-risk drinking in subcritically injured trauma patients. *Journal of Trauma* **61**, 805–814.

O'Brien CP, McLellan AT (1996). Myths about the treatment of addiction. *Lancet* **347**, 237–240.

Saitz R et al. (2006). Challenges applying alcohol brief intervention in diverse practice settings: populations, outcomes, and costs. *Alcohol: Clinical and Experimental Research* **30**, 332–338.

Schappert SM, Burt CW (2006). Ambulatory care visits to physician offices, hospital outpatient departments, and emergency departments: United States, 2001–02. *Vital and Health Statistics* **13**(159).

Schermer CR et al. (2006). Trauma center brief interventions for alcohol disorders decrease subsequent driving under the influence arrests. *Journal of Trauma* **60**, 29–34.

Spirito A et al. (2004). A randomized clinical trial of a brief motivational intervention for alcohol-positive adolescents treated in an emergency department. *Journal of Pediatrics* **145**, 396–402.

Substance Abuse and Mental Health Services Administration. (2006) Results from the 2005 National Survey on Drug Use and Health: National Findings. NSDUH Series H-30, DHHS Publication No. SMA 06-4194). Office of Applied Studies, Rockville, MD.

Zatzick D et al. (2004). A randomized effectiveness trial of stepped collaborative care for acutely injured trauma survivors. *Archives of General Psychiatry* **61**, 498–506.

SECTION V :
APPLICATION AND IMPLICATIONS OF FINDINGS FROM EMERGENCY DEPARTMENT STUDIES
INTRODUCTION

Norman Giesbrecht - Centre for Addiction and Mental Health | Toronto, ON CANADA

The papers in this section examine the applications and implications of findings from emergency department studies. Several themes emerge and they are summarized briefly in the following paragraphs. The implications of five case studies are summarized in Chapter 14. Borges points to the challenges of converting the results of emergency room (ER) studies in Mexico into public health measures. Cremonte indicates that ER research in Argentina highlights the need for a comprehensive national health program focusing on alcohol, and improvement of social conditions for research. The paper by Swiatkiewicz, focusing on Poland, highlights the sharp contrast between the burden from alcohol found in ER studies and official perceptions in the national health care assessment that the alcohol burden is not significant. Similarly, Sovinova's and Csemy's analysis of lessons from an ER study in the Czech Republic found a significant role of alcohol in the overall injury rate, but a policy and prevention response that was inadequate. Focusing on India, Benegal points out that the heavy burden of alcohol-related violent assaults is not adequately recognized, and there is a need to integrate regular screening for alcohol into the ER procedure. Several themes, thus, emerge from these five country descriptions: alcohol is a significant risk factor in cases presenting in emergency departments; in contrast, alcohol-related problems are not perceived to be a significant health and safety issue; and, there is no routine monitoring or screening for alcohol in ER settings.

The next two papers highlight dimensions of finding a more effective and synergistic interface between the ER and the community. As noted by Giesbrecht and Moskalewicz, both ER research initiatives and community action projects can benefit from a perspective that looks outside the projects' parameters. For example, ER studies can be used to inform community-based prevention practice and monitor the impact of policy initiatives; and community action projects can use ER data as a valuable resource in identifying local problems, tracking and evaluation of impacts. These latter themes are illustrated and elaborated in the paper by Holder.

The last three papers examine the ER experience in three different contexts: the United States, Europe and globally. Greenfield and Cherpitel indicate how policy decisions have had a bearing on how ER studies are conducted, and, in turn, how ER data is a valuable resource for policy deliberations and decisions. Hope also makes the case that ER studies can be an essential resource, in connection with other methods and resources, in monitoring the impact of alcohol policies. This section concludes with a paper by Giesbrecht, Cherpitel, Room and Stockwell, who note that while ER studies are useful in surveillance, documenting the scope of the problem and identifying causal relationships between alcohol use and injuries, the goals of the initiative and methods used must be sensitive to local contexts and resources. Nevertheless, while there are many conceptual and practical challenges in conducting these studies and interpreting the findings, ER-based research is relevant to advocating for better health responses, training and resource development, and stimulating more effective policy responses to the burden from alcohol at the local, national and international levels.

CHAPTER 14 :
PRACTICAL EXPERIENCES IN FIVE DIVERSE CULTURAL CIRCUMSTANCES

14.1 – Application and Implications of Findings from Emergency Department Studies in Mexico

Guilherme Borges - National Institute of Psychiatry, Metropolitan Autonomous University | Mexico City, MEXICO

Main Findings of ED Studies in Mexico

Since 1986 the National Institute of Psychiatry "Ramón of the Muñiz" (INPRFM) has conducted research on the relationship between alcohol consumption and injuries/medical emergencies involving representative samples of patients who attend the emergency department (ED). The main goals of these studies have been to measure the prevalence of alcohol use among emergency department patients, to study their risk factors and consequences, and to propose interventions to reduce alcohol use and related problems. During the past 20 years there have been 15 emergency departments surveyed in the country by the National Institute of Psychiatry. Data from all of these EDs were collected using a similar methodology developed by Cherpitel (1989).

A sample of adult patients, 18 years and older, admitted to the emergency department and reporting an injury or medical emergency was drawn from ED admission forms which reflected consecutive patient arrival in the ED over a several-weeks period. All eligible patients seen in the ED during a 24-hour period were approached and asked consent to participate. Patients with severe psychiatric disorders and follow-ups of previous consultation were excluded. Patients were approached to be breathalyzed as an estimate of blood alcohol content (BAC) and interviewed as soon as possible after admission to the ED. Patients who were too severely injured or ill to be interviewed in the ED were followed into the hospital and interviewed after their condition had stabilized. A cadre of trained interviewers at each site obtained the BAC estimate and administered a standard 25-minute questionnaire. Patients were interviewed regarding the reason for the ED visit, drinking in the six hours prior to the injury or illness event, quantity and frequency of usual drinking during the last year, and demographic characteristics, among other items.

The first of these studies was carried out in Mexico City in 1986. The second took place Acapulco, Guerrero in 1987. The third study was carried out in Pachuca Hidalgo, in 1996-97 and, finally, the fourth study was conducted in 2002, in a hospital located in the southern part of Mexico City (Tlalpan), as part of the WHO Collaborative Study on Alcohol and Injuries. The results of these studies included 4,950 patients from 15 different emergency department in three cities in Mexico (see Table 1).

Table 1. Characteristics of 15 emergency departments surveyed by "Instituto Nacional de Psiquiatría", 1986-2003

	Hospital	Study	Principal Investigator	Year	Duration	N	% Trauma	Response rate	Type of Service
1	XOCO [c]	**Mexico City**	H. Rosovsky	1986	1 week p/hospital	352		88%	Public
2	La Villa [c]					331			Public
3	Balbuena [c]					203			Public
4	Ruben Leñero [c]					314			Public
5	Cruz Roja					293			Public
6	A-B-C					69			Private
7	A. López Mateos [d]					352			Social Security
8	Lomas Verdes [e]					274			Social Security
Subtotal						**2188**	**74.0**		
9	General-SSA [c]	**Acapulco, Gro**	G. García	1987	5 weeks	242		87%	Public
10	Cruz Roja					275			Public
11	Magallanes					123			Private
Subtotal						**640**	**53.6**		
12	General-SSA [c]	**Pachuca, Hgo**	G. Borges	1996-97	12 weeks	460		93%	Public
13	General- ISSSTE [d]				4.5 weeks	351			Social Security
14	General-IMSS [e]				5 weeks	606			Social Security
Subtotal						**1417**	**47.4**		
15	Gea González (trauma only)	**Tlalpan**	G. Borges/L. Mondragón	2002	6 ½ weeks	705	100	94.9%	Public
TOTAL						**4950**			

[c] Secretaria de Salud y Asistencia (SSA).
[d] Instituto de Social Security al Servicio de los Trabajadores del Estado (ISSSTE).
[e] Instituto Mexicano de Social Security (IMSS).

Basic results on several measures of alcohol prevalence are presented in Table 2. In all studies, it is clear that alcohol was a prevalent factor among these patients. The episodic and intoxicated use of alcohol consumption appear frequently in traumatic emergencies, and chronic and heavy consumption, sometimes with symptoms of alcohol dependence, is present among the medical emergency patients as well. Both trauma and medical emergency patients who report drinking within six hours prior to the event are very likely to attribute their emergency to the use of alcohol, and may therefore be willing to change their behavior. These results suggest that alcohol plays a prominent role among patients who seek emergency department care.

Table 2. Main results of 15 emergency departments surveyed by the Instituto Nacional de Psiquiatría, 1986-2003, by trauma and medical emergency

	STUDY	N	PERCENTAGE (%) TRAUMA									
			Violence	Accident traffic	Males	<30	BAC+	Alc. 6 hrs	HD	Depend. (3+)	5+drinks	Attribution*
1	Mexico City	1620	28.1	14.3	70.7	55.6	21.3	27.5	4.5	12.5	22.8	62.9
2	Acapulco, Mexico	343	26.8	9.6	70.8	57.9	21.2	29.2	3.4	10.4	Na	74.5
3	Pachuca, Mexico	672	17.9	16.8	66.8	53.9	15.9	14.9	3.1	6.3	17.6	78.6
4	Tlalpan, Mexico	705	15.6	8.3	60.0	48.8	15.5	17.4	2.2	7.0	19.3	68.0

	STUDY	N	PERCENTAGE (%) MEDICAL EMERGENCIES									
			Coronary Emerg.	Respiratory Emerg.	Males	<30	BAC+	Alc. 6 hrs	HD	Depend. (3+)	5+drinks	Attribution*
1	Mexico City	568	8.0	9.2	47.0	35.2	6.4	11.4	6.9	12.6	17.4	77.0
2	Acapulco, Mexico	297	6.7	10.1	51.2	44.8	5.4	11.9	7.2	13.7	Na	71.9
3	Pachuca, Mexico	745	5.2	17.9	38.7	37.2	3.4	4.2	2.6	5.0	8.5	86.7
4	Tlalpan, Mexico	Na										

Na: Not applicable.
BAC+: Blood alcohol content ≥0.01
Alc. 6 hrs Reported drinking within 6 hours prior to the injury or medical problem
HD: Heavy Drinker is one who drinks at least weekly and reports 12 or more drinks on at least one occasion during the last year.
Depend (3+) positive on three or more dependence symptoms; for Tlalpan, positive on for alcohol dependence according to DSM-IV.
5+ drinks: had five or more drinks on one occasion in the last 12 months.
*Attribution: Believe the event would not have happened if he or she had not been drinking

ED Studies and Cultural Aspects of Alcohol Use in Mexico

Two main aspects of these studies reflect a more general pattern of alcohol consumption in Mexico. First, comparatively speaking, the rates of chronic alcohol involvement (i.e., alcohol abuse or dependence) are relatively low in the Mexican EDs, ranging from 5% to 13.7%. In contrast, these rates are higher than rates in general population surveys in the country (Merikangas et al, 2000; Medina-Mora et al, 2005), lower than in other EDs in the US, but higher than in EDs in Poland (Cherpitel, 2006). Secondly, the rates of acute and intoxicating levels of alcohol involvement are, on the contrary, very large in Mexican EDs, with a range of 4.2% to 29.2% for self-reported alcohol consumption. The combination of infrequent drinking, but consuming a large number of drinks on a single occasion, has been noted as a characteristic of the Mexican drinking pattern – fiesta drinking (Medina-Mora et al, 2000) that leads to acute alcohol related-problems, including injuries, and especially violence-related injuries, seen in the ED.

Have These Findings Informed Community-Based Prevention or National Alcohol Policies?

Converting these results into public health measures has proven difficult in the country. This is not related just to ED studies, but to a more general problem related to the role of scientific inquiry and communication of findings to public health agencies in Mexico. There has been a divorce in Mexico between researchers/academicians and policy-making administrators. The first group is concerned that research findings are not used to formulate policies, while the latter group is concerned that research findings are not easily accessible and understandable. Findings from the series of Mexican ED studies appear understandable and straightforward: alcohol is a main cause of the injury burden in the country. Programs for detecting, treating and referring patients in this setting are necessary. Making these simple messages available to all involved in managing the problem is a next aim in dissemination of our research findings in Mexico.

References

Borges G et al. (1998). Alcohol consumption in emergency room patients and the general population: a population-based study. *Alcohol: Clinical and Experimental Research*, **22**(9),1986-1991, 1998a.

Borges G et al. (2004). Episodic Alcohol Use and Risk of Nonfatal Injury. *American Journal of Epidemiology*, **159**, 565-571.

Borges G et al. (1994). Casualties in Acapulco: Results of study on alcohol use and emergency room care. *Drug and Alcohol Dependence*, **36**,1-7.

Cherpitel CJ (1989). Study of alcohol use and injuries among emergency room patients. In: *Drinking and Casualties: Accidents, Poisonings and Violence in an International Perspective*. Giesbrecht N et al., eds, pp. 288-299, Routledge-Tavistock New York.

Cherpitel CJ, Rosovsky H (1990). Alcohol consumption and casualties: a comparison of emergency room populations in the United States and Mexico. *Journal of Studies on Alcohol*, **51**(4),319-26.

Cherpitel CJ (2006). *Alcohol and injuries: A review of emergency room studies since 1995*. World Health Organization, Geneva

Medina-Mora ME, Borges G, Villatoro J (2000). The measurement of drinking patterns and consequences in Mexico. *Journal of Substance Abuse*,**12**(1-2),183-196.

Merikangas KR et al. (1998). Comorbidity of substance use disorders with mood and anxiety disorders: results of the International Consortium in Psychiatric Epidemiology. *Addictive Behaviors*, November-December, **23**(6),893-907.

Medina-Mora ME et al. (2005). Prevalence, service use, and demographic correlates of 12-month DSM-IV psychiatric disorders in Mexico: results from the Mexican National Comorbidity Survey. *Psychological Medicine*. **35**(12),1773-1783.

14.2 – Practical Experiences and Lessons with Emergency Room Studies in Argentina

Mariana Cremonte - National University of Mar del Plata | Mar del Plata, ARGENTINA

Introduction

The socio-political processes in Latin America have had a dramatic impact on the history and development of research in social sciences in this region. In Argentina, following a promising beginning, research in the social sciences was virtually paralyzed during the 1930s, in what is now remembered as the anti-positivist reaction; all future attempts to pursue developments in social science were severely disrupted and compromised with each successive military government (Vilanova, 2003). As is the case in some neighboring countries, the absence of a research culture is still felt after nearly 20 years of democratic governments. In this context it is not surprising to find epidemiological studies on alcohol to be extremely scarce.

Social Science Research on Alcohol

In Argentina there have been only two national studies evaluating alcohol consumption in the general population (Míguez, 1999; SEDRONAR, 2004); these, along with Argentina's participation in the WHO GENACIS project (WHO, 2005) provide the only available evidence of drinking habits in the general population. From these studies a high prevalence of drinking in a Mediterranean pattern was observed, with many adults consuming alcohol on a daily basis. Argentina has been traditionally a wine producing country, with a wine drinking culture originating from the influence of Spanish and Italian immigrants. Within this cultural tradition, alcoholic beverages have been primarily associated with their nutritional value, consumed at the family table along with the meal, and not strongly associated with their psychoactive effects. Although the consumption of wine still surpasses the consumption of other alcoholic beverages, there has been a rapid and sharp increase in beer consumption (Tendencias Económicas, 2002). This increase is probably associated with a change in the drinking pattern emerging in the younger generations, including, for example, switching towards a more high-quantity/low-frequency pattern, and involving the consumption of large quantities of beer on weekends (Míguez, 2003). A similar transformation of drinking habits has also been observed in other developing countries and has been attributed to new global strategies of alcohol production and marketing (Room et al., 2002). In concert with the rather limited research on drinking patterns and habits in Argentina, it is not surprising that there has been very little research attention paid to the consequences of drinking and related problems. Furthermore, in sharp contrast with high-income countries, until very recently Argentina had no surveillance of the consumption of alcohol or other substances. However, in 2002 a National Drugs Observatory was established, which was directly dependent on the national government to gather and disseminate reliable data on the access, distribution and consumption of psychoactive substances, including alcohol. To date, the Observatory has implemented, among other initiatives, two national studies on the consumption of psychoactive drugs in the general population among high school students and among prison inmates.

Emergency Room Studies

These two dimensions – the lack of data regarding alcohol consumption and a restricted academic research tradition – constitute a significant context in the execution of emergency room (ER) studies in Argentina. On one hand these types of studies pose a series of difficulties: there is lack of funding and material resources, there is limited institutional support from hospital and university authorities, and there are no previous studies which could serve as a model or provide practical experience to build on. On the other hand, these studies can add worthwhile knowledge, given the study's historical value and uniqueness in Argentina. Thus, these limitations can turn research and data shortage into a very strong source of motivation. Two ER studies were conducted in Argentina. One was funded by Conicet (National Science and Technology Council) through a fellowship stipend to the author and allowed Argentina's participation in the Emergency Room Collaborative Alcohol Analysis Project (ERCAAP) (Cherpitel, 2003); the other was funded and implemented by WHO as part of the Collaborative Project on Alcohol and Injuries (Cherpitel, 2005). Both were conducted during 2001 on representative samples of patients 18 years and older admitted to ER of a large publicly funded hospital located in the city of Mar del Plata, a coastal city in the State of Buenos Aires. The WHO study sample was restricted to injured patients, only, who arrived at the ER within six hours following the injury event. In both studies patients were breathalyzed to estimate blood alcohol level (BAC), and asked a series of questions related to their injury and alcohol use. For the ERCAAP study, data were collected from 779 patients, representing a 92% completion rate. Interviews were not obtained due to: patients refusing consent (4%), leaving the ED before completion of the interview (3%), or being unable to understand the questions due to cognitive impairment or language barriers (1%) (Cremonte and Cherpitel, 2008). Results indicated that 28% of all injured patients reported drinking prior to the event (Cherpitel et al., 2005) and that alcohol constituted a risk factor for an injury (OR=4.52), specially among males and those under 30 years of age (Borges et al., 2006).

Context and Implications

When conducting ER research in Argentina, the variability intrinsic to the legal, institutional and social-cultural context lead to some modifications regarding the practical implementation of the study design. For example, it was not necessary to inquire regarding ethnic background, since the area where the study was conducted had, at that time, a very homogenous ethnic composition, although this has recently changed due to immigration from neighboring and Asian countries and emigration of young Argentineans to Europe. Since drinking is widespread within the culture, refusal to either participate in the study or to acknowledge one's own drinking was very rare (4% of the sample and 1% of those having a positive BAC, respectively). However, it remains to be explored how social desirability may have affected responses regarding intoxication and alcohol-related problems. The ability to implement the study and patients' high level of participation were likely facilitated by a social environment which has not yet been strongly dominated by a legal orientation to such interaction; i.e., patients and hospital personnel did not modify their behavior to avoid possible legal repercussions. Moreover, the application of the breathalyzer was not associated with law enforcement or the police. Although Argentina passed a law regulating BAC level while driving in 1997, the law is seldom enforced and most Argentineans had never experienced being breathalyzed before participating in the ER study. While most public hospitals are understaffed and overloaded with patients, participation of ER personnel in the study was outstanding. Even staff that was not formally assigned to the project found ways to collaborate. For most of the personnel this may have been their first and only opportunity to be part of a research project, thus motivating their enthusiastic participation in the study. Although the results did not reflect any changes at the institutional policy level, staff attitudes and handling of alcohol problems in the ER have changed following their participation in the study.

Findings from the study reflected the particular social context in which the study was conducted, with some types of injuries observed differing from those found in the literature. For example, there was a comparatively larger number of injuries related to bicycle and horse-drawn vehicles, both of which are common means of transportation in Argentina. Results also indicated an extremely high prevalence of motor vehicle injuries, accounting for one fourth of all injuries, and higher than that found in many other countries. The percentage of intentional injuries (15%) was also higher than that described in the literature for other countries.

In Argentina external causes constitute the fourth most common cause of death among all age groups and the first most common among children, adolescents and young adults. Violent-related deaths also prevail among young people (INE, 2001). In this context, ER studies provide unique evidence of alcohol involvement in injuries. Although much additional effort is needed in order to transform the knowledge acquired from these studies into health politics and policy development, the results of this ER study, and other research on alcohol in Argentina, highlight the need for a comprehensive national health program addressing alcohol use. In this regard, two dimensions require specific attention: collective endeavors aimed at improving social conditions for research, and the promotion of scientific-resilience. The strengths and advantages of these ER studies can be found even in an adverse milieu for research, and the experience provided to be a very rewarding one.

References

Borges G et al. (2006). Acute alcohol use and the risk of non-fatal injury in sixteen countries. *Addiction*. **101**(7), 993-1002.

Cherpitel CJ et al. (2003). A cross-national meta-analysis of alcohol and injury: data from the Emergency Room Collaborative Alcohol Analysis Project (ERCAAP). *Addiction* **98**, 1277-1286.

Cherpitel CJ et al. (2005). Multi-level analysis of alcohol-related injury among emergency Department patients: a cross-national study. *Addiction*. **100**(12),1840-50

Cremonte M, Cherpitel CJ (*2008*). Performance of screening instruments for alcohol use disorders In Emergency Department patients in Argentina. *Substance Abuse and Misuse*. **43**(1), 125-139

Instituto Nacional de Epidemiología "Dr J. Jara" (2001) *Causas de mortalidad por grupo de edad en base a estadísticas vitales*. Información Básica. Mar del Plata, Argentina.

Míguez H (1999). *Estudio Nacional sobre Consumo de Sustancias Adictivas en la República Argentina*. Subsecretaría Nacional de Prevención del Uso indebido de Drogas. Resultados Generales. Buenos Aires, Argentina.

Míguez H (2003). Epidemiología de la alcoholización en Argentina. *Revista Vertex*, XIV, Sup. II., Buenos Aires, Argentina.

Room R et al. (2002). *Alcohol in developing societies: a public health approach*. Finnish Foundation for Alcohol Studies, vol. 46, in collaboration with WHO. Helsinski.

SEDRONAR (2004) *Informe preliminar. Segundo Estudio Nacional sobre el consumo de sustancias psicoactivas Argentina 2004*. Sedronar. Buenos Aires, Argentina.

Tendencias económicas (2002) *La economía Argentina*. Tendencias Económicas. Buenos Aires, Argentina.

Vilanova A (2003). *Discusión por la psicología*. Departamento de Servicios Gráficos, Universidad Nacional de Mar del Plata, Mar del Plata, Argentina.

WHO (2005). *Alcohol, Gender and Drinking Problems: perspectives from low and middle income countries*. Obot I, Room R, eds. World Health Organization. Geneva.

14.3 – Prospects for Emergency Room Studies and Their Impact on Alcohol Policy in the Polish Context

Grażyna Świątkiewicz - Institute of Psychiatry and Neurology | Warsaw, POLAND

Introduction

Poland joined the Emergency Room Collaborative Alcohol Analysis Project (ERCAAP) in 2001. The studies were carried out in two large hospitals located in very different urban centers: Warsaw and Sosnowiec. Warsaw, the capital of Poland, represents a relatively rich and culturally heterogeneous place. Sosnowiec is located in the large industrial mining region of Silesia, which currently suffers from an economic depression. As in other ERCAAP studies, probability samples of patients were selected, breathalyzed to estimate blood alcohol concentration, and interviewed, with a translated version of the same instrument used in the US Santa Clara emergency room (ER) study (Cherpitel 1998).

In Poland emergency care delivery is different from that in the US and many other countries. There are no emergency rooms or department, as such, offering multidisciplinary medical help for emergent cases. Instead the specialist clinics (eg. orthopedic, ophthalmology) of public hospitals offer 24-hour emergency services a number of days per month. Each specialist clinic is expected to offer from one to a dozen emergency days per month, and this service is coordinated at the given community level across different hospitals. Given these circumstances, the procedure of obtaining a probability sample was more complicated than in other ERCAAP studies. In both hospitals emergency days for each specialty were first sampled, and than, according to the numbers of patients usually admitted to each special service procedure, the sampling frame for patients was defined. In total, in each of the two hospitals about 750 patients were interviewed, with completion rates ranging between 65-68%. Almost 20% of the eligible patients refused to be interviewed; other main reasons for non-interviews were the patient's condition and failure to locate the patient.

Before the ERCAAP studies were initiated, there was very little information on the prevalence of alcohol-related disorders in emergency services. Two unpublished reports were produced in the mid-1990s for local authorities, one in Warsaw and one in a small city located in southern Poland. The Warsaw study was carried out on the ambulance service, and showed that about 10% of medical interventions were alcohol-related. The study carried out in southern Poland reported that one-fifth of general hospital admissions consisted of patients with alcohol-related problems (Świątkiewicz et.al 1997). Additional information on alcohol-related disorders was available from studies carried out in primary health services. A screening study using the AUDIT was conducted in 1997 (involving 42 000 patients), and showed that more than 18% in the primary care population could be classified as high-risk drinkers. Among men the proportion was much higher, reaching 35% (Święcki 2001). Another study was conducted in 20 primary clinics and involved 4,373 adults aged 18-80. This study found that 12% were problem drinkers and 19% were assessed as alcohol dependent (Maxwell, Ignaczak & Czabała,2002).

In Poland alcohol treatment is commonly perceived as the primary method of addressing alcohol-related problems. The concept of the preventive paradox, which focuses prevention efforts on the general population and their drinking behavior, is not popular. This preoccupation with treatment, and the down-playing of population-based prevention, would partially explain why there have been so few studies on the impact of alcohol on the health services burden, and also why no standards and procedures exist on how to treat emergency service patients with alcohol-related problems.

Our studies in Warsaw and Sosnowiec represented the first attempts in Poland to assess the extent and nature of problem drinking among emergency service patients. Findings from the ERCAAP studies provide evidence-based arguments for those who are interested in stimulating public discussion on how to change the medical system, in order to make it more responsive to alcohol-related problems at an early stage, and before the problems are chronic and specialist treatment for addiction is required.

Main Findings from the ERCAAP Studies in Poland

In this short article it is only feasible to offer some general conclusions from the Polish studies that are cited here. Additional information is described elsewhere (Cherpitel et.al 2004; Moskalewicz et.al 2006), including a more detailed description of the study methodology, application of standard questionnaires and statistical procedures and models used for data analyses.

In both Polish hospitals a substantial and visible proportion of males were noted whose drinking patterns represented risky behavior. In the Warsaw hospital more than 16% of the male respondents were classified as heavy drinkers. Their consumption of absolute alcohol (100% alcohol equivalent) exceeded 12 liters annually. In Sosnowiec almost every fourth (24.9%) male patient reached this level of consumption. Almost 16% of male emergency service patients in Warsaw and 20% in Sosnowiec reported drinking prior to the injury which brought them to the emergency service. Among those who visited the emergency service due to medical conditions, the proportions were much less (6% in Warsaw and 10% in the Sosnowiec hospital). The statistical analyses examined several drinking patterns including quantity and frequency of drinking and frequency of drunkenness. These analyses confirmed that injured males were significantly more likely to be heavier drinkers than those males presenting with other health problems or conditions. Among males, injury status was significantly associated with alcohol-related problems in the year preceding the interview (Cherpitel at. al., 2004).

Applicability of ERCAAP Findings to Polish Circumstances

The Polish data analyzed along with other data from the international Emergency Room Collaborative Alcohol Analysis Project and the World Health Organization (WHO) Collaborative Study on Alcohol and Injuries showed a strong relationship between alcohol and casualties, similar to that found in ER settings of the other countries that participated in these studies. Furthermore, differences found between the two hospitals, which operated in culturally and economically diverse regions in Poland, partially supported earlier findings suggesting that cultural context may affect the magnitude of the alcohol-injury relationship (Cherpitel 1998; Cherpitel et.al 2005a). More recent analysis of the ERCAAP and WHO studies across 13 countries found that cultural context also influences the level of performance of screening tests. Data from that study was a basis for the hypothesis that RASP4 screening instrument for alcohol problems performed better in countries higher on societal-level detrimental drinking patterns than those countries lower on detrimental patterns of drinking (Cherpitel et. al 2005b).

The emergency services questionnaire administered in both the Warsaw and Sosnowiec hospitals included a number of screening instruments (CAGE, AUDIT, and RAPS4) (Cherpitel et al. 2005a). Although the Polish population is relatively homogenous in terms of ethnicity and culture, the study found that despite the lack of heterogeneity, no one screening instrument or cut point was optimal for identifying alcohol-related problems across region and demographic sub-groups. These findings suggested that screening instruments for identifying patients who require intervention should be very carefully tested before being implemented on a wider scale.

The extend of problem drinking among emergency patients in Polish hospitals seems to be important in making the case that emergency services are potentially a promising site for a brief intervention (Cherpitel et. al., 2005a). This conclusion gave rise to a pilot study conducted in Sosnowiec and aimed at determining the feasibility of whether a brief intervention is acceptable to emergency service patients.

The aim of the first stage of the study was to determine the proportion of patients who, on the basis of a short interview, screened positive for eligibility for a brief intervention and agree to be contacted again in six months for a similar interview. The screening questionnaire was short, containing only ten questions in addition to basic socio-demographic information. The RAPS4 instrument was selected as the screening criterion since it had been found to perform better for identifying problem drinking than other screening instruments in the previous emergency services study in Sosnowiec (Cherpitel et.al., 2004). It was assumed that one positive response to the RAPS4 qualified a patient for a brief intervention. The results of the first stage of the feasibility study were very promising. Among RAPS4 positive patients, 95% gave their consent to be contacted again.

In the second phase of the feasibility study, patients who gave their consent were contacted again after six months to determine how many of those who had agreed could be reached and were willing to be interviewed again. The interviewer successfully contacted 73% of these patients, and no one contacted refused to give a short interview. The major causes of attrition were a wrong telephone number and wrong address. We found it surprising that in almost 25% of cases interviewers had to visit respondent's places of living because they could not be reached by telephone.

In summary, this feasibility study in Sosnowiec demonstrated that almost all the emergency service patients who screened positive on the RAPS4 as problem drinkers agreed to be contacted again, and that more than 70% could subsequently be reached for a six-month follow-up. Results of this feasibility study encouraged us to plan a larger investigation in order to assess the effectiveness of a brief intervention in the emergency services setting in Poland.

According to the last report prepared for WHO (Anderson and Baumberg 2006), Europe is the heaviest drinking region in the world. Poland is one of the European countries which, since the beginning of the 1990s, has undergone deep social and economic transformations. One of the negative aspects of these social changes has been increased availability and consumption of alcohol followed by a rising prevalence of alcohol-related disorders. In Poland alcohol-related disorders constitute a heavy burden on mental health services. The proportion of male patients with such disorders reached more than 40% of all male hospitalisations (Moskalewicz and Świątkiewicz 2004). There is also evidence of growing alcohol-related mortality, particularly due to liver cirrhosis. In the last 15 years mortality due to liver cirrhosis among males age 20-64 has increased twofold (Brodniak et.al 2002).

The findings from the two Polish ERCAAP studies, and from primary care studies noted above, show that alcohol-related admissions constitute a visible burden on general medical services. The evidence from these studies provides a basis to formulate recommendations to encourage a wider participation of general health services in the early identification of high-risk drinking and in responding to harmful alcohol consumption. However, a challenge is determining to whom such recommendations should be addressed.

It is a common belief that the health services system and its personnel are natural partners interested in public health improvement. Paradoxically, their interests are at times contradictory to public health interests. In most countries service providers, even those from public units, function like other personnel in trying to maintain their position in the world where free market economy rules dominate. In Poland the bigger the hospital's patient burden, the better their position to be able to negotiate a favourable contract with the National Health Fund. This is likely one reason why the health services organisers and providers are not interested in implementing routine procedures (e.g. brief intervention) aimed at reducing the number of admissions. Health care administrators, the Ministry of Health and the National Health Fund could better represent public health interests. These institutions should focus on looking for effective options that would systematically decrease the alcohol-related burden on health services and reduce the economic burden.

Since the beginning of the 1990s, the Polish health care system has experienced some significant reforms. The most important was enforced on January 1st 1999 and involved a general health insurance act which created a new insurance-based budgetary model for health care funding. Prior to that, the funding of health services came from the state budget. In spite of numerous attempts to improve the system, it received a negative evaluation by the general public and among health service personnel. In June 2006 the Ministry of Health prepared a document entitled "Information for the Parliament of the Polish Republic on the Status in Health Protection" for discussion in the parliament. This extensive document (126 of pages) contained only one paragraph focusing on alcohol issues. The authors noted that in general overall alcohol consumption is not high and the only problem highlighted was the unfavourable structure of alcohol consumption – referring to the relatively high proportion of spirits in total consumption (Ministry of Health 2006).

From this document it appears that according to the health care administrator's perspective, the alcohol-related burden to the system is not perceived as a problematic issue. More visible public debate is needed to change this perspective. The ERCAAP studies which have been conducted in Poland to date provide evidence that public health advocates can use to draw attention to the large alcohol-related burden in the health care system.

References

Anderson P, Baumberg B (2006). Alcohol in Europe: a public health perspective. European Commission

Brodniak W, et al. (2002). Mortality among treated alcoholics In Poland. Paper presented at the 28th Annual Alcohol Epidemiology Symposium, Paris 3-7 June 2002

Cherpitel CJ (1998). Drinking patterns and problems and drinking in the injury event: an analysis of emergency room patients by ethnicity. *Drug and Alcohol Review*, **17**, 423-431.

Cherpitel CJ, Moskalewicz J, Świątkiewicz G (2004). Drinking patterns and problems in emergency services in Poland. *Alcohol and Alcoholism*, **39** (3), 256-261

Cherpitel CJ et al. (2005a). Screening for alkohol problems in two emergency samples In Poland: comparison of the RASP$, CAGE and AUDIT. *Drug and Alcohol Dependence* **80**, 201-207.

Cherpitel CJ et al. (2005b). Cross-national performance of the RAPS4/RAPS-QF foe tolerance and heavy drinking: data from 13 countries. *Journal of Studies on Alcohol*. **66**(3), 428-432.

Maxwell LB, Ignaczak M, Czabała JC (2002). Prevalence of tobacco and alcohol use disorders in Polish primary care settings. *European Journal of Public Health* **12**(2), 139-144.

Ministry of Health (2006). Information for Parliament of Polish Republic on the status in health protection. The Cabinet document no 662, Warsaw 31 May, 2006

Moskalewicz J, Świątkiewicz G (2004). Rozpowszechnienie związanych z alkoholem zaburzeń zdrowia w Polsce w latach dziewięćdziesiątych. (prevalence of alcohol-related health disorders in Poland In the 1990s) in: Kiejna A., Rymaszewska J. (red) Epidemiologia Zaburzeń psychicznych. (Mental Disorders Epidemiology) Biblioteka Psychiatrii Polskiej, Warszawa, 39-51.

Moskalewicz J et al. (2006). Results of two emergency room studies. *European Addiction Research* **12**, 169-175.

Świątkiewicz G, Sierosławski J, Stępień E (1997). Obciążenie szpitala ogólnego, pogotowia ratunkowego i posterunków policji w Kędzierzynie Koźlu interwencjami związanymi z nadużywaniem alkoholu. (Alcohol burden on general hospital, ambulance services and the police in Kędzierzyn Koźle) Institute of Psychiatry and Neurology – unpublished report submitted also to State Agency for Solving Alcohol Related Problems.

Święcki P (2001). Audit. Terapia Uzależnienia i Współuzależnienia **2**, 16-18

14.4 – Alcohol in Czech Society and Lessons from the Prague Emergency Department Study

Hana Sovinova, Ladislav Csémy - National Institute of Public Health | Prague, CZECH REPUBLIC

Alcohol Consumption in the Czech Society

The Czech Republic, part of the European Union (EU), is one of the EU countries highest in alcohol consumption. According to the "WHO Health For All" database, registered alcohol consumption per capita in the Czech Republic was 13.6 liters of absolute alcohol in 2003. It was notably the highest in consumption among the EU candidate countries as well as among the 15 EU member states, with the exception, only, of Luxembourg. Alcohol consumption has not changed markedly since 1990, increasing only slightly by 2003. Consumption of alcohol declined in 2004, which, according to the Czech Statistical Office, was caused by forward buying before the expected increase in consumer tax for alcohol on January 1, 2004 (Český statistický úřad, 2004). In addition to a high level of total alcohol consumption, composition appears to be particularly harmful, with spirits representing one-third of the alcohol consumed (Table 1), and this has been stable over the last 10 years.

Table 1   Registered per capita (total population) consumption of alcohol beverages in the Czech Republic since 1990

Year	Total registered consumption in litres of 100% alc.	Spirits (litres of 40% alc.)	Beer (litres)	Wine (litres)
1990	8.9	7.2	155.2	14.8
1991	9.1	8.3	146.9	14.8
1992	9.4	8.0	163.3	15.0
1993	9.2	7.8	153.6	15.3
1994	9.4	7.9	156.7	15.4
1995	9.4	7.9	156.9	15.4
1996	9.5	8.0	157.3	15.8
1997	9.8	8.3	161.4	15.9
1998	9.8	8.2	161.1	16.0
1999	9.9	8.3	159.8	16.1
2000	9.9	8.3	159.9	16.1
2001	9.9	8.2	156.9	16.2
2002	10.0	8.3	159.9	16.2
2003	10.2	8.4	161.7	16.3
2004	9.8	7.6	160.5	16.5

Source: Czech statistical office (Statistical yearbook of the Czech Republic (1960-2005))

An extensive population survey was performed in 2002, when the Czech Republic participated in the "Gender, Alcohol and Culture – An International Study" (GENACIS). This survey provided the most recent data describing the patterns of alcohol consumption in different segments of the population. Within the study, 2,551 adults from the Czech Republic, aged 18 to 64, were randomly selected and interviewed. The average daily alcohol consumption calculated on the basis of answers to questions concerning the frequency and quantity of alcohol beverage intake was 16.6 g of pure ethanol equivalent. Compared to the average consumption for females (7.2 g), the consumption for males was higher (26.4 g). Alcohol consumption varied by age group (see Table 2), and was highest for middle-aged males between the ages of 35 and 44. The lowest alcohol consumption was for males aged 18 to 24, who represent the youngest males in the sample. Among females, however, the differences in alcohol consumption across age groups were not as pronounced. The lowest reported alcohol consumption in age and gender subgroups was among females over the age of 54, while the highest alcohol consumption for females was among those aged 36 to 44 (Csémy, Sovinová, 2003).

Table 2. Average daily alcohol consumption in grams according to age group and gender

	Age categories				
	18 to 24	25 to 34	35 to 44	45 to 54	55 to 64
Males	20.7	23.0	35.2	27.1	24.8
Females	7.8	7.2	8.9	7.1	5.0

Source: Czech statistical office (Statistical yearbook of the Czech Republic (1960-2005))

Alcohol and Health Problems

Interest in the alcohol issue is determined in part by the position of alcohol in modern society, and particularly by the complex interactions of alcohol in relation to health which has implications for the public health system. There is evidence that alcohol plays a substantial role in overall mortality – not only with regard to its obvious connection with injuries, but also with regard to a number of diseases in which alcohol contributes to the development and progression. Those of particular concern included oncology diseases, diseases of the gastrointestinal tract and cardiovascular diseases. Using the data from 25 European countries, Her and Rehm (1998) showed that an increase in average alcohol consumption per inhabitant was associated with an increase in overall mortality rate, and a decline in consumption was associated with a decline in mortality. Nemtsov (2002) has published data on the human loss caused by alcohol in Russia, including an increased suicide rate. Evstifeeva et al. (1997) compared mortality due to oncology diseases over 20 years in three Western European and three Eastern European countries (including the former Czechoslovakia). In all six countries examined an increase in mortality was found, due to oncology diseases related to changes in smoking and alcohol consumption rates.

Alcohol and Injuries

Publications based on large epidemiological investigations generally report that a substantial share of injuries can be linked to alcohol use (Rehm et al., 2003; Cherpitel et al., 2005). It also appears that patterns of alcohol consumption are associated with the incidence of injuries. Drinking a large amount of alcohol in a short period of time is highly risky (Gmel et al, 2006). In 2000, 22% of the Czech population was treated for injuries/accidents: 8.7% traffic accidents, 14.0% work and school injuries, 22.2% sports injuries and 55.1% miscellaneous. However, alcohol was reported in only 2.5% of all registered injuries (Institute of Health Information and Statistics, 2000).

The participation of Prague in the WHO Collaborative Study on Alcohol and Injuries provided the first opportunity in the Czech Republic to investigate the association between injuries and alcohol intake in patients in emergency department (EDs). The main results from the Prague study are summarized below.

The Main Results of the Prague Study

This research was performed in the emergency department at one of the largest university hospitals in Prague. In total, 511 persons who experienced an injury within six hours before the first visit for the emergency condition were included in the study. The proportion of females was slightly greater (55.6%), and female participants were older than males (average age 44.8 and 35.5, respectively). Relatively minor differences were found in the level of education between males and females. A full work load, i.e. 30 or more hours per week, was reported more often by males than females (77% and 47%, respectively).

Injuries and Testing for Alcohol

The presence of alcohol was examined using several methods. In addition to alcohol testing by means of an alcometer (breath alcohol analyzer), a clinical assessment using the ICD-10 Y91 codes was performed. Other sources of information were also used, including a list of signs of intoxication and, lastly, a personal interview with the patient concerning alcohol intake within six hours before the injury. The ascertained range of the presence of alcohol before the injury varied on the basis of these indicators (see Table 3). The alcometer detected a positive blood alcohol concentration (BAC) in 70 persons, i.e. 15.2% of the people in whom the alcometer was used (n=462). Mostly low concentrations of BAC were detected and a value of 0.05 or higher was measured in only 14 persons (3%). Based on clinical observations (Y91 codes), the presence of alcohol in the blood was reported in 28 persons (5.5%). At least one of the signs of intoxication was identified in 54 participants (10.6%), and 40 (7.8%) admitted drinking alcohol before the injury. In relation to the indicators used, it appears that the real estimate of the role of alcohol in injuries in the Czech Republic is around 10% – as detected in this study of an emergency service.

Table 3. Indicators of alcohol involvement in injury

Alcohol involvement based on:	Males	Females	Total
Observational assessment (Y91.x)	23	5	28 (5.5%)
Observational assessment – clinical signs	44	10	54 (10.6%)
Breath alcohol analysis (any alcohol detected)	51	19	70 (15.2%)
Breath alcohol analysis (BAC ≥ 0.05)	12	2	14 (3%)
Alcohol consumption within period of 6 hours prior to the injury (based on self- report)	32	8	40 (7.8%)

Agreement of Assessment Based on Y91 Codes and BAC Measured by the Alcometer

The level of consistency between the results from the clinical observational assessment (Y91 codes) and the values measured using the alcometer was not high. Of the 14 persons in whom the measured BAC was greater than 0.05, the alcohol influence (intoxication) was recorded using the Y91 codes in only 7 cases. In contrast, alcohol influence or slight intoxication was recorded in five persons in whom no alcohol was measured using the alcometer. Somewhat better consistency was found between clinical observation of intoxication and self-reported level of intoxication. A total of 11 persons stated that they had been intoxicated (from mild to very severe) at the time of the event, and 7 of these were classified in the mild to severe category of intoxication according to the clinical observational assessment.

Injuries and Drinking Patterns

Information on drinking behavior in the patient sample can be regarded as a significant by-product of this research. The data from this ED study complements the broader epidemiological information on the overall high rates of alcohol use in the adult Czech population. In our ED study, we detected regular alcohol consumption (twice weekly or more often) in 63.6% of all males and 26.4% of all females. The average typical alcohol consumption recalculated per 100% alcohol was 57 ml and 43 ml in males and females, respectively. Patients with a BAC of more than 0.05 in this study also had a significantly higher typical daily average alcohol consumption (91 ml of ethanol) than those with a lower or negative BAC.

Perception of Alcohol Problems in Emergency Departments Practice

We also obtained qualitative information from key informants on their perceptions of alcohol problems in the emergency department. The proportion of injuries in which the role of alcohol is noted is influenced, in part, by the ED team's perceptions of the effects of alcohol on injuries. An essential fact is that health insurance coverage of medical expenses is not influenced by alcohol intoxication in the Czech ED. Therefore, regardless of whether ED staff is interested in assessing the role of alcohol, alcohol involvement or intoxication does not have to be reported in relation to the provision of treatment, even if this information would have a bearing on appropriate medication and treatment.

The attitude of injured patients is another factor which may influence knowledge concerning the effects of alcohol on injuries. ED workers indicated that among patients presenting in the ED for treatment of minor injuries related to alcohol intoxication, many do so only when the effects of alcohol have largely worn off. This is consistent with our own research experience when many patients were not enrolled in the study because they had come to the ED more than six hours after being injured. Whatever the reasons for the postponement of treatment (e.g. feeling shame due to drinking or for some other reason), if these patients are more likely to be drinking prior to injury than those who arrive sooner at the ED following injury, the prevalence of alcohol involvement in the injury event will be underestimated.

Implications for Health Promotion and National Alcohol Policy

Results of the Czech ED study were published (Sovinova et al., 2002) and disseminated among experts engaged in the problems of health promotion and prevention (National Network of Institutes of Health). As far as we can assess, the results of this work have been accepted with interest by the professional community, and have highlighted the problems of alcohol and its risks on the health of the population. Priorities of the alcohol policy within the national health care system are set by the program Zdraví 21 ("Health 21"), which has been approved by the government and which includes issues related to reducing alcohol-related injuries.

Current alcohol legislation defines a number of standards and criteria (e.g., availability of alcohol products only to those above the age of 18, regardless of the type of beverage, and zero tolerance for drinking and driving); however, thorough enforcement of all statutory provisions is insufficient, which consequently reduces the efficacy of the alcohol policy. Current provisions make it feasible to achieve a decrease in damage related to alcohol consumption by emphasizing better enforcement of the current statutory provisions. In addition, implementation of specific interventions is also required, such as brief intervention in general practice and the ED, which has, to date, been rarely used.

Conclusions

This investigation of the role of alcohol in injury occurrence indicates that the share of alcohol in the overall number of injuries is around 10% in the Czech Republic. This value is approximately the same as in other developed countries. Various indicators used for the determination of alcohol involvement in patients presenting to the ED do not show high consistency, and, in the context of Czech medical practice and tradition, it is hardly conceivable that a method of determination of alcohol, other than that based on an objective measurement of BAC, could be put into practice. Additionally, since neither current guidelines for medical practice nor health insurance companies require blood alcohol determination, physicians in emergency departments have little, if any, reason to systematically and routinely monitor the presence of alcohol among patients presenting with injuries. Therefore, given the significant role of alcohol in the overall injury rate, we cannot be satisfied with the current status of dealing with this problem in the Czech Republic. However, data from this research, including a comparison with foreign experience, has contributed to improved knowledge among public health professionals, and has increased a focus on alcohol-related issues within health promotion programmes.

References

Český statistický úřad (2004) Vyjádření ČSÚ k interpretaci údajů týkajících se statistiky spotřeby lihovin (http://www.czso.cz/csu/redakce.nsf/i/vyjadreni_csu_k_interpretaci_udaju_tykajicich_se_statistiky_spotreby_lihovin)

Cherpitel CJ, Ye Y, Bond J (2005). Attributable risk of injury associated with alcohol use: cross-national data from the emergency room collaborative alcohol analysis project. *American Journal of Public Health* **95**(2), 266-72.

Csémy L, Sovinová H (2003). Spotřeba alkoholu v České republice. In: Kouření cigaret a pití alkoholu v České republice. Sovinová H, Csémy L. (Eds.) Státní zdravotní ústav, Praha

Evstifeeva TV, MacFarlane GJ, Robertson C (1997). Trends in cancer mortality in central European countries: The effect of age, birth cohort and time-period. *European Journal of Public Health*, **7**(2),169-176.

Gmel G et al. (2006). Alcohol-attributable injuries in admissions to a swiss emergency room-an analysis of the link between volume of drinking, drinking patterns, and preattendance drinking. *Alcohol: Clinical and Experiental Research* **30**(3), 501-509.

Her M, Rehm J (1998). Alcohol and all-cause mortality in Europe 1982-1990: A pooled cross-section time-series analysis. *Addiction* **93**(9),1335-1340.

Nemtsov AV (2002). Alcohol-related human losses in Russia in the 1980s and 1990s. *Addiction* **97**(11),1413-25.

Rehm J et al. (2003). Alcohol-related morbidity and mortality. *Alcohol Research and Health* **27**(1),39-51.

Rehm J et al. (2003). The relationship of average volume of alcohol consumption and patterns of drinking to burden of disease: an overview. **Addiction 98**(9),1209-28.

Sovinová H et al. (2002) Alkohol a úrazy. Státní zdravotní ústav, Praha.

14.5 – Alcohol and Injuries: India

Vivek Benegal - National Institute of Mental Health and Neurosciences | Bangalore, Karnataka INDIA

Background

Alcohol has been identified as an important risk factor in injury occurrence. The problem of alcohol-related injuries is particularly alarming in developing countries, like India, where increasing rates of alcohol consumption are coupled with hazardous patterns of drinking, injury rates are extremely high, and appropriate public health policies have not been implemented.

The Burden of Injuries in India

An examination of 'years of potential life lost' in India, indicated that injuries are the second most common cause of death after the age of 5 years (Mohan and Anderson 2000). Data estimated for the year 2005 suggested that injuries contributed to nearly 850 000 deaths, and nearly 17 000 000 persons were hospitalized. Further, nearly 42 500 000 persons had minor injuries, incapacitating them for shorter or longer periods. Nearly 70% of these deaths and injuries occurred among men 15–44 years of age. Eighty per cent of these deaths and injuries occurred in rural areas, where health care is poor and deficient. One-third of disabilities were due to injuries with an estimated 7 million persons suffering from various disabilities. If no systematic efforts are introduced and implemented, it is estimated that the number of deaths due to injuries is likely to increase to 1.1 million by 2010 and 1.2 million by 2015 (Gururaj, 2005).

Patterns of Alcohol Use in India

India is generally regarded as a traditional 'dry' or 'abstaining' culture. The prevalence of alcohol use is low; estimated at 21% among adult males (Ray et al, 2004), and less than 5% among women (Benegal et al, 2005). The per capita consumption is 2 litres of absolute alcohol equivalent per adult per year, and adjusting for undocumented consumption (illicit beverages and tax evaded products account for 45-50% of total consumption), this is likely to reach 4 litres (Benegal et al, 2003; Singh, 1986).

'Dry' cultures are known to predispose to deviant, unacceptable and anti-social behavior related to alcohol use as, well as chronic disabling alcoholism (Blum and Blum, 1969). Repeated observations have documented that more than 50% of all drinkers in India satisfy criteria for hazardous drinking. The typical consumption pattern is one of heavy solitary drinking, involving predominantly spirits and usually more than 5 standard drinks per occasion (Gaunekar et al, 2004). Among drinkers there is surprisingly little difference between amounts consumed by men and women. A large proportion of drinkers of both genders drink daily or almost daily. The dominant drinking expectancies favor drinking to intoxication, and alcohol use is strongly associated with expectations of disinhibition, especially among males, which 'legitimizes' male drunkenness and violence (Benegal et al, 2005; Gupta et al, 2003). Needless-to-say, this translates into substantial rates of alcohol-related morbidity involving a high social cost. Alcohol-related problems account for over a fifth of hospital admissions in India, but are under recognized by primary care physicians. Alcohol misuse has a disproportionately high association with deliberate self-harm, high-risk sexual behavior, HIV infection, tuberculosis, esophageal cancer, liver disease and duodenal ulcer, and alcohol consumption has been implicated in over 20% of traumatic brain injuries (Benegal, 2005).

The impact of globalization and economic liberalization (exposure to satellite television, rapid socioeconomic transition and growing disposable income) has influenced a widespread attitudinal shift to greater normalization of alcohol use which is reflected in a steady rise in alcohol beverage sales over the last 20 years, with an annual growth rate of 8-10% for spirits and 35% for beer. This has opened a vast emerging market for trans-national alcohol companies. Concurrently, there has been a significant lowering of age at initiation of drinking, with a drop from a mean of 28 years to 20 years, between the birth cohorts of 1920-30 and 1980-1990 (Benegal, 2005).

The preoccupation with prohibition-centric alcohol policies and the general public perception of India as a 'dry' culture, has worked against a rational and more balanced examination of the impact of alcohol-related health consequences on public health. The focus has been on supply reduction, brief attempts at prohibition, and volume-based taxation encouraging spirits consumption relative to beer. The dependence of most state governments on alcohol taxes, which provide 20% of their annual tax income, makes a mockery of most attempts at supply limitation.

The Indian Motor Vehicle Act mandates a legal limit of 30 mg / 100 ml and recommends fines and/ or imprisonment for transgression. Implementation is poor, however, and the little enforcement that takes place is non-random in geographical coverage, non-visible, and non-uniform. There has been very little attention given to the aspect of early detection and brief intervention at the level of primary health care providers, emergency room personnel or the police.

Attempts at modifying individual behavior by increasing public awareness through media campaigns, are often non-systematic, with minimal scientific input, are poorly focused, and have not been evaluated systematically. There have been few estimates of the contribution of alcohol to injury causation in India. The WHO Collaborative Study on Alcohol and Injuries assumes special significance, in this regard, since it has been the first systematic assessment of the impact of alcohol on emergency department (ED) attendees in the country.

The WHO Collaborative Study on Alcohol and Injuries – India

The India site for the WHO Collaborative Study on Alcohol and Injuries was in the ED of Victoria hospital, located in the city of Bangalore in the state of Karnataka in southern India. Victoria hospital is the largest general hospital in the state and serves a large catchment area comprising the city market, the city railway station and several densely populated working class neighborhoods. Referrals, especially in case of accidents and injuries, are also received from the entire urban agglomeration of Bangalore city (which has a population of 5 686 844) and its rural hinterland. In the calendar year Jan – Dec 2000, 32 485 patients were seen in the ED, of which 42% were referrals treated for injuries of various kinds.

The objectives of the study were threefold: 1) describe the prevalence of alcohol-related injury; 2.) test the validity of the Y91, ICD-10 code, by comparing observational ratings of ED personnel with breath alcohol measurements, and 3) explore the feasibility of using the ICD-10 alcohol codes as a credible and valid alcohol-related injury data source. Subjects with injuries presenting to the Victoria hospital ED were surveyed continuously from May to July 2001. ED attendees were assessed, by trained study personnel, for the impact of alcohol on their injuries using the WHO questionnaire, and a breathalyser to estimate breath alcohol levels, following an observational rating of alcohol intoxication by the casualty medical officer on duty at the ED.

High Proportion of Alcohol-Related Injuries

A very high proportion of the 658 injuries seen during the study period were alcohol-related. Almost a fourth (24%) of all persons presenting with injuries to the ED (30% of male injured and 4% of female injured) had consumed alcohol within six hours prior to the occurrence of their injury. A further 12.8% of the injured (14% of male injured and 10% of female injured) definitely implicated alcohol use by another person as a contributing cause of their injuries (alcohol use by the perpetrator of the injury and not by the injured patient). "Possible" alcohol use by the perpetrator of the injury was recorded in 22.5% of the injured (20% of male injured and 29% of female injured); here the information about alcohol intoxication in the perpetrator was given by secondary sources (relatives, bystanders etc.) and not by the primary injured patient (who was often afraid of indicting a close family member). Pooling the injuries 'primarily linked' to alcohol use (injuries resulting from the patient's own alcohol use) and the injuries 'secondarily linked' to alcohol (those resulting from someone else's alcohol use) raised the proportion of alcohol-related injuries to 59% of all the injuries (64% of male injured patients and 43% of female injured patients) treated at the ED over two calendar months.

This is somewhat higher than that reported in previous international studies, where between 10% and 35% of injury cases were found to be alcohol-related (although these figures were based on only those injuries 'primarily linked' to alcohol) (Cherpitel, 1993; Maio et al. 2000; Pickett et al. 1998), but similar to figures from some low-income countries (Cherpitel et al, 2005). The high rate of injuries attributable to alcohol, as recorded in this study, supplements previous observations from India on the inordinately high association between alcohol use and health consequences in studies from general hospitals (Sri et al, 1997), road traffic accidents and suicide (Gururaj, 2005).

Hazardous Patterns of Alcohol Use in ED Attendees

The patterns of drinking observed among the ED population reflected the prevailing pattern in the general population. A low prevalence of alcohol use in both males and especially in females was found, contrasted with frequent heavy use in a large proportion of users with harmful consequences of drinking. Drinking was restricted to spirits (rather than beer), marked by bingeing (>5 drinks per drinking occasion) and one out of two alcohol users scored above the cut-off for hazardous drinking on the AUDIT. In this study the current alcohol-related injury was in most cases not an aberration caused by an occasional binge; similar amounts of alcohol had been consumed during the same time period the week before, which was typical of the patient's pattern of consumption. Almost a tenth of the regular alcohol users had prior ED visits for an injury during the last year.

Spectrum of Injuries

The largest proportion of injuries consisted of violent and intentional injuries (including various forms of assault), and accounted for a third of all injuries and more than half of all alcohol -related injuries. Road traffic accidents accounted for less than 20% of all injuries and only 12% of alcohol-related injuries. While much attention has been paid to driving while intoxicated in other countries, in India, and probably in other developing countries as well, the spectrum of alcohol-related injuries is different. Various socio-cultural and economic phenomena likely account for these differences. For example, factors like overcrowding, poverty and unemployment, coupled with the keen competition for scarce resources (all of which are persistent in most develop-ing economies), often provide a catalyst for interpersonal violence. This is not to suggest that adopting anti-drinking and driving measures is not an urgent concern, but, overall, as a means of reducing the burden of preventable injury, this focus will clearly not be enough. The problem foci of interventions will need to be much wider.

Gender and Injuries

There is an overwhelming male preponderance (30% male, 4% female) among injuries resulting from one's own drinking, which is understandable in the light of the predominantly male usage of alcohol in India. The gender balance shifts dramatically for injuries attributable to others' use of alcohol, however (34% of males injured and 39% of females injured). A strong gender differ-ence in the spectrum of injury is also evident, with injuries due to burns, hanging, poisoning and assault over-represented among women, while road traffic accidents and injuries due to assault were more common among men.

Absence of Weekly Variation in Incidence of Alcohol-related Injuries

No weekly variation was observed in the incidence of injuries reporting to the ED. A high preva-lence of alcohol-related injuries on weekends is highlighted in most international reporting on injuries, attributed to the influence of weekend drinking binges. However, the observed pattern in India was one of frequent (nearly every day) heavy drinking, and not weekend bingeing.

Detection of Intoxicated Persons

Medical officers in the ED were able to reliably distinguish intoxicated from non-intoxicated patients, identifying 80% of patients with breath alcohol concentration (BAC) over 30 mg/dL, (the legal limit for driving in India), with a false positive rate of 6%. There was much poorer agreement, however, between the medical officers' clinical assessment ratings of the level of intoxication (using the ICD-10 Y91 categories), and the level of intoxication based on BAC (using the ICD-10 Y90 categories) (kappa around 26%). This validated one objective of the study. With minimal training, it is possible to screen for alcohol-related injuries in primary care settings and emergency depart-ments, without the necessity for costly equipment. However, with this level of training/exposure, it is difficult for ED personnel to clinically differentiate levels of severity of intoxication, which is better achieved using breath-analyzers.

Unfortunately, it has been difficult to convince ED personnel to continue screening for alcohol use, since the only purpose they envisage for this screening is to fulfill forensic and legal requirements. The health and legal system does not have a process which automatically refers those patients screening positive for alcohol to substance abuse treatment services.

Impact of the WHO Study in India

The study received ample coverage in the media, albeit in the English language press. The results were also successfully used to advance public health guidelines for alcohol-related problems, in the Health Policy Document of the State of Karnataka. This was the first time in India that such a policy document recognized alcohol as a public health problem, and measures to address alcohol-related injuries formed an important part of the recommendations. The findings have subsequently influenced three major prevention initiatives in the country: a) the "Suraksha Sanchar ", a multi-sectoral programme against drinking and driving launched in Bangalore during 2000; b) an initiative to reduce drinking and driving in the Indian capital, New Delhi, and several urban centers around the country, by the Indian Association for Alcohol Policy and the Indian Medical Association in 2006; and c) deliberations on strategies for early detection of alcohol problems and brief intervention by primary care and emergency department physicians throughout the country, by the Ministry of Health & Family Welfare, Government of India in 2005.

Lessons for Future Initiatives

The health burden due to injuries in Indian society is very high and a large proportion of that appears to be alcohol-related, however, this is not adequately recognized. The greatest portion of such injuries is unrelated to drinking and driving, but due to violent assault by intoxicated individuals victimizing non-drinkers. Measures aimed only at drinking and driving, while urgently required, will not address the full spectrum of alcohol-related injuries. In this context, emergency departments are a potential window for early detection and appropriate intervention for alcohol-related problems.

There is thus a need to integrate regular screening for alcohol into ED procedures. Health planners in India are becoming aware of the need to institute and legally mandate a requirement that hospital-based injury surveillance systems record the alcohol intoxication status in the injury victim or in the perpetrator of an injury. This could be either a clinical assessment based on a simple checklist, or preferably, a record of breath/blood alcohol levels. Clear protocols need to be instituted to refer patients with alcohol-related injury to relevant treatment providers. This requires training of medical and police personnel regarding alcohol-related harm, early detection and brief intervention. Properly planned public awareness campaigns, backed by stricter implementation of existing drinking and driving laws and regulations relating to alcohol availability, need immediate attention.

The findings from this survey highlight the need to conduct more research on the effect of alcohol use on injuries; opportunities should be explored to do this across the country as well as in urban and rural settings. Data on the social cost of such injuries is likely to be more persuasive for law-makers and policy planners than data on damage. Initial work on social cost of alcohol misuse in India (Benegal et al, 2000) suggests that the cost of treating alcohol-related problems far outweighs the profits that the state accrues from the production, sale and taxation of alcoholic beverages. Above all, the findings from this and similar studies need to be widely publicized in order to influence health policy and planning, so that more effective initiatives are undertaken to reduce preventable injuries and alcohol-related problems in India.

References

Benegal V (2005). India: alcohol and public health. *Addiction*, **100**(8), 1051-6.

Benegal V et al. (2005). Women and alcohol in India. In *Alcohol, Gender and Drinking Problems: Perspectives from Low and Middle Income Countries.* Obot IS, Room R, eds, pp. 89-124. World Health Organisation, Geneva.

Benegal V, Gururaj G, Murthy P (2003). *Report on a WHO Collaborative Project on Unrecorded consumption of Alcohol in Karnataka, India* [monograph on the Internet]. (Available at http://www.nimhans.kar.nic.in/Deaddiction/lit/UNDOC_Review.pdf Accessed on 31st May, 2006)

Benegal V, Velayudhan A, Jain S (2000). *Social Costs of Alcoholism: A Karnataka Perspective.* NIMHANS Journal, **18** (1&2) 67.

Blum RH, Blum EM (1969). *A Cultural Case Study*, pp. 188-227. In: Blum RH et al. Drugs I: Society and Drugs, Jossey-Bass, San Francisco, 1969, pp. 226-227.

Cherpitel CJ et al.; Emergency Room Collaborative Alcohol Analysis Project (ERCAAP) and the WHO Collaborative Study on Alcohol and Injuries. (2005) Multi-level analysis of alcohol-related injury among emergency department patients: a cross-national study. *Addiction*,**100**(12),1840-50.

Cherpitel CJ (1993). Alcohol and injuries: a review of international emergency room studies. *Addiction*, **88**, 923-937.

Gaunekar G et al. (2004). Drinking Patterns of Hzardous Drinkers: A Multicenter Study in India. In *Moonshine Markets: Issues in Unrecorded Alcohol Beverage Production and Consumption.* Haworth A, Simpson R, eds, pp. 125-144. Brunner-Routledge, New York.

Gupta PC et al. (2003). Alcohol consumption among middle-aged and elderly men: a community study from western India. *Alcohol and Alcoholism* **38**, 327-31.

Gururaj G (2005). Injuries in India: A national perspective. In: NCMH Background Papers – *Burden of Disease in India National Commission on Macroeconomics and Health* eds, pp. 325-347. Ministry of Health & Family Welfare, Government of India, New Delhi

Maio R et al. (2000). Adolescent injury in the emergency department: opportunity for alcohol interventions. *Annals of Emergency Medicine*, **35**(3) 252-7

Mohan D, Anderson R (2000). Injury prevention and control: International course on injury prevention and control. TRIPP, New Delhi.

Pickett W et al. (1998). Surveillance of alcohol-related injuries in two Canadian emergency department settings: an analysis and commentary. *Contemporary Drug Problems*, **25**, 441-451

Ray R et al. (2004). The Extent, Pattern and Trends of Drug Abuse in India: National Survey. United Nations Office on Drugs and Crimes & Ministry of Social Justice and Empowerment, Government of India. New Delhi.

Singh G (1986). Epidemiology of alcohol abuse in India. *In: Proceedings of the Indo US Symposium on alcohol and drug abuse.* Ray R, Pickens RW NIM, eds, HANS Publication No. 20, Bangalore.

Sri EV, Raguram R, Srivastava M (1997). Alcohol problems in a general hospital–a prevalence study. *Journal of the Indian Medical Association.* **95**:505-6.

CHAPTER 15 :
COMMUNITY CONTEXT AND EMERGENCY ROOM RESEARCH: TWO SOLITUDES OR OPPORTUNITIES FOR COLLABORATION?

Norman Giesbrecht - Center for Addiction and Mental Health | Toronto, ON CANADA |
Jacek Moskalewicz - Institute of Psychiatry and Neurology | Warsaw, POLAND

Introduction

The chapter examines the potential for closer synergy between community action projects (CAP) and emergency room (ER) studies focusing on alcohol and trauma. Despite major differences in agendas, methods, actors and ideological backgrounds, there is significant potential for synergy. However, the potential for collaboration across these two traditions is under-developed and under-utilized. Three main questions are addressed: How does the community context of the ER have a bearing on local initiatives to reduce alcohol-related injuries? What are the potential community-oriented applications of ER data on alcohol and injuries? What challenges are likely to emerge in the community-ER interactions? The practical aim for such collaboration is to facilitate active exchange across community action and ER projects in order to reduce alcohol-related harm.

Community Context of ER Studies

The emergency room (ER) does not operate as a universal, purely technical response to acute health problems of individuals. ER operation depends on a number of mainly external character-istics such as the following: population (age, health status, longevity); health system in general and its capacity; organization of emergency health services; natural environment (climate, winds, floods, terrain); man-made environment; and socio-cultural environment.

ER operation is also affected by alcohol-specific characteristics, such as the following: overall consumption in the host population; per session consumption, including binge drinking; frequency of drinking; place of consumption; time of consumption; consumption context (leisure, working, driving, companions, occasion); alcohol availability; alcohol promotion/advertising; drunkenness control (formal); social control (informal); stigma associated with heavy drinking and/or drunkenness; and the social definition of alcohol and alcohol-related problems.

Interaction of external and alcohol-specific factors

The interaction between alcohol and external characteristics affects the risk of injury, and there-fore may mediate or reinforce demands for emergency services and their potential in prevention and reduction of acute health problems. Individual medical complications, therefore, become a public health problem, and ER studies may emerge as a significant partner in community action projects on alcohol.

To contribute more to this partnership, ER studies should address all of the above characteristics instead of focusing exclusively on the direct relationship between individual drinking and risk of injury, mediated by a few socio-demographic features of an individual. On the other hand, community action projects could more intensely involve ER services and research as a source of information, as well as the ER as a significant agent in ameliorating alcohol-related problems, and a valuable intervention setting.

– ER and general population

It is well known that risk of injury is higher at younger ages, nevertheless, this relationship may vary from community to community. In some communities the likelihood of injury is very high among teenagers and young adults and then declines rapidly among those aged 30+, while in other cultures a high risk of injury is found in older cohorts up to 50 years old. It therefore can be expected that the burden on emergency services will be higher in developing countries of relatively low life expectancy with a high proportion of youth or young adults and a substantially smaller proportion of elderly. It can also be assumed that in communities where health in general is poor, risk of injuries may also be on the rise as unhealthy people are more likely to be victims of accidents and other traumatic situations.

Alcohol magnifies these risks. Its almost unrestricted availability in a number of developing countries (except those of Islamic tradition) is followed by an increase in alcohol consumption and an increasingly negative impact on health indicators, particularly among younger generations (Room et al 2000). In all European countries, alcohol is a main cause of death for males aged 15 to 29, but the magnitude of its role varies from a prevalence of 41% of all deaths in that age group in so called Europe C (Baltic States, Belarus, Hungary, Kazakhstan, Moldova, Russia, Ukraine) to around 25% in the remaining European countries (Rehm et al., 2006). This variation can be attributed to different consumption levels that vary in response to alcohol availability and level of binge drinking.

– ER and health system in general and its capacity

There is a considerable variation in health care systems throughout the world, from ones offering free health care for all citizens, and those with mandatory health insurance systems that may cover either the total population or some sectors, to entirely privatised care. There is also a substantial variation in the organization of emergency care. Often emergency services are offered free of charge, regardless of the financial or insurance status of the patient. Where this is the case the demand for emergency services may be very high as these services tend to be utilised by those who are deprived of more regular care and whose less serious medical problems may develop into acute conditions. Alcohol-related cases can also be over-represented in the ER as people excluded from subsidized support (e.g. homeless, migrant workers) tend to drink in less secure environments and are more likely to become intoxicated due to bad nutrition and poor health, as well as due to consuming a lower quality of alcohol, including non-beverage alcohol.

A crisis of the welfare state, including the high costs of delivery of health and social services, may produce contradictory results with regard to the demand for emergency services. On the one

hand there may be pressures to extend services to those who are high-risk drinkers or dependent on alcohol, and, on the other hand, initiatives to save costs may become a high priority.
"Blaming the victim" has become an issue of public debate in relation to costs of health care. As a
consequence, those who can be blamed for their own disease or injury may be expected to pay
for their treatment. Similar policies are often implemented by car insurance companies, which do
not pay compensation if the driver was under the influence of alcohol at the time of the accident.
The potential deterrent effect of this policy has not been sufficiently studied. Negative side effects
can be predicted for health services, including corruption, under-reporting, and serious health
problems among those who postpone seeking help while drinking heavily until they are sober.

– Natural environment (climate, winds, floods, terrain)

The type and frequency of emergency interventions may be affected greatly by weather, climate
and natural disasters. In climates with hot summers and severe winters, annual distribution
of emergency interventions will be different compared to countries with more stable climatic
conditions. Natural disasters are usually accompanied by peaks in emergency admissions. The
probability of injury might be higher in hilly terrain compared to flat areas.

Alcohol consumption is very likely to reinforce the negative effects of extreme weather or natural
disasters. In Poland, several hundred people die annually from being exposed to low tempera-
tures in the wintertime. It is said that 80% of these individuals are under the influence of alcohol.
No data have been collected as to the proportion that freezes to death after using alcohol to
obtain a sense of warmth in unheated houses or other cold shelters. Respective figures for Russia,
and particularly for the northeast region, suggest that the proportion must be very high indeed,
considering the combination of a severe climate, high alcohol consumption and a high level of
social isolation. There is also no evidence regarding the proportion of the elderly French who
died in the thousands during the hot summer of 2003 who abused alcohol in search of its thirst-
quenching properties. Finally, heavy drinkers in a natural disaster are at higher risk of injury, and
are less co-operative during rescue efforts, contributing to increased chaos and potentially also
to an increase in the risk of casualties among others.

– Man-made environment

Human beings shape the environment where they live, work, move and enjoy their leisure.
Cultures differ in their potential and efforts in making this environment safe in terms of accidental
casualties including falls, fires, drownings, and accidents associated with driving a car or operating various types of machinery. All of the following environmental factors may contribute to an
increased risk of unintentional injury, as examples: uneven side-walks, slippery floors, dilapidated
stairs, unprotected stair-cases, insufficient protection from machinery in motion, lack of side-walks
or pedestrian pathways providing sufficient separation from traffic. In addition, the probability of
intentional injury – for example, attempted homicide or other assault – may also be on the rise
due to a high population concentration at risk in inner city ghettos, availability of firearms, and
insufficient lighting in public spaces, among other factors. Alcohol outlets and their density and
time of operation constitute a crucial network in the human environment. The effective utilisation
of an unsafe space requires sobriety and vigilance to avoid unintentional injuries, and if an unsafe
environment is additionally complemented by a high density of alcohol outlets, then the overall
risk of injury related to alcohol may be extremely high (Gruenewald, Millar & Roeper 1996).

– Socio-cultural framework

There are several socio-cultural factors that affect the risk of injury in a community. One of these is the position of safety in the value system of the population. If safety issues are of low status this can result in not only negligence of safety aspects in daily life leading to risky behaviour, but also to negligence in constructing and maintaining apartments, offices, factories, streets and machinery, resulting in an increase in unintentional injury as noted above.

Social definitions of alcohol and alcohol-related problems are well integrated into the socio-cultural framework in a community, and reflect prevailing value systems. If health, safety, and self-control are not priority issues, then drinking will tend to increase risks for health and safety, including accidental casualties and intentional injury. Social definitions are not universal and may differ from community to community, from one gender to another, across social classes, and from one institution to another. They have paramount impact on the position of alcohol among other urgent issues in a community as a whole, as well as on the priority given to medical and public health domains and the priority given to alcohol in a particular ER or hospital. All of these factors will influence whether or not there will be opportunities for ER studies to inform community action, or for a community initiative to use ER-based data and the ER as an important resource.

Community action projects on alcohol and emergency room studies

There is an extensive and growing research literature on these projects, and evaluations have shown that they have had an impact on orientation to alcohol policy, drinking and drinking-related harm (e.g., Casswell and Gilmore, 1989; Holder et al., 2000; Wagenaar et al., 2000). However, it is uncommon to find the ER as a setting for the data collection and monitoring and/or the intervention component of these projects. Nevertheless, there is a potential to use the ER in the community action project context in order to achieve one or more of the following goals:

- ER data can be used to increase awareness of the damage from alcohol;

- ER data can be used as an indicator of the range of problems in the community and the need for effective interventions and alcohol policies;

- the ER can be a setting for an intervention among the range of prevention strategies used in a community-based project; and,

- ER data can be used in combination with other data to monitor progress and evaluate the outcome and impact of a community action project.

ER Studies and Injuries – Potential Applications

There is an extensive, and ever expanding, research literature on the presenting conditions, drinking experiences, blood alcohol concentration (BAC) levels, and demographic characteristics of patients in emergency rooms. Occasionally, one sees reference to prevention applications at the end of a paper, but this is not common practice, with the important exception of the research on screening and brief interventions (SBI) in ER contexts. While SBI has received a good deal of recent attention in this regard, it is, nevertheless, somewhat baffling or puzzling, given the volume of research in the ER, that there appears to be relatively infrequent reference of the implications of this research for interventions in the ER or the community setting.

A preceding section of this chapter has discussed the ER in the community context, and more specifically, the potential interaction between community action research projects and the ER. The following paragraphs offer a preliminary exploration of the same general topic, but from the other side, namely looking at the potential prevention applications from the context of the ER. The basic question is the following: How might the ER and ER data on alcohol and injury play a role in prevention strategies in the community?

The potential applications can be organized into three general levels: individual, group and community. Furthermore the first two might be either in the ER or hospital setting or outside. This is illustrated in Table 1; however, this table assumes that there is a clear conceptual and institutional boundary between the ER/hospital setting and the community, which may not be the case. There are numerous hospitals that have extensive community out-reach programs that seek to prevent or better manage both acute and chronic diseases, including trauma.

Table 1: ER and Prevention/Intervention Activities on Alcohol and Injuries: Examples by Setting and Level

Level Setting	Individual	Group	Community
Within the ER and Hospital	■ Screening and brief interventions ■ Inpatient treatment for presenting problem	■ Group counselling; ■ AA meetings	
Outside the ER and Hospital	■ Out-patient programs, e.g. coordination home visits with public health staff	■ Group counselling; ■ AA meetings ■ Workers in specific plants receive prevention ■ DUI programs based on ER data	■ Drinking locals identified for prevention & control ■ Work with police or street workers to reduce harm ■ ER data to monitor and evaluate alcohol policy initiatives and community action projects

Individual level interventions

A prime example of an individual-level intervention is that of the development of screening and brief treatment. This has been shown to be effective in reducing consumption and harm in other settings, such as primary care physician offices, and has potential for expanded implementation in ER settings (Babor et al. 2003). The application of SBI protocols, unfortunately, has tended to be project specific rather than institutionalized as routine practice.

Group level interventions

Group level interventions can also benefit from ER data on alcohol and injuries. Patients presenting with a history of alcohol problems can be referred to counseling. There is also the potential to use aggregate ER data to make the link to the outside context, but this is uncommon; for example, providing information to union or management leadership about certain work sites that demonstrate a high volume of alcohol-related injuries. Also, if specific on-premise sales venues turn up frequently as the place of last drink related to either intentional or unintentional injuries, this would be useful information for prevention campaigns and interventions.

Population level and the community context

There are several potential contributions of ER studies to prevention at the community level. These studies can contribute to a more comprehensive profile of alcohol-related damage and raise awareness of alcohol as a risk factor. These studies can also identify drinking contexts/settings where risks of alcohol-related damage are elevated – such as districts, streets, and spaces. The ER-based information can be used to inform choices among prevention options, and offer projections of the impact of possible policy changes on alcohol and trauma. Finally, ER data are a useful tool in monitoring and evaluating the impact of a community-based intervention.

Challenges

There are a number of challenges in bringing together the worlds of ER research and community action projects; it is feasible to briefly mention eight.

- Different protocols: There are different protocols involved. Community-based prevention may be more ad hoc vs. a high level of standardization in ER operations.

- Ethical considerations vary: An ER intervention may have implications of "blaming the victim", involve a compromise of privacy, and involve research during a time of suffering in the ER setting. By focusing on alcohol this may shift attention from other causes of problems.

- Management issues: When the cultures of community action and ER practice come together there is likely not a transparent or obvious protocol for determining who owns the project and runs it. There may be competition and envy for resources and divergent priorities with regard to research, service delivery, and the community-based prevention agenda, as well as no clear guidelines for setting priorities.

- Setting priorities – alcohol-related problem: This may be influenced by epidemiology, emerging 'popular' priorities and special interests. For example, is priority for a prevention trial given to the most common cases of alcohol-related trauma, the most willing patients, or the most serious presenting cases?

- Setting priorities – level of activity: Decisions about prevention will likely also need to consider whether there are resources to focus on all three – individual, group and community level activities, and, if not, which should receive priority.

- Institutionalization: The transition from a research project to routine prevention practice seldom involves an easy or obvious route. Both the ER management and staff and the community might be willing to tolerate a temporary intrusion as part of the prevention protocol, but a long-term strategy will generate unique dynamics. Institutionalization will involve a substantial increase in resource commitments, as well as ethical issues, for example, routinely collecting alcohol/other drug data on all adult ER patients.

- Different ideological backdrops: The community action project (CAP) and the ER research study likely involve different ideological orientations. The CAP will likely emphasize collective responsibility and solidarity, whereas the ER project may emphasize individual responsibility, individual identification (potential stigmatization), and individually oriented interventions.

- Different orientations: Communities seek to manage aggregate-level alcohol-related harm while ERs are oriented toward treating individuals.

Conclusions

The two fields of community-based action projects and ER studies of alcohol and trauma can co-exist, either competing for scarce resources, or seeking opportunities for co-operation. So far the potential for collaboration and co-operation has been under-utilized. There are some risks and challenges involved, however. Two major challenges need to be emphasized. The ER may run the risk of stigmatizing the drinker and his or her behaviour, through identification and labeling. The community action project can lead to unduly focusing on alcohol as a single cause and thereby shift attention from other structural problems in the community that contribute to alcohol-related problems.

Both approaches to prevention, with admittedly different foci and protocol, can be modified to work in a coordinated way that will produce a substantial added value. ER data can inform and monitor the impact of a community-based intervention, while a community action project can serve to raise the profile of hospital-based interventions focusing on alcohol and trauma.

References

Babor T et al. (2003). *Alcohol, No Ordinary Commodity: Research and Public Policy*. Oxford University Press, Oxford

Casswell S, Gilmore L (1989). An evaluated community action project on alcohol. *Journal of Studies on Alcohol*, **50**, 339-346.

Gruenewald PJ, Millar AB, Roeper P (1996). Access to alcohol: geography and prevention for local communities. *Alcohol Health and Research World* **20**, 244-251.

Holder HD et al. (2000). Effects of community based interventions on high risk drinking and alcohol related injuries. *Journal of the American Medical Association*, **284**, 2341-2347.

Rehm J, Taylor B, Patra J et al. (2006). Volume of alcohol consumption, patterns of drinking and burden of disease in the European region, 2002. *Addiction*, **101**, 1086-1095.

Room R et al. (2000). Alcohol policies in developing societies: perspectives from a project. *Journal of Substance Use* **5**(1), 2-5.

Wagenaar AC et al. (2000). Communities Mobilizing for Change on Alcohol: Outcomes from a randomized community trial. *Journal of Studies on Alcohol*, **61**, 85-94.

CHAPTER 16 :
COMMUNITY PREVENTION OF ALCOHOL-INVOLVED INJURIES: THE ROLE OF EMERGENCY ROOM STUDIES

Harold D. Holder - Prevention Research Center | Berkeley, CA US

Introduction

The papers in this collection confirm that alcohol is a major contributing factor to injuries, based upon emergency room (ER) studies. Emergency room studies, mortality data, and population surveys have found that the main areas of such alcohol-involved trauma are motor vehicle crashes (the leading cause of accidental death in the U.S.), falls, drownings, and burns (the second, third, and fourth leading causes of accidental death) (see Saltz, Gruenewald and Hennessy, 1992). Nearly 70% of young adult (aged 20-24) deaths in traffic crashes involve alcohol. Traffic crashes are the leading cause of death for young people under 25, and nearly 70% of young adult (aged 20-24) deaths in traffic crashes involve alcohol (Fell and Nash 1989; Zador 1989). Not often recognized, but also of significance for understanding alcohol-related injuries and fatalities, is the role of alcohol in assaults (which produce injuries of varying severity), homicides, and suicides. In the case of assaults and homicides, estimates of alcohol involvement range from 40% to 50% (Pernanen, 1991; Parker, 1993). In the case of both homicide (Parker and Rebhun, 1995) and suicide (Rossow, 1993; Gruenewald et al., 1995), there is increasing evidence of a causal involvement of alcohol. Thus alcohol-involved trauma is an important prevention target.

Prevention – Reduction of Alcohol-involved Problems at the Population Level

The goal of alcohol prevention is to reduce the future incidence of specific alcohol-involved problems. Community-level alcohol problem prevention is most effectively based upon a public health perspective and is concerned with reducing alcohol-involved problems at the population level, i.e., where the entire community is at risk. Alcohol-involved problems which are most sensitive to heavy and/or high-risk drinking in the moment are typically acute and immediate. For example, impairment by a drinking driver resulting in a traffic crash is an immediate event.

A public health perspective for prevention of alcohol problems considers the community as a dynamic system (See Holder, 1998). No single prevention program, no matter how good, can sustain its impact, particularly if system-level changes are not accomplished, e.g., increased enforcement of alcohol sales or drink driving or training alcohol servers to prevent over-serving. At the local level, policy makers can establish the priorities for community action to reduce risky behavior involving drinking which, in turn, can reduce the number of alcohol-involved problems. Changes in the environment effect changes in drinking, which can reduce acute alcohol-involved problems, such as injuries.

Determining the Effectiveness of a Prevention Intervention: ER Studies as a Resource for Community-Based Prevention Projects

A key requirement of any scientifically-based community prevention trial is a strong evaluation that (a) clearly and operationally defines the population-level alcohol-involved problems that are to be targeted and (b) provides for measurement of each problem over time, i.e., at least pre and post intervention. Two acute alcohol-involved problems that have been shown to be sensitive to prevention interventions are: (1) alcohol-involved traffic crashes; and (2) trauma treated in emergency rooms and hospital inpatient care. Both are incomplete but necessary indicators that overlap (i.e., victims of traffic crashes often end up in hospital emergency rooms).

While traffic crash records provide important data for community prevention evaluation, they are insufficient in at least two ways. First, these data provide information only about alcohol-involved crashes resulting from driving. While important, these data cannot document non-traffic alcohol-involved injuries. Second, the typical consequences of crashes, e.g., non-fatal injuries to drivers and passengers, are not systematically recorded in police reports, and there is great inter-country variation in whether and how this is recorded.

Across the U.S., the level of blood alcohol concentration (BAC) for acute death cases resulting from an either intention or unintentional traumatic event is often, but not universally, available. Thus, it is theoretically possible to determine the number of acute death cases over time in which alcohol was present. Such data are not always systematically collected from autopsies in many areas, as the determination of BAC for injury deaths can be at the discretion of the local coroner.

Another serious problem in using mortality data for evaluation of a local prevention trial is the small number of acute deaths that occur in small population communities. These small numbers often lack statistical power to detect a statistically significant change that might be attributed to a community prevention trial. Thus, while such data could be useful, if available on a systematic basis, the prevention effort may lack a sufficient number of cases on which to carry out appropriate statistical analyses of the potential prevention effect.

On the other hand, non-fatal injury cases occur at a much higher frequency than fatal injuries. Even in relatively small population communities, the number of these cases can often be sufficient for statistical analyses. Thus, while ERs in community hospitals provide excellent sites to monitor the level of alcohol-involved injuries in a community, the available ER clinical data are unlikely to be adequate for evaluation. There are several potential problems that distract from this alternative for community prevention evaluation. Rarely do ERs collect BAC data on injury patients in a systematic manner. Such data are obtained only when there is a medical reason. As a result, the available BAC information is substantially influenced by medical decision-making and selection bias.

Even if BAC data were collected on all ER injury patients, hospitals are typically reluctant to release such data for fear of litigation or criminal prosecution. For example, if a death or serious injury can be attributed to a driver, or even a non-driver, who was seriously impaired by alcohol (e.g., showed a high BAC level on blood tests), then this situation could trigger either a criminal action by the police or civil liability litigation on the part of other victims, or both. In many situations, obtaining access to BAC data on all ER injury cases from hospital records alone is quite problematic, if not impossible.

An emergency room study, initiated as a purposeful means to evaluate a community prevention trial, provides a viable alternative that can be used to compare the incidence of alcohol-related injuries (and their secular trend through the implementation years), both over time and between experimental and comparison sites, if part of the research design. Emergency room data can be collected from all target and comparison sites throughout the course of any prevention project through personal interviews with a systematic sample of patients, and by collecting BAC levels using breathalyzers.

ER injury data would ideally be collected from all ERs in the local communities involved in the prevention trial every day of the week and across all hours of the week over the life of the prevention trial. While ideal, this data collection schedule may not be possible due to the high costs associated with continuous data collection. Less frequent data collection schedules can be used appropriately. Alcohol-involved injuries are much more likely to occur at night (typically after 8 pm) and on Friday and Saturday nights, and a data collection schedule that systematically and consistently collects data during such time periods can be used.

ER studies, as a part of an evaluation of a community prevention trial, are not necessarily a means to obtain proper epidemiological estimates of the incidence and prevalence of alcohol-involved injuries. Rather, the purpose of ER studies within a community trial is to document if any changes have occurred in alcohol-involved injuries in the community such that if a change is detected, then this change can be correctly attributed to the effects of the prevention effort itself in the 'intervention' community. This assumes that other explanations for any observed change can be ruled out. Therefore, valid baseline measurement provides a standard against which any subsequent data can be compared. For purposes of evaluation, the requirement is that subsequent ER data be collected in an identical fashion, during the same hours and days as the baseline data, and identical patient sampling procedures (if any) are also used. This assists in reducing measurement bias into the ER study. Further, a similar ER data collection protocol should be implemented in the comparison or 'control' community where there is no special prevention effort.

It is necessary that sufficient baseline (pre-invention) ER data have been collected to provide a valid "pre" measure of alcohol-involved injuries. At a minimum, sufficient "post" intervention data should also be collected. If resources are available, then routine collection of ER data during the period of prevention intervention creates the opportunity for a longer time series than a simple pre and post measurement. Time series ER data are much preferred over a simple pre and post design, since ER data are often subject to natural trends that are independent of the prevention interventions themselves. Such trends need to be statistically controlled in analyses in order to reduce the risk of Type I errors. For example, a natural downward trend in alcohol-injury data may occur independent of the prevention intervention, but be inappropriately interpreted as a prevention effect. An adequate time series of ER data increases the ability of the statistical analyses to detect a significant change in alcohol-involved injuries, and to accurately attribute any observed changes to the community trial itself. For example, for most tests of an intervention within a time series, i.e., interrupted time series analyses, it is suggested that there are at least 50 time observations.

A long series of observations significantly enhance confidence in inferring causal relationships from observed differences by reducing a number of threats to the internal validity of the design, such as maturation, selection, testing, and regression-to-the-mean effects (Cook and Campbell, 1979). Statistical models can be developed for any time series following the general analytic approach outlined by McCleary and Hay (1980). With adequate time series data, it can be possible to utilize ARIMA techniques associated with Box and Jenkins (1976) in conjunction with Box-Tiao intervention analysis (Box and Tiao, 1975). The superiority of the Box-Jenkins strategy to regression analysis and other analytic approaches on this issue has been demonstrated – see Newbold and Granger (1974), and Vigderhous (1978).

In summary, the advantage of using ER data collection for evaluating effectiveness of a community prevention trial is the potential of obtaining alcohol-involved injury data over time, i.e., longitudinal measurements, for evaluation. Such data collection, if undertaken systematically and over the life of the prevention trial, provides real-time information about the actual effectiveness of the overall interventions of the community trial. No need to await archival data to be released (often a delay of 12 months or more). If desired, these data can be used by the community prevention trial to assist in mid-course corrections of the prevention interventions, and to increase potential prevention trial impact.

Limitations of ER studies for Evaluating Community Prevention Efforts

ER studies as part of an evaluation of a community prevention trial face some important challenges. Routine data collection by non-hospital staff can be expensive if undertaken with sufficient frequency to provide data useful for evaluation. The typical ER study to collect epidemiological data of the relationship of alcohol to injury is unlikely to provide adequate data in support of prevention evaluation. The demands of evaluation are at a minimum for a pre and post measurement. To increase statistical power more frequent measurements during the intervention period are necessary.

Another challenge is that in many local areas, especially urban areas, trauma centers exist to handle the more serious injury cases and as such can by-pass the local ERs in community hospitals. Serious injury cases (often the types of injuries that have alcohol involvement) may not be treated at a local hospital ER. Rather as a result of triage or decisions made by emergency medical teams, the most serious cases can be routed to specialized trauma treatment centers and thus ER data alone may not represent all of the alcohol-involved injuries in a community. This limitation of general hospital ER settings is not a fatal flaw in the data collection as long as all data are collected in identical fashion over the life of the project. Therefore, while incomplete, the ER alcohol-involved injury data can provide a valid indicator of overall changes in such injuries over time and thus be used appropriately for prevention evaluation.

It is also likely that within the prevention target area, there is more than one ER operating. The challenge to the prevention evaluation is to obtain permission to collect data in all of these emergency settings, a significant barrier in many situations. Obtaining access to any ER for purposes of data collection is always a challenge since data collection staff must be present over an extended time. It is an important challenge for any community prevention trial to obtain permission to be present at a regular schedule to collect data. The presence of data collection staff can be perceived by ER staff as disruptive or invasive. This concern may be a less serious barrier for a one-time ER study that is undertaken for epidemiological purposes. However, for longitudinal data collection for purposes of prevention evaluation, the data collection process is present over a much longer period, and hospital and medical staff objection can increase.

Example of a Community Prevention Trial That Utilized ER Data for Evaluation

The Community Trials Project (Holder, et al., 2000) is an example of a prevention project that had a goal of reducing alcohol-involved injuries, and that utilized ER data collection as part of its evaluation. This study evaluated a five-year community-based environmental preventive intervention implemented in three U.S. communities over three years, with a one-year baseline and one-year follow-up. A multiple time series design with three pairs of intervention and matched comparison communities was used to evaluate the efficacy of the prevention programs. The three comparison communities were matched to the experimental communities on the basis of common state regions and similar demographic compositions.

To achieve the goal of reducing overall alcohol-involved trauma, the Community Trials Project implemented and evaluated five broad types of prevention activities referred to here as components. Each component addressed one or more intermediate variables, had its own set of prevention activities, and was designed to be mutually reinforcing with other components. The five interacting components included: (1) a "Community Knowledge, Values, and Mobilization" component to develop community organization and support for the goals and strategies of the project; (2) a "Responsible Beverage Service Practices" component to reduce the risk of intoxicated and/or underage customers in bars and restaurants; (3) a "Reduction of Underage Drinking" component to reduce underage access; (4) a "Risk of Drinking and Driving" component to increase enforcement efficiency regarding Driving While Impaired, and reduce drinking and driving; and (5) an "Access to Alcohol" component to reduce overall availability of alcohol. The interventions changed the drinking environment by mobilizing community support, raising the standards for responsible beverage service in bars and restaurants, increasing local enforcement of drinking and driving laws, curtailing the retail sale of alcohol to minors, and using zoning and other municipal powers to reduce the number and density of outlets selling alcohol.

Outcome measures included data of alcohol-related traffic crashes, alcohol-involved injury and assault cases treated at local emergency rooms, hospital admissions for treatment of injuries with high likelihood of alcohol-involvement (Treno, Cooper, and Roeper, 1994; Treno and Holder, 1997), and self-report measures of drinking and drinking and driving. Emergency room data were collected from all target and comparison sites every other Friday and Saturday night from 8 am to 4 pm. The ER data collection was undertaken over 12 months prior to prevention intervention, and then continuously over the next four years of the trial. These data were collected through personal interviews with all patients, and by collecting BAC levels with breathalyzers, using the protocol developed and tested by Cherpitel (1989).

This trial presented clear evidence that environmental interventions can reduce alcohol-involved traffic crashes and other acute trauma. These changes produced a 10% reduction in nighttime injury crashes and a 6% reduction in crashes in which the driver had been drinking in the intervention communities relative to comparison communities. A 43% reduction in assaults directly observed in emergency rooms between two matched sites (i.e., intervention and comparison communities) was obtained as well as a 2% reduction in serious assault cases requiring hospital admission. A reduction of 68 assault cases per 100 000 adult population per year between the two matched sites was obtained.

Observations and Conclusions

ER studies of alcohol-involved injuries have been successfully applied in a large number of countries as discussed in this collection in order to obtain epidemiological estimates of the relationship of alcohol to injuries. ER surveys as part of an evaluation of a community prevention trial have occurred less frequently and are a more recent application of this type of survey. The appropriate and consistent application of such data collection as part of an evaluation design can provide valuable and unique data, i.e., data that are not available for prevention evaluation in any other fashion.

Such use of ER surveys requires a careful consideration of the challenges that are inherent in an evaluation of a community prevention project. Regular and intensive use of ER-based data collection can be expensive, and could increase resistance of medical staff to the routine presence of ER data collectors. The number of observations necessary for adequate evaluation will likely exceed the requirements for a single epidemiological study. However, ER surveys provide unique data about alcohol-involved injuries that are not currently available via any other means.

The long history of ER studies, and the validated techniques for such data collection, is a valuable gift to the field of alcohol prevention. Until a time when ERs consistently collect BAC data on injury patients and make these data readily available anonymously for research purposes, researcher-sponsored ER studies for evaluation will be essential. There is currently no other way to document the level of alcohol-involvement in acute intentional and unintentional injuries in order to evaluate community prevention trials seeking to reduce such injuries without ER studies.

References

Box GEP, Jenkins GM (1976). *Time Series Analysis: Forecasting and Control*. Holden-Day, San Francisco.

Box GEP, Tiao GC (1975). Intervention analysis with applications to economic and environmental problems. *Journal of the American Statistical Association* **70**, 7079.

Cherpitel CJ (1989). A study of alcohol use and injuries among emergency room patients. In: *Drinking and Casualties: Accidents, Poisonings and Violence in an International Perspective*. Giesbrecht N et al., eds, pp. 288-299. Tavistock, London.

Cook T, Campbell D (1979). *Quasi-experimentation*. RandMcNally, Chicago, Illinois.

Holder HD (1998). *Alcohol and the Community: A Systems Approach to Prevention*. Cambridge University Press, Cambridge.

Holder HD et al. (2000). Effect of community-based interventions on high risk drinking and alcohol-related injuries. *Journal of the American Medical Association* **284**(18), 2341-2347.

McCleary R, Hay R (1980). *Applied Time Series Analysis for the Social Sciences*. Sage Publications, Beverly Hills.

Newbold P, Granger CW (1974). Experience with forecasting univariate time series and the combination of forecasts. *Journal of the Royal Statistical Society*, Series A 137, 131165.

Parker RN (1993). The effects of context on alcohol and violence. *Alcohol Health and Research World*, **17**, 117-122.

Parker RN, Rebhun L-A (1995). Alcohol and Homicide: A Deadly Combination of Two American Traditions State University of New York Press, Albany, NY. Pernanen K (1991). *Alcohol in Human Violence*. Guilford Press, New York.

Rossow I (1993). Suicide, alcohol, and divorce; aspects of gender and family integration. *Addiction*, **88**, 1659-1665.

Treno AJ, Cooper K, Roeper P (1994). Estimating alcohol involvement in trauma patients: Search for a surrogate. *Alcoholism: Clinical and Experimental Research* **18**(6), 1306-1311.

Treno AJ, Holder HD (1997). Measurement of alcohol-involved injury in community prevention: The search for a surrogate III. *Alcoholism: Clinical and Experimental Research* **21**(9), 16951703.

Vigderhous G (1978). Forecasting sociological phenomena: Application of Box-Jenkins methodology to suicide rates. In: Sociological Methodology. Schuessler KF, ed, 20-51. Jossey Bass, San Francisco.

Zador P (1989). *Alcohol-related Risk of Fatal Driver Injuries in Relation to Driver Age and Sex*. Insurance Institute for Highway Safety, Washington, DC.

CHAPTER 17 :
ALCOHOL POLICY AND PUBLIC HEALTH IMPLICATIONS IN THE U.S. CONTEXT

Thomas K. Greenfield, Cheryl J. Cherpitel - Alcohol Research Group | Emeryville, CA US

Recent Policy Developments Bearing on Injuries and Emergency Services

This chapter considers, first, recent developments in U.S. alcohol policy having a bearing on injuries and the ED, especially those designed to target drinking driving; second, policies specifically related to the ED, such as those promoting or deterring adoption of routine screening, brief intervention and referral to counseling; and, third, it comments on the role of ED data in relation to policy development and legislation at various jurisdictional levels – federal, state and local.

Recent policy issues

One of the key accomplishments in U.S. federal alcohol policies in recent years has been the adoption of the .08 BAC definition for driving while intoxicated (DWI) as a standard. The legislation was passed by huge majorities in both houses (senate and representatives) in October 2000 as part of a federal transportation appropriations bill. As with the earlier federal minimum drinking age (MDA) legislation, also attached as an incentive to a transportation spending bill, the measure would withhold federal highway funds to states not adopting the .08 standard by 2004, however the incentive was staged to increase the percentage of funds to be withheld over time, by up to 8% by 2006. States changing their DWI laws to comply with the lower limit by 2007 would have the withheld funds returned. At the time of the legislation's passage, the District of Columbia and 18 states already had put in place the .08 BAC or lower (Colorado had set it at .05) standard for drunk driving, but most remaining states had a .10 standard. Opponents cited the state prerogative for setting blood alcohol levels. By the end of 2002, a Los Angeles Times article noted that 17 states, including Minnesota, North Carolina, Nevada and Delaware, were resisting the federal push to lower the BAC for DWI (Vartabedian, 2002). For example, the Minnesota state senate with support from the alcohol industry, which argued the .08 standard would trap "social drinkers," initially refused to pass the law despite strong support from law enforcement and medical groups like the Minnesota Medical Association (MMA). It is significant in relation to ED data that the MMA cited the following in making its case for the reduced BAC: "According to statistics from the Minnesota Brain Injury Association, 22 percent of the people that show up in hospital emergency rooms have alcohol impairment" (Minnesota Medical Association, 2001). Although other studies were also noted by the MMA, such ED statistics are often effective at mobilizing legislative and public support for stiffer drunk driving laws, since drinking driving contributes greatly to admissions for injuries. By July 2003, when Iowa adopted the .08 BAC legal definition, 38 other states had adopted the .08 or lower DWI limit, according to an editorial in the Iowa State Daily (Paseka et al., 2003). The Iowa editorial also noted that the boating laws had not been brought into compliance and still retained the higher limit, which it said "just doesn't add up", noting boating accidents were just as serious as highway ones. One of the problems with federal inducements to states to enact laws that may reduce alcohol-related admissions to the ED is that only the targeted specific law may be changed. The editorial evidences, however, how such a change can become a rallying point for evidence-based enactments of related laws such as legal BAC limits for boaters.

Other broad policy approaches such as increases in price, often driven by changes in taxation (Babor et al., 2003), are also effective in reducing alcohol-related morbidity and a range of problems, including the rate of fatal car accidents and drinking driving, particularly in special population subgroups such as youth (Chaloupka et al., 1998). Price of alcoholic beverages has been implicated in a variety of outcomes including various crimes, assaults, domestic and other forms of violence (Cook and Moore, 1993; Cook and Moore, 2000; Grossman et al., 1998), that may in turn lead to admissions to the ED and trauma centers. Despite the excellent evidence base for alcohol tax and price increases as an effective measure to reduce casualties among other alcohol problems, unfortunately, federal excise tax on alcoholic beverages has not been raised since the early 1990s. Congress has not been inclined to raise alcohol taxes for preventive purposes; when alcohol excise taxes have been raised – as when the federal taxes on beer and spirits were raised in 1991 – it has been based on the logic of a general revenue measure (Giesbrecht et al., 2004; Giesbrecht and Greenfield, 2003). If anything, the present Congress has been more responsive to pressure to *lower* taxes on beer and spirits, apparently related to beer and spirits industry lobbying efforts. Taxes on alcoholic beverages in the U.S. have not been linked to the consumer price index, and so the effect of them in real dollars is weakened over time (Babor et al., 2003).

Another policy approach found to reduce many alcohol problems, such as those resulting in ED admissions, is regulating physical availability through such policies as retail monopolies (in 18 states), and state regulations from the departments of alcohol beverage control limiting outlet density. Spatial location and characteristics of bars and other on-premise outlets have been found to be related to drinking driving (Gruenewald et al., 1999) and violent crime (Gorman et al., 2001). A number of studies have shown evidence that bunching of drinking places (increased density) can result in motor vehicle crashes, pedestrian injury collisions and violent assaults (Babor et al., 2003). Thus in summary, a variety of national, state and local policies affecting alcohol problems, and in particular those resulting in injuries often brought to the ED, are known to be effective. Because the ED addresses injuries that may involve alcohol, it represents a mechanism that in effect selects for alcohol involvement. Thus, it represents an ideal opportunity for alcohol screening, intervention and referral, the topic addressed next.

Screening and Brief Intervention in the ED

Screening and brief interventions in the ED have policy relevance in a number of ways. First, from a secondary prevention standpoint, the ED may provide a critical venue for targeted interventions for high-risk individuals, a policy priority of the National Institute of Alcohol Abuse and Alcoholism (NIAAA). Second, for this to be accomplished, policy barriers to screening will need to be addressed. Here we cover both of these issues.

National surveys in the U.S. have repeatedly found that almost half of the trauma centers do not routinely obtain a blood alcohol concentration (BAC) on their patients (Soderstrom et al., 1994), and BACs often miss those patients meeting criteria for hazardous, harmful or abusive drinking or for alcohol dependence. While two-thirds of BAC positive patients in a trauma center met criteria for dependence, almost half of the BAC negative patients also met diagnostic criteria (Soderstrom et al., 1994). Another study in the ED found that a positive BAC identified only 20% of those with a current alcohol use disorder (Cherpitel, 1995). In an effort to reduce alcohol-related visits in the ED, the U.S. Department of Health and Human Services (2000) included as one of the objectives of Healthy People 2010 the adoption of screening, brief intervention and referral to treatment (SBIRT) in the ED by emergency service providers.

The National Alcohol Screening Day (NASD) initiative which annually promoted a day of national screening for alcohol problems across a variety of settings, recently provided an opportunity to demonstrate that it is possible to conduct emergency department screenings for at-risk and alcohol dependent drinking and thereby increase the likelihood of intervening via brief interventions and referral to appropriate treatment. This study, which was funded by NIAAA and the Substance Abuse Mental Health and Services Administration (SAMHSA), was conducted at 14 academic-based EDs (Emergency Medicine Research Collaborative (SBIRT), 2004) and included developing methods to effectively implement a model of screening, brief intervention and referral to treatment into the practice of emergency medicine, identifying ED patients with alcohol use disorders via screening, and developing and testing methods to facilitate referral of patients to specialized treatment (Murray, 2005). The study also included the development of a SBIRT curriculum for use in the ED setting, and training of ED service providers across the 14 sites to facilitate adoption of SBIRT as a standard of care in the ED. A model SBIRT curriculum was initiated and tested by measuring changes in responsibility and confidence in providing SBIRT, perceived barriers to SBIRT and rates of SBIRT adoption among ED service providers receiving the training (Academic ED SBIRT Research Collaborative, 2007). While provider changes were all in the positive direction, changes from baseline to 12 month post-training were only modest, with the largest change an 11% increase in actually providing SBIRT. Unfortunately, these findings, although promising, do not seem overly encouraging.

A meeting of ED clinicians and researchers on identification and intervention with alcohol problems among emergency department patients (2001) was supported by the Centers for Disease Control and Prevention (CDC) in conjunction with a number of other federal agencies. Recommendations included an increase in support from funding agencies for research in screening and interventions for alcohol problems among ED patients (Hungerford and Pollock, 2002), and a more recent conference, also supported by CDC, recommended that SBIRT should become routine practice in trauma centers and an essential component of trauma care (Centers for Disease Control and Prevention, 2005). SBIRT has been recently mandated, as well, by the American College of Surgeons Committee on Trauma, commencing in 2007 (Committee on Trauma, 2006). In due course it will be of interest to evaluate the effectiveness of such a mandate from a professional credentialing organization.

Despite these recommendations, legal barriers to alcohol screening in EDs and trauma centers presently exist in many U.S. jurisdictions. In 1947, the National Association of Insurance Commissioners (NAIC) developed a model law entitled the Uniform Accident and Sickness Policy Provision Law (UPPL). The model law proposed the following language to be adopted in State laws: *The insurer shall not be liable for any loss sustained or contracted in consequence of the insured's being intoxicated or under the influence of any narcotic unless administered on the advice of a physician* (National Association of Insurance Commissioners (NAIC), 1947).

Subsequently, as many as 38 states adopted this law retaining the model law's formulation. A number of more enlightened states, like South Dakota, have enacted statues which specifically prohibit denial of insurance benefits on this basis. For further information on the UPPL, including tables of states where these statutes exist, see the Alcohol Policy Information System (APIS) developed for NIAAA (2006). (http://www.alcoholpolicy.niaaa.nih.gov/ ; accessed 24/10/09)

In spite of the fact that the National Association of Insurance Commissioners in 2001 rescinded its original model law recommendation, it remains true that in a majority of states, this relatively little-known law, the UPPL, is still on the books (Chezem, 2004/2005). In such states the UPPL statute explicitly permits third-party payers to deny reimbursement for medical services for patients who are alcohol positive at the time of injury and it is unnecessary to establish either the level of intoxication or a causal link between alcohol and the injury event. Fortunately, even in many states with UPPL statutes, insurers do not in fact deny payment for alcohol-related claims. But this may, paradoxically, in part be because of ED and trauma center physicians' unwillingness to identify alcohol or other drug involvement. The chilling effect of the UPPL is seen, for example, in a national survey of members of the American Association for the Surgery of Trauma, which found that among those who do not routinely measure BAC (over 50%), most (82%) cited insurance barriers as their reason for failure to screen routinely (Gentilello et al., 2005a). Thus, regardless of U.S. federal alcohol initiatives supporting screening and brief intervention for alcohol problems in the ED and trauma centers, the likelihood of routinely implementing SBIRT seems diminishingly small. As long as state insurance laws are in place in many jurisdictions which imply or may inflict upon patients potentially devastating financial burdens, there is likely to result an understandable reticence on the part of emergency service provides to identify problem drinking. It is clear that policy development in these states needs to focus on removing these legal barriers and reimbursement constraints that interfere with appropriate screening for alcohol problems.

Repeal of the UPPL is a cause that has been taken up by a variety of professional organizations. As one example, the American Society of Addiction Medicine (ASAM) has issued a public policy statement making a detailed case for repeal. The logic of ASAM's policy position seems persuasive. To summarize the ASAM statement's argument, it notes that currently less than 15% of injured patients in U.S. hospitals are screened for alcohol or other drugs, and referred for counseling. Without identification of such problems by screening, even where the UPPL is in place, insurers typically cannot identify alcohol or drug involvement and so end up covering the health costs anyway. Absent screening, treatment and referral, re-offending with alcohol-related vehicular violations and incurring further substance-related medical costs will be common. Studies have indicated that very substantial health cost savings may follow from screening and brief intervention at these so called "teachable moments" when the patient is admitted for injuries (American Society of Addiction Medicine, 2005). The policy statement cites a cost-benefit analysis at the University of Texas Southwestern Medical School and the University of Washington. This study "demonstrated that routine ER and trauma center alcohol screening and intervention would result in an estimated three-year net national savings of $1.82 billion in direct medical costs, all of which go to payers of health care (insurers, state and federal governments). Direct medical costs are estimated to comprise only 15% of total costs, with the balance attributable to property damage, lost wages and other losses. The study found that nearly $4 in direct medical costs are saved for every dollar invested in ED and trauma center screening and intervention" (Gentilello et al., 2005b).

Studies of alcohol policy development in the U.S. (Greenfield et al., 2004a) have suggested a kind of inertia in existing alcohol-related legislation making it difficult for even strong evidence-based reforms to be implemented (Johnson et al., 2004). This may be particularly the case when the governing image (Room, 1987) of alcohol problems that comes most readily to the minds of both legislators and the public suggest that the alcohol abusing individual is morally responsible for his or her difficulties by choosing to drink irresponsibly and thus "has it coming to him or her self". Despite the stakeholders who may appear to benefit from maintaining this governing image of alcohol problems even when lip service is paid to alcoholism as a disease (e.g., the insurance companies for whom 'moral hazard' is to be minimized), there appears to be a gathering momentum for legal reform in this area. Informed opinion is recognizing that screening and intervention may be preferable to the de facto obfuscation and problem denial resulting from UPPL. In the alcohol policy process, as in other policy areas, what political scientists have dubbed "windows of opportunity" have been noted to occur when events in various policy streams create openings that are crucial for enacting or revising laws. At such times, concerted efforts can result in political realignments that may rapidly affect legislative possibilities (Greenfield et al., 2004b). It will be of considerable interest to observe whether overturning these problematic health insurance laws gains traction across the states, clearing the way for implementation on a wider scale of serious SBIRT translational and knowledge transfer efforts. Political scientists have also noted that such windows do not stay open for a long time, so the opportunity must be seized rapidly and with zeal, or it may be lost (Kingdon, 1995).

Need for Marshalling Data from EDs

Finally, data derived from the ED is important for both a policy enactment and a policy evaluation standpoint, which we discuss in conclusion. As the Minnesota Medical Association instance mentioned in the first section showed, data regarding alcohol-related admissions for injuries in ED and trauma centers can be effectively used in campaigns to enact evidence-based laws to reduce alcohol-related problems. Studies of alcohol problem prevention relying on community organizing have illustrated the importance of establishing such data collection both for outcomes monitoring, and for mobilizing community action (Holder, 1998; Treno and Holder, 1997; Voas et al., 1997). For example, alcohol-related injuries, and specifically motor vehicle injuries, treated in EDs were used as baseline data and to evaluate program effectiveness in a community action trial to prevent alcohol-related problems (Holder et al., 2000) (See Chapter 16 in this volume).

Critical to such efforts to employ data from EDs, as with any evidence accumulation, studies should follow scientific desiderata to allow confirmation, refutation and where possible, causal inference, so as to provide a sound basis for policy development (Babor et al., 2003). As the Emergency Room Collaborative Alcohol Analysis Project (ERCAAP) and the WHO Collaborative Study on Alcohol and Injuries cross-national studies show, systematically gathered ED data using standard protocols has the potential to inform policy makers in many countries regarding the extent of alcohol-related injuries seen in the ED (Cherpitel et al., 2003; Macdonald et al., 2005; Young et al., 2004). Ongoing surveillance of alcohol-related injuries in trauma centers and the ED (not routinely carried out in the U.S.) would be important for monitoring the impact and burden of alcohol in the ED and in the community, and has important potential for informing alcohol policy. The chapters in the current volume also reflect the way in which the ED has importance as a policy-relevant fulcrum. Hospital and ED admissions can be a sensitive barometer for assessing other environmental and policy interventions. To increase the policy relevance further, it may be important, in addition to carefully screening for consumption prior to ED admission, to begin to ask about the type of locations and settings in which the immediately prior drinking took place (data which have been collected by the ERCAAP and WHO collaborative studies). Such information may be used in conjunction with community surveys or archival data to develop a risk profile of types of settings or even specific establishments giving rise to an unusually high level of alcohol-related injuries, for example (Mcleod et al., 1999; Stockwell et al., 2002). In sum, the ED is well placed to provide important data on the epidemiology, intervention and prevention of alcohol-related injury for informing policy development.

Beyond the Babor et al. (2003) general dictum regarding characteristics of studies that will best further evidence-based alcohol policies, more thought will need to go to the following questions: What types of ED-related information will most facilitate enactment of good alcohol policies? We believe evidence of the following kinds will be most helpful: prevalence and incidence of damage from alcohol; evidence of high-risk drinking situations that bring casualties to the ED; high-risk drinkers seen in the ED; and lastly, socio-economic and other broad-based social conditions that intensify injuries that can be observed when monitoring occurs systematically over time and across jurisdictions. This brief chapter has outlined some of the ways that both current and enhanced ED data may be used to inform alcohol policies.

References

Academic ED SBIRT Research Collaborative (2007). An evidence-based alcohol screening, brief intervention and referral to treatment (SBIRT) curriculum for emergency department (ED) providers improves skills and utilization. *Subst Abuse* **28**(4), 79-92.

American Society of Addiction Medicine (2005) *Public Policy Statement on Repeal of the Uniform Accident and Sickness Policy Provision Law (UPPL)* (www.asam.org/1REPEAL%20OF%20UPPL%20 7%2D05.pdf), accessed 15 May 2006). American Society of Addiction Medicine, Inc., Chevy Chase, MD.

Babor TF et al. (2003). *Alcohol: No Ordinary Comodity. Research and public policy.* Oxford University Press, New York, NY.

Centers for Disease Control and Prevention (2005). Recommendations for trauma centers to improve screening, brief intervention, and referral to treatment for substance use disorders. *Journal of Trauma* **59**(Suppl.), S37-S42.

Chaloupka FJ, Grossman M, Saffer H (1998). The effects of price on the consequences of alcohol use and abuse. In: *Recent Developments in Alcoholism*, vol. 14. Galanter M, ed, pp. 331-346. Plenum Press, New York, NY.

Cherpitel CJ (1995). Screening for alcohol problems in the emergency department. *Annals of Emergency Medicine* **26**(2), 158-166.

Cherpitel CJ et al. (2003). Alcohol-related injury in the ER: a cross-national meta-analysis from the Emergency Room Collaborative Alcohol Analysis Project (ERCAAP). *Journal of Studies on Alcohol* **64**, 641-649.

Chezem L (2004/2005). Legal barriers to alcohol screening in Emergency Departments and Trauma Centers. *Alcohol Research and Health* **28**(2), 73-77.

Committee on Trauma, American College of Surgeons. (2006). *Resources for Optimal Care of the Injured Patient.* Chicago, Illinois: American College of Surgeons.

Cook PJ, Moore MJ (1993). Taxation of alcoholic beverages. *Economics and the prevention of alcohol-related problems.* Hilton ME, Bloss G, eds, pp. 33-58. National Institute on Alcohol Abuse and Alcoholism, Rockville, MD.

Cook PJ, Moore MJ (2000). Alcohol. In: *Handbook of Health Economics*, vol. 1. Culyer AJ, Newhouse JP, eds, pp. 1629-1673. Elsevier Science, Amsterdam, Neth.

Emergency Medicine Research Collaborative (SBIRT) (2004) [Editorial] Screening, Brief Intervention and Referral to Treatment (SBIRT): new funding opportunities for emergency departments. *Emergency Medicine News* **26**(12).

Gentilello LM et al. (2005a). Effect of the uniform accident and sickness policy provision law on alcohol screening and intervention in trauma centers. *Journal of Trauma* **59**, 624-631.

Gentilello LM (2005b). Alcohol interventions in a level 1 trauma center. *Annals of Surgery* **241**(4), 541-550.

Giesbrecht N et al. (2004). Changing the price of alcohol in the United States: perspectives from the alcohol industry, public health, and research. *Contemporary Drug Problems* **31**(Winter), 711-736.

Giesbrecht NA, Greenfield TK (2003). Preventing alcohol-related problems in the US through policy: media campaigns, regulatory approaches and environmental interventions. *Journal of Primary Prevention* **24**(1), 63-104.

Gorman DM, Speer PW, Gruenewald P J (2001). Spatial dynamics of alcohol availability, neighborhood structure and violent crime. *Journal of Studies on Alcohol* **62**, 623-636.

Greenfield TK et al. (2004a). A study of the alcohol policy development process in the United States: theory, goals, and methods. *Contemporary Drug Problems* **31**(Winter), 591-626.

Greenfield TK, Johnson SP, Giesbrecht NA (2004b). The alcohol policy development process: policy makers speak. *Contemporary Drug Problems* **31**(Winter), 627-654.

Grossman M, Chaloupka FJ, Sirtalan I (1998). An empirical analysis of alcohol addiction: results from the Monitoring the Future panels. *Economic Inquiry* **36**(1), 39-48.

Gruenewald P et al. (1999). Beverage sales and drinking and driving: the role of on-premise drinking places. *Journal of Studies on Alcohol* **60**, 47-53.

Holder HD (1998). *Alcohol and the community: A systems approach to prevention.* Cambridge University Press, Cambridge, MA.

Holder HD et al. (2000). Effect of community-based interventions on high-risk drinking and alcohol-related injuries. *Journal of the American Medical Association* **284**(18), 2341-2347.

Hungerford DW, Pollock DA, eds (2002). A*lcohol Problems among Emergency Department Patients: Proceedings of a Research Conference on Identification and Intervention, Arlington, VA,* March 19–21, 2001. National Center for Injury Prevention and Controls, Centers for Disease Control and Prevention, Atlanta, GA.

Johnson S et al. (2004). The role of research in the development of U.S. federal alcohol control policy. *Contemporary Drug Problems* **31**(Winter), 737-758.

Kingdon JW (1995). *Agendas, Alternatives, and Public Policies.* Harper Collins, New York, NY.

Macdonald S et al. (2005). The criteria for causation of alcohol in violent injuries based on emergency room data from six countries. *Addictive Behaviors* **30**, 103-113.

Mcleod R et al. (1999). The relationship between alcohol consumption patterns and injury. *Addiction* **94**(1), 1719-1734.

Minnesota Medical Association (2001). Senate votes down new.08 blood-alcohol level, July 2, 2001. Minnesota Medical Association, Minneapolis, MN.

Murray M (2005). Academic emergency medicine alcohol screening and intervention collaboration. *International Conference on Alcohol and Injuries: New Knowledge from Emergency Room Studies.* October 3–6, Berkeley, CA.

National Association of Insurance Commissioners (NAIC) (1947). Uniform Individual Accident and Sickness Policy Provision Law (UPPL). In *Accident and Health Insurance Rate and Policy Standards*, Vol. II, pp. 180-185. National Association of Insurance Commissioners, Kansas City, MO.

National Institute on Alcohol Abuse and Alcoholism (2006) *National Alcohol Policy System* (www.alcoholpolicy.niaaa.nih.gov, accessed 15 May 2006). National Institute on Alcohol Abuse and Alcoholism, National Institutes of Health, Rockville, MD.

Paseka N et al. (2003). Editorial: New blood-alcohol limit may save lives. *Iowa State Daily* **July 3, 2003**.

Room, R. (1987) Governing images and self-control: a comment on Drew's 'Beyond the disease concept of addiction'. *Australian Drug/Alcohol Review* **6**, 51-54.

Soderstrom CA, Dailey JT, Kerns TJ (1994). Alcohol and other drugs: An assessment of testing and clinical practices in U.S. trauma centers. *Journal of Trauma* **36**(1), 68-73.

Stockwell T et al. (2002). Alcohol consumption, setting, gender, and activity as predictors of injury: a population-based case-control study. *Journal of Studies on Alcohol* **63**(3), 372-379.

Treno AJ, Holder HD (1997) Measurement of alcohol-involved injury in community prevention: the search for a surrogate III. *Alcoholism: Clinical and Experimental Research*(9), 1695-1703.

U.S. Department of Health and Human Services (2000) *Healthy People 2010: Understanding and improving health*, vol. 1 & 2. U.S. Department of Health and Human Services, Washington, DC.

Vartabedian R (2002). A Spirited Debate Over DUI Laws. The government's effort to compel states to lower blood-alcohol limits encounters resistance. A senator in Iowa calls the policy 'blackmail'. *Los Angeles Times* **December 30, 2002**, A.1.

Voas RB, Holder HD, Gruenewald PJ (1997). The effect of drinking and driving interventions on alcohol-involved traffic crashes within a comprehensive community trial. *Addiction* **92**(Supplement 2), S221-S236.

Young DJ et al. (2004). Emergency room injury presentations as an indicator of alcohol-related problems in the community: a multilevel analysis of an international study. *Journal of Studies on Alcohol* **65**, 605-612.

CHAPTER 18 :
IMPLICATIONS OF EMERGENCY DEPARTMENT STUDIES FOR ALCOHOL POLICY IN A EUROPEAN CONTEXT

Ann Hope - Trinity College | Dublin, IRELAND

European Region and Alcohol Policy

The European Region is most broadly defined by the World Health Organisation (WHO), covering 53 countries with a combined population of 877 million people. The WHO European Region stretches from Western Europe to the Russian Federation and the Central Asian Republics, and from Iceland in the north to Cyprus in the South. Within the WHO region, the European Union (EU), a smaller geographical block, now has 27 Member States and a population of nearly half a billion (Health for All WHO database). Both European institutions have important implications for alcohol policy in the region.

WHO European Region

The WHO European Region has the highest per capita alcohol intake in the world, twice as high as the world average. Alcohol is the third most serious risk factor for European ill health and premature death. The disease burden from alcohol in the region, calculated at 10.8% for 2002, is also twice as high as the world average. In 2002, an estimated 600 000 people died prematurely from alcohol-related causes in the region (Rehm & Taylor 2005). Injuries, unintentional and intentional, account for the largest portion (43.5%) of the alcohol attributable burden of disease, followed by neuro-psychiatric conditions. These alcohol-related injuries include motor vehicle accidents, drownings, house fires, falls, poisonings, assaults, homicide and suicides. The acute hospital emergency department is where many of these alcohol-related injuries present for initial health care treatment. However, only recently has alcohol research focused on the emergency department (ED) at an international level to provide a greater understanding of the nature and extend of alcohol-related injuries. The WHO Collaborative Study on Alcohol and Injuries, set in the emergency room and the Emergency Room Collaborative Alcohol Analysis Projects (ERCAAP) have undertaken research with capacity for international comparisons (Cherpitel et al. 2003, WHO 2005).

The WHO Regional Office has adopted many alcohol policy initiatives to prevent and reduce alcohol-related harm, starting in 1992 with a region-wide action plan of alcohol, followed by the European Charter of Alcohol (1995) and the Declaration on Young People and Alcohol (2001). Reducing alcohol related injuries has been a central part of the two European Alcohol Action Plans. A resolution was adopted by the WHO Regional Committee for Europe in 2005, which endorsed a framework for alcohol policy in the WHO European Region (EUR/RC55/RI). The framework sets out a long-term strategy, emphasising the importance of alcohol policy based on the best scientific evidence and includes tackling alcohol-related injuries through a joined up approach with the WHO strategic area on injuries and violence (WHO, 2006).

European Union

The European Union (EU), now encompassing 27 Member States, was established in 1957 to have a "common market" where trade barriers would be removed and economic development would prosper. Through a succession of treaties, new forms of cooperation between member state governments have been introduced. The EU now deals with many other subjects such as regional development, environmental protection, citizens' rights, consumer protection and public health. In fact, public health only received an independent mandate in 1993 to promote health and inter-state co-operation (Maastricht Treaty) which was further strengthened in 1998 by ensuring all EU community policies achieve a high level of health protection and compliment national policies (Amsterdam Treaty). However, despite the expanding role of the EU, trade and economics continue to be the driving force today with the overriding goal of a single market in which goods, services, people and capital can move around freely (Lisbon Agenda).

During the first thirty years of the EU, alcohol was primarily seen as an agricultural product and as a trade commodity (Österberg & Karlsson, 2002). The late arrival of public health on the EU agenda and the limited powers given to it, militated against the implementation of effective alcohol measures to protect public health (Sutton & Nylander, 1999). In addition, many effective alcohol policy measures at the national level have been weakened by the application of EU trade and economic directives to alcohol products and services, such as liberalization of the rules governing cross-border shopping for alcohol which forced some countries to lower their alcohol taxes (Anderson & Baumberg 2006), and abolition of the wholesale alcohol monopoly systems in Finland, Sweden and Norway, which were deemed to impede the 'free flow' of trade within the EU (Holder et al 1998). Alcohol marketing practices at a trans-national level have been acknowledged as having a significant impact on Member States' ability to frame effective health policies (DG SANCO 2005). The challenge is to find different ways to monitor the impact of relevant EU policy measures on alcohol-related harm.

In the European Union, there have been several important public health alcohol policy initiatives in recent years, including a Council Recommendation of 5 June 2001 on the drinking of alcohol by young people, in particular children and adolescents (*Official Journal of the European Communities*, L161, 16 June 2001), and Council Conclusions of 5 June 2001 on a Community strategy to reduce alcohol related harm (*Official Journal of the European Communities*, C 175, 20 June 2001). The Council recommendation on alcohol stemmed from a call for action from the European Parliament in response to the introduction of alcopops into Europe in 1995 (Sutton & Nylander 1999). The Council of Health Ministers acknowledged that the Council Recommendation represented a first step towards the development of a more comprehensive approach by means of a Community strategy to reduce alcohol-related harm. The Council Conclusion, adopted at the same time as the Council Recommendation, was again repeated by the Council of Ministers in June 2004. In response, the Commission prepared and adopted in October 2006 a Communication on an EU strategy to support Member States in reducing alcohol related harm where alcohol related injuries from alcohol-related road traffic accidents have been identified as priority theme for action (COMMISSION OF THE EUROPEAN COMMUNITIES, 2006).

Emergency Department Studies and Their Role in Alcohol Policy

Alcohol related injuries are a major source of death and disability. The data base is well established at the macro level and at the individual level in specific areas such as drinking and driving, alcohol and violence and alcohol and suicide (Anderson & Baumberg 2006). The new emergency department (ED) studies (the WHO Collaborative Study of Alcohol and Injuries) have provided a unique focus in a setting where a broad range of alcohol related injuries are funnelled. The ED links the macro with the micro level. Therefore, ED studies could have a crucial role to play in helping to inform and implement effective alcohol policies in three main areas;

a) creating a common knowledge base in Europe,

b) monitoring alcohol policy effectiveness at international, national and community levels and

c) developing effective interventions in the ED to reduce alcohol-related harm.

Creation of a Common Knowledge Base in Europe

The importance of a solid knowledge base to help inform alcohol policy has been highlighted by WHO and the public health section of the EU (DG SANCO). Much of the evidence on alcohol and injuries has been established at the macro level (Rehm & Taylor 2005). However, strengthening the evidence base at the micro level through emergency room studies could expand the research base and the understanding of the prevalence of alcohol and injuries in the ED, as well as the patterns of attendance and types of injuries. Emergency department studies help to explore the contextual circumstances of when, where and how alcohol-related injuries occur. They can also identify demographic characteristics of high risk drinking groups presenting to the ED. While alcohol and violence has traditionally had a public safety or social harm focus, (police enforcement) ED studies highlight the additional negative health consequences experienced as a result of alcohol related injuries by the drinker and by others beside the drinker (third party harm).

A number of European countries, Belarus, Czech Republic, Ireland, Italy, Poland Spain and Sweden have carried out ED studies (Cherpitel et al 2003; Hope et al 2005; Moskalewicz et al 2006; WHO 2005). These studies identified the number of patients presenting with alcohol-related injuries in the ED in acute hospitals. The proportion of patients who consumed alcohol within 6 hours prior to their injury ranged from a high of 42% in one hospital in Ireland to a low of 8% in the Czech Republic. However, the majority of studies reported between 15% and 30% presenting with alcohol-related injuries (Table 1).

Table 1. European ED studies, Alcohol-related injuries (the proportion of participants who consumed alcohol within 6 hours prior to their injury)

	Drinking prior to injury %	Place of injury on street/road %	Un-intentional %	Intentional self-inflicted %	Intentional by someone else %
Country					
*Belarus	30	47			
*Czech Republic	8	23			
*Sweden	15	33			
**Ireland					
Total	23	51	58	6	35
Mater	42	56	55	7	38
Beaumount	16	44	62	6	32
Waterford	16	48	59	7	33
Galway	19	51	62	6	32
Sligo	24	47	65	0	35
Letterkenny	24	51	51	7	42
***Poland					
Warsaw	16				
Sosnowec	20				
****Spain					
Barcelona	16				

* WHO Collaborative Study Group protocol; **Hope et al 2005 using WHO protocol; *** Moskalewicz et al 2006 using ERCAAP protocol; **** Young et al 2005 using ERCAAP protocol

In most of the countries studied, patients with alcohol-related injuries were most likely to be injured on the street or road, with the exception of Czech Republic where the patient's own home was the most likely location of injury occurrence. One of the most recent and largest of these ED studies was in Ireland, where six hospitals and over 2,000 patients participated. The main factors that significantly predicted alcohol-related injury attendance in the ED, while controlling for demographics and other relevant variables, were ED attendance between midnight and 6 am, hazardous drinking (5-11 drinks on one occasion) at least monthly, positive score for alcohol problems (2 or more on the RAPS4), ED attendance on weekends, injured on the street or road, lower socio-economic status and ED attendance in an inner city hospital (Hope et al 2005). It is evident that such ED studies provide a rich source of information and strengthen the base for policy makers in decision making. Such information allows for the prioritizing of policy measures relevant for national and local alcohol policy development. Given this research area is relatively new from an international perspective, it would be beneficial to build on these studies. Member States should be encouraged and supported to replicate such studies to establish a broad comparative data base across Europe, as well as to provide information for national alcohol policy development. In co-ordinating future ED studies, consideration should be given to having a core comparable protocol and design, and core interview questions (as has been undertaken with the Emergency Room Collaborative Alcohol Analysis Project and the WHO Collaborative Study on Alcohol and Injuries), with specific studies each having the option to add unique supplementary questions to address local issues and situations. This data base could be used to develop and consolidate agreed key alcohol indicators at the international level.

Monitoring the Impact of Alcohol Policies

The changing landscape of alcohol policy across Europe creates a major challenge in the wave of economic growth and market liberalisation with the expanding European Union. ED studies could provide a base line, in tandem with other methods, for monitoring key indicators of alcohol-related harm at the international, national and community levels, such as alcohol and violence, the drinking environment, alcohol and attempted suicide, drink driving injuries, family violence and workplace injuries. Such studies could identify the hot spots through mapping where and when alcohol-related injuries occur, and compare ED attendance for injuries in areas where there is high overall consumption and high prevalence of heavy drinking in contrast to areas where overall consumption and prevalence of heavy drinking is substantially lower. In examining the situational context of alcohol-related injuries presenting in the ED in Ireland, a clear picture emerged, in which half of the alcohol-related injuries occurred on the street or road, in comparison to only one-quarter of non-alcohol related injuries, with pubs as the primary drinking place prior to the event. Over half of the alcohol-related injuries were unintentional while one third were intentional and inflicted by someone else, with the vast majority of perpetrators consuming alcohol at the time they inflicted the injury. This contrasted significantly with non-alcohol-related injuries, of which almost all (94%) were unintentional injuries (Hope et al 2005).

On a European-wide basis ED studies could help monitor the impact of policy changes in availability and pricing practices, serving practices in bars and drink driving enforcement, in particular the cross-border effect (between Member States). ED studies could also monitor the drinking environments that give rise to alcohol related injuries and monitor international and national sporting or cultural events where heavy drinking can occur, resulting in increased risk of injuries. Such information would be of assistance in developing enforcement policy strategies, as well as in building the evidence base for strengthening alcohol control measures if appropriate. Community mobilisation projects have successfully targeted alcohol-related injuries (Holder et al 2000; see also Chapter 16 in this volume). However, few projects have used the emergency department as a source of data for monitoring effectiveness. Therefore, a useful application of the ED studies could be, in tandem with other measures, monitoring the impact of community interventions. The systematic ongoing screening in the ED of hazardous and harmful drinking patterns could monitor treatment outcome measures such as a reduction in repeat visits by problems drinkers to the emergency department.

Emergency Department – A Place to Intervene for Injury Prevention

The prevalence of alcohol-related injuries in the emergency department can have a negative impact on the delivery of health services in the ED in terms of increased numbers in attendance, the range of presenting injuries and the high level of repeat visits that tend to occur with problem drinkers (Hope et al 2005; WHO 2005). The fact that two thirds of those presenting with alcohol-related injuries reported that the injury would not have happened if they had not been drinking is an indication of the additional burden that alcohol-related injuries place on the emergency department. (Hope et al 2005). There is also an increased health and safety risk for staff working in the ED , due to the level of intoxication among patients. Taking a proactive approach, the ED can be seen as an opportunity for providing a setting for intervention to address alcohol-related problems. Taking the 'learnable moment' with the patient to link the presenting injury with their hazardous and harmful drinking pattern can be carried out using different protocols and with good effect (Dinh-Zarr et al 2004). The emergency department has been identified as a potential setting for early intervention across the European region (Anderson & Baumberg 2006); however, two important steps are needed. First, further studies are necessary to answer such questions as what are the most appropriate screening tools, who should be targeted, who should deliver the intervention, when and where should it take place and what are the measures for success? Second, building capacity within the emergency department, through information and training of health professionals, is essential for delivery of effective and efficient interventions to reduce alcohol-related harm while recognizing that the priority of emergency department personnel is treating the presenting injury, whether alcohol is involved or not.

Conclusion

Alcohol-related injuries carry a high proportion of disability and premature death across Europe, especially among the 15 -29 age group. In the European Region, political commitment to reduce alcohol-related harm has been demonstrated by the adoption of the WHO resolution, the New Framework on alcohol policy and by the EU Communication on reducing alcohol related harm in Europe which send a clear signal of the level of commitment for effective alcohol action across Member States in the EU to reduce alcohol-related problems.

Gathering relevant information and monitoring policy effectiveness are part of an overall strategy to tackle alcohol-related harm. Extending the evidence base by means of ED studies could play an important role as both an anchor and a linkage to other international, national and community strategies. To date, the ED has been under-utilized for a variety of reasons, including resources, capacity, work pressure, perception of roles, differing priorities and proven effectiveness in reducing repeat visits. As policy changes take place across Europe, the ED should be utilized to monitor the impact of such changes on alcohol-related harm. While there have been some successes in addressing problem drinking in the emergency department environment, further research is needed to fine-tune the most appropriate, effective and efficient ways to handle alcohol-related injuries. A strong knowledge base and an effective policy mix provide the way forward to reduce alcohol-related injuries.

References

Anderson P, Baumberg B (2006). Alcohol in Europe, a public health perspective. Instutute of Alcohol Studies, London.

Cherpitel CJ et al. (2003). A cross-national meta-analysis of alcohol and injury: data from the Emergency Room Collaborative Alcohol Analysis Project (ERCAPP). *Addiction*, **98**, 1277-1286.

COMMISSION OF THE EUROPEAN COMMUNITIES (2006). Communication from the Commission to the Council, the European Parliament, the European Economic and Social Committee and the Committee of the Regions. An EU strategy to support Member States in reducing alcohol related harm. COM (2006) 625 final.
(http://eur-lex.europa.eu/LexUriServ/site/en/com/2006/com2006_0625en01.pdf)

Dinh-Zarr T et al. (2004). Interventions for preventing injuries in problem drinkers. The Cochrane Database of Systematic Reviews, Issue.3. Art No.: CD001857. DOI: 10. 1002/14651858. CD001857.

Holder HD et al. (2000). Effect of community-based intervention on high-risk drinking and alcohol related injuries. *Journal of the American Medical Association*, **248**, 2341-7.

Holder H et al. (1998). *European Integration and Nordic Alcohol Policies; Changes in alcohol controls and consequences in Finland, Norway and Sweden, 1980-1997.* Aldershot, England.

Hope A et al. (2005). Alcohol and Injuries in the Accident and Emergency Department: A National perspective. Dublin: Health Promotion Unit, Department of Health and Children.

Moskalewicz J et al. (2006). Results of two emergency room studies. *European Addiction Research*, **12**(4), 169-175.

Österberg E, Karlsson T (2002). Alcohol policies at the European Union level. In E Osterberg and T Karlsson (Eds.) *Alcohol Policies in EU Member Sates and Norway: A collection of Country Reports.* Helsinki, STAKES.

Rehm J, Taylor B (2005). Volume of alcohol consumption, patterns of drinking and burden of disease in the European Region – implications for alcohol policy. WHO Regional Office for Europe meeting on Alcohol Policy in the WHO European Region, Stora Brannbo, Sweden, 13-15th April.

Sutton C, Nylander J (1999). Alcohol policy strategies and public health policy at the EU-level: The case of alcopops. **Nordic Studies on Alcohol and Drugs**, English Supplement 16, 74-91.

WHO (2005). WHO Collaborative Study on Alcohol and Injuries. First Report. WHO Collaborative Study Group, Department of Mental health and Substance Abuse, Department of Injuries and Violence Prevention, Geneva: World Health Organisation.

WHO (2006). Framework for alcohol policy in the WHO European Region. World Health Organisation Regional Office for Europe, Copenhagen.

Young DJ et al. (2004). Emergency injury presentation as an indicator of alcohol-related problems in the community: a multilevel analysis of an international study. *Journal of Studies on Alcohol*, **65**, (5), 605-12.

CHAPTER 19 : ALCOHOL POLICY AND PUBLIC HEALTH IMPLICATIONS IN A GLOBAL PERSPECTIVE

Norman Giesbrecht - Centre for Addiction and Mental Health | Toronto, ON CANADA

Cheryl J. Cherpitel - Alcohol Research Group | Emeryville, CA USA

Robin Room - Turning Point Alcohol and Drug Centre and University of Melbourne | Melbourne, AUSTRALIA

Tim Stockwell - Centre for Addictions Research of BC | Victoria, BC CANADA

Introduction

Given the evidence of alcohol as a major risk factor for disease, disability and mortality both globally and regionally (WHO 2002), it is timely to examine alcohol-related trauma by using the 'window' of the evidence gleaned from studies conducted in emergency room settings. Emergency rooms provide a useful setting for monitoring, testing hypotheses and exploring new topics about the relationships between alcohol consumption and trauma and chronic disease.

As this volume illustrates, the field has grown dramatically in the over 20 years since the 1st international symposium on this topic (Giesbrecht et al. 1989). There have been substantial changes in the methods and analysis since the mid-1980s, and the number of countries where emergency room (ER) studies have been conducted is now in the dozens. Last but not least, this research has contributed to raising the profile of alcohol as a contributor to trauma. It has provided policy advocates and analysts with important tools and resources for prevention activities, and supported the work of clinicians in the development, implementation and evaluation of brief interventions.

Context

However, there is substantial local, intra-country and international variation in the accessibility of ER services and the populations that use them. There is also considerable variation across countries in the ER capacity – e.g. volume of patients seen at ERs per 10 000 population. Also, health care financing, and the availability of alternatives to ER treatment, may greatly influence the likelihood of an injury being treated in an ER. Even in developed countries it is likely that there is considerable variation in the percentage of accidental injuries that show up in an ER. For example, in isolated workplace settings such as mines or other primary industries, many accidents may be treated on-site. Violent incidents are likely to be under-represented in ER settings, given that perpetrators (e.g., gang violence) and victims (e.g., sexual assault) may consider it more threatening to be exposed, and have their identity registered by seeking professional medical attention, than to seek ways to treat their injuries informally and in a less transparent manner. Where there is armed conflict, especially for extended periods, there are likely to be high rates of trauma, but these may not be represented in the ER statistics, since hospital services may be disrupted during this time, and the armed force will likely have some treatment capacity in the conflict zone. Nevertheless, there are numerous contexts and periods of economic stability and relative consistency in health care delivery, where it may be assumed that ER-based data are representative of the trauma and other acute morbidity that took place in a jurisdiction at a given time interval.

Utility

ER studies are useful in surveillance, in documenting the scope of alcohol problems, in identifying causal relationships between alcohol and injuries, and as a resource for developing prevention initiatives and policies and monitoring their impact. Furthermore, they offer an opportunity to examine interactions between alcohol and other drugs and their combined contributions to casualty experiences and statistics. In the past two decades there has been a substantial elaboration in ER-based research, and also in the awareness of and application of this research along the lines noted above.

In the broader context, ER studies can serve a critical "canary in the mine" function. The findings and statistics that they generate highlight emerging alcohol problems, trends in drinking-related damage and causal processes. This is especially critical during an era where many countries are experiencing a combination of deregulation of alcohol controls combined with an increasing globalization of alcohol production and enhanced marketing of alcoholic beverages. The initial impacts of these developments on drinking patterns, high risk drinking and trauma may be first detected in ER studies, well before their impacts show up in chronic disease statistics.

ER studies are particularly important in environments where no data exist on alcohol-related health problems, where data are sketchy, or when other approaches to study alcohol involvement in health problems, such as surveys or interviews by clinicians, face considerable difficulties, as is the case, for example, in countries where beverage alcohol is banned. For example, it may be feasible to obtain consent for urine or blood samples, whereas direct questions about drinking may be considered too intrusive in a culture where the official position is abstinence from alcohol.

Studies using ER data can also contribute both to monitoring of alcohol-related conditions in a jurisdiction and to evaluation of the prevention activities and programs. However, there are a number of noteworthy challenges. It may be difficult to find a good match between the agenda and foci of the ER study and goals of more generic monitoring initiatives. Also, good monitoring involves a combination of standardized protocols with the aim of collecting data using the same procedure over an extended time period. Most ER studies are time-limited, involving a few weeks or months of on-site data collection. Thus an ER study may point the way to establishing a monitoring system or include piloting of a monitoring protocol, but typically it is not designed to also serve as a monitoring tool.

Nevertheless, where protocols are established to collect data routinely from ER settings, then this becomes a powerful tool for tracking trends in the type and severity of injuries where alcohol plays a role, and also the changes over time in the demographic characteristics or drinking patterns of those so injured. Thus, the ER data might serve as an 'early warning system', illustrating, for example, a rise in alcohol-related problems before the change is evident in in-patient morbidity statistics or mortality data. Despite this clear potential, it appears that there is a long way to go to establishing ER-based monitoring systems that include alcohol tracking in a number of countries.

There is also substantial potential for ER studies to be linked to prevention activities either through stimulating community-based or nationally-based prevention activities or by using the ER study protocol as a resource (not necessarily the primary or only one) to track the impact of a prevention-oriented intervention. Both of these potential applications and linkages to prevention have been under-utilized, but there are signs that this might change. Chapter 14 provides several illustrations of how data from ER studies have been a resource for stimulating local or national prevention activities and assessing its impact – see papers by Borges, Cremonte, Swiatkiewicz, Sovinova and Csemy, and Benegal in this volume.

Looking outside the alcohol arena, it is also clear that ER studies provide a resource in the injury surveillance area. The evolving protocol of ER studies focusing on alcohol (see Cherpitel – Chapter 7 in this volume) – having been used in dozens of countries – provides a substantial resource for injury surveillance – even in those instances where the surveillance system does not yet include an alcohol tracking dimension. All major aspects of a high quality surveillance system are addressed in the higher quality ER studies focusing on alcohol, and thus provide a rich, but often under-utilized, resource for injury surveillance specialists to borrow from and implement.

ER studies have also contributed to broadening perceptions of the range of alcohol problems. To the non-specialist who is not involved in alcohol studies, alcohol problems may be perceived as involving drinking and driving, liver cirrhosis, alcohol dependence and possibly alcohol poisoning. In developing countries the range of perceived links between alcohol and health problems is likely narrower. However, alcohol has been identified as a contributing cause to over 60 disease conditions or types of trauma (e.g., Babor et al. 2003), and the list is growing with further research. ER studies provide a rich resource for the policy planner, advocate and analyst by illustrating the role of alcohol in a wide range of trauma as well as chronic diseases. These include, for example, falls, fires, water accidents, motor vehicle crashes and various types of violence including self-inflicted injuries (see especially Chapter 5 by Roizen).

Last but not least, ER studies also make a substantial contribution to identifying alcohol-injury causal relationships. This is a major theme of a number of chapters in this book – see especially papers by Ye and Cherpitel (Chapter 1), Borges et al (Chapter 2), Bond and Macdonald (Chapter 3), Rehm, Popova and Patra (Chapter 4), and Rehm and Room (Chapter 6). Nevertheless there remain opportunities for further work, including that of developing a better understanding of the causal relationships between alcohol use and trauma in developing countries where the place of alcohol in society and contexts of drinking and drinking-related activities is quite different from the patterns in western societies with established drinking cultures.

Challenges

ER studies provide a nexus between epidemiology, treatment, prevention practice, policy development and evaluations of interventions. The challenges of conducting this research and translating it into clinical practice or policy advice are considerable and they have already been explored in some depth in sections II, III and IV of this publication. However, it should be noted that in developing countries the challenges might take on added dimensions and intensity. A few are highlighted below.

A major challenge internationally is to develop the capacity to do the studies and to disseminate the findings so as to inform practice and policy-making. Conducting an ER study requires a combination of capabilities, including the following: obtaining funding, conducting fieldwork research, facility in project and data management, and skills in data analysis. If the findings are to be translated into policy initiatives, this requires at least two parties: persons knowledgeable about the research who have links to policy makers and can provide policy-relevant synopses; and persons in the policy-making arena who are interested in receiving the research findings and using them to inform policy development. At the systemic level it requires a 'window of opportunity' where there is the convergence of problem recognition, the formation and refining of policy proposals and political support (Kingdon, 1984). These windows of opportunity cannot be fully managed, but their potential will be enhanced where ER clinicians and researchers are involved in capacity building on alcohol issues and highlighting their implications for effective alcohol control policies.

Another challenge is that of finding areas of convergence and comparability when the research and its applications are drawn from institutions that are quite different. There is a diverse mosaic of settings, contexts and cultures evident under the 'emergency room' umbrella. In many countries other institutional arrangements may substitute for emergency departments. And in developed countries there is no clear consistency in the range of services provided in an ER, the type of clients or presenting problems that are dealt with in these settings, and the response capacity of an institution. Studies based in ER settings continually face the limitations and challenges of generalizing from institutions that are not identical on these key dimensions.

There are also challenges related to variations in the drinking culture where the ER studies were conducted. Drinking styles, characteristics of heavy drinkers and activities typically conducted during heavy drinking episodes vary substantially by country and within regions of a jurisdiction. The normative dimensions combined with perceived access to services, may encourage minor injuries to be taken for professional attention in one setting and discouraged in others. These variations in drinking behaviour and informal 'rules' about use of services will impact the scope of injuries that come to ER attention and what policy interventions can attain priority.

Roles for WHO and its National Partners

The World Health Organization plays a key role in many aspects related to the areas of ER studies and drawing out their implications for practice and policy development. This includes providing a horizontal connection among national and regional initiatives, and also making the links between injuries and alcohol, and mental health issues and alcohol. The WHO has been a co-sponsor of several initiatives related to ER research, including conferences in Toronto in August 1985 and in Berkeley in October 2006, and undertaking the multi-site Collaborative Study on Alcohol and Injuries, which was an epidemiologic study of the association of alcohol and injury, as well as an examination of the concordance of clinical observation of alcohol intoxication (ICD-10 Y 91 codes) with estimated blood alcohol concentration using breath analysis (ICD-10 Y90 codes) in emergency rooms in 12 countries.

Future Initiatives

There are a number of topics and themes for future initiatives, some of which are summarized briefly below.

Epidemiological studies

While we know that alcohol consumption is associated with injury, further studies are needed of the magnitude of alcohol involvement in injury occurrence and alcohol involvement by type and cause of injury. These can include secondary analyses of existing data, and aggregations of different datasets. Issues to be investigated in these studies include: the distribution of level of alcohol impairment of those with a particular injury; dose-response relationships; the existence and extent of co-factors which contribute to the occurrence of the particular injury (e.g., other drug use, icy conditions or other environmental factors); distribution in terms of the context of occurrence of the particular type of injury; distribution of usual drinking patterns (volume and heavy episodic consumption) and injury; and, hangover effects associated with drinking in the event. Contextual variables affecting associations at the study level are also an important consideration. Particularly for intentional injuries, studies are needed which measure and analyze cultural variations in alcohol expectancies and in the excuse value of alcohol. The impact of alcohol involvement by severity of injury and extent of disability, and alcohol dependence and abuse as predictors of severity and disability are also important to examine.

Clinical assessment of alcohol intoxication

Research is needed on optimal ways of recording estimated BAC, degree of intoxication and other clinically significant information for use in replacing or reformulating the alcohol codes (Y90 and Y91) in ICD-10, looking towards their adoption in ICD-11.

Methodological comparisons of control subjects for estimating risk of injury in ER patients

Studies of general populations concerning injuries and alcohol involvement in the injury are needed. Analyses of such population surveys and of ER samples in the same catchment area can be used to highlight the effect of alcohol on whether and when an injured person presents to an ER, and can illuminate factors that distinguish injury-prone from other heavier drinkers. Furthermore, studies are needed with designs which allow for estimation of the relative risk of injury from a given blood-alcohol level in particular contexts and with other particular co-factors. These can include case-control, case-crossover and other designs that yield a control condition.

Methods of obtaining BAC estimates

Development and testing of methods of obtaining BAC estimates with the least possible intrusion is needed. The forensic significance of accurate measurements makes them much more problematic for clinical use, and unlikely to actually be collected on a routine basis. Given this, the emphasis should be on methods that yield estimates that are sufficient for monitoring and epidemiological research purposes, but not accurate enough to be useful forensically.

New controlled study methods to obtain relative risks (RRs)

For ER studies to be used in estimating the population burden of alcohol in injuries, there is a need to obtain measures in control populations of comparable drinking events without injury. Two main methods have been used: control-group respondents matched on various characteristics with the ER clients, and the case-crossover method where the ER client serves as his/her own control, e.g. in terms of drinking exactly one week before the injury. Both methods have their drawbacks as well as their advantages, and there is a need for further development of methods in this area.

Comparisons of individual-level risk of injury estimates from ER studies with aggregate-level data

Comparative analysis is needed of individual-level estimates of risk of injury (RRs) obtained from ER studies with those estimates obtained from aggregate-level data. These analyses will contribute data on the magnitude of alcohol's causal role in injury as well as inform and improve the estimates of the global burden of disease for injury attributable to alcohol consumption.

Studies of capacity development

While the focus of alcohol-specific ER studies is on damage from alcohol, the work presented in this volume raises awareness of response capacities, systems and gaps in the response network for ER systems and monitoring generally. Thus, future work should include a combination of retrospective analyses as well as pilot studies on how these gaps can be narrowed, and what is required to increase capacity. This line of investigation will likely also involve cost analyses as well as an examination of knowledge exchange models and their relevance to different cultural settings.

Brief intervention studies

In the area of screening and brief intervention (see section IV) the promising findings are mainly limited to studies in developed countries. Further work should explore several themes: who delivers the intervention, the type of ER settings most amenable to this protocol, recidivism outcomes, and the feasibility of on-going implementation. There are opportunities for further work in developing countries that involve fine-tuning of protocols, capacity development and training. This line of intervention and evaluation also has implications for development and application in other settings, such as community health facilities, college health offices, and employee assistance programs. These analyses might be particularly useful in developing a transportable brief intervention package that can be easily distributed across cultures in a multitude of ER settings.

Dissemination of research and policy development

Studies are needed of the best approaches for disseminating major research findings from recent ER studies to inform prevention activities and to encourage consideration of the ER as an epidemiologic resource and a resource to inform brief intervention and policy development.

Continued work on alcohol and injuries in ED

An on-going project for continued work in the area of alcohol and injuries in ER studies is the Emergency Room Collaborative Alcohol Analysis Project (ERCAAP) and the WHO Collaborative Study on Alcohol and Injury which has combined data using the same instrumentation and study protocols (see Chapter 7 by Cherpitel) on nearly 17 000 injured patients and over 10 000 non-injured patients across 46 ER sites in 17 different countries. These data have been used for further establishing the alcohol-injury relationship (see Chapters 1 – Yu and Cherpitel, 2 – Borges et al., 3 – Bond and Macdonald, and 9 – Room), as well as for determining the lifetime risks of injury associated with alcohol consumption to inform drinking guidelines development for Canada (Taylor, Rehm, Room et al., 2008) and Australia (Rehm, Room & Taylor, 2008), and for informing the formulation of revised criteria for alcohol use disorders. Future planned work on this project, with continued funding from the U.S. National Institute of Alcohol Abuse and Alcoholism (Cherpitel PI) includes the addition of similar data from 20 countries with a doubling of the ER sites for, among other research aims, updating estimates of the alcohol attributable fraction of injury morbidity to alcohol, in order to inform comparative risk assessment for the Global Burden of Disease 2005.

A Sense of Urgency

Measuring and monitoring the role of alcohol in injury is an important area for further development. The alcohol policy arena is still under-developed and under-resourced in most countries. Alcohol is the 5th leading contributing risk factor to the global burden of disease and disability of the 26 examined (WHO 2002). More than half of the alcohol burden is estimated to be in the form of injury, intentional or unintentional. In light of the increase in consumption in some of the more populous countries it is expected that alcohol-related damage will increase.

ER studies continue to draw attention to the associations and causal links between alcohol consumption and trauma. They also provide a significant resource for those who wish to advocate for greater attention to alcohol issues, and implementation of sound alcohol policies so that the global harms from alcohol can be more effectively curtailed. Further work on the alcohol dimension in emergency services is thus a public health priority.

References

Babor T et al. (2003). *Alcohol, No Ordinary Commodity: Research and Public Policy*. Oxford: Oxford University Press.

Giesbrecht N et al., eds (1989). *Drinking and Casualties: Accidents, Poisonings and Violence in an International Perspective*. London: Tavistock/Routledge.

Kingdon JW (1995). *Agendas, Alternatives, and Public Policies, 2nd edition*. New York: Harper Collins College Publishers.

Rehm J, Room R, Taylor B (2008). *Method for moderation: measuring lifetime risk of alcohol-attributable mortality as a basis for drinking guidelines*. Toronto, Canada: Centre for Addiction and Mental Health.

Taylor B, Rehm J, Room R (2008). Determination of lifetime injury mortality risk in Canada in 2002 by drinking amount per occasion and number of occasions. *American Journal of Epidemiology*, **168**, 1119-1125.

WHO (2002). *World Health Report 2002: Reducing Risk, Promoting Health Life*. Geneva: World Health Organization.